After-sales Service of Engineering Industrial Assets

Vicente González-Prida Díaz
Adolfo Crespo Márquez

After-sales Service of Engineering Industrial Assets

A Reference Framework for Warranty Management

Springer

Vicente González-Prida Díaz
Adolfo Crespo Márquez
Department of Industrial Management
Universidad de Sevilla
Sevilla
Spain

ISBN 978-3-319-34679-3 ISBN 978-3-319-03710-3 (eBook)
DOI 10.1007/978-3-319-03710-3
Springer Cham Heidelberg New York Dordrecht London

Printed on acid-free paper

Springer is part of Springer Science+Business Media (www.springer.com)

*I dedicate this book to my daughters
Pilar and Paloma, for whom the effort
entailed in the development of a research
in a field as complex (and at time arduous)
as engineering, can one day serve them
as an example of self-development,
perseverance and discipline, qualities
that are often devalued nowadays
in our society*

Vicente

*I dedicate this book to my wonderful
family: my wife Lourdes and my sons:
Gonzalo, Lourdes and Adolfo*

Adolfo

Foreword

I am delighted and honored to write a foreword to this useful book. Indeed, I have personally known Adolfo Crespo Márquez for many years and in consequence Vicente González-Prida Díaz as Adolfo's Ph.D. student. Our collaboration has been initiated from complementary skills on maintenance engineering by investigating innovative work related to new forms of maintenance strategies (i.e., proactive strategy), organizations (i.e., e-maintenance) and engineering (i.e., system engineering). This work was also achieved in close partnership with industrial community of maintenance and asset management taken into account Adolfo involvement in Spanish industrial society called INGEMAN. A lot of collaboration results exist, but the main one is the contribution to the creation of the working group Advanced Maintenance Engineering, Service and Technology (A-MEST) within the IFAC TC 5.1. The WG is a founded repository defending new advanced maintenance/asset management challenges and issues. Adolfo and his team were able to take the opportunity of A-MEST to broaden the scope of their activities while upgrading their initial skills in Maintenance Management. In that way, they are now key representative actors in Europe and beyond on maintenance/asset management engineering. This position is proven by their high scientific credibility supported with relevant articles and books.

This new book entitled "After-sales Service of Engineering Industrial Assets. A Reference Framework for the Warranty Management" is fully in line with this position by considering a specific item of maintenance/asset management engineering related to warranty management—after-sales service which is a field actually not well formalized. In the book, an asset is considered as product or service launched in the market, that presents a certain level of complexity (complex industrial asset), and that after the sale requires an assistance (warranty) leading to an organization of resources supporting the buyer technical services (after-sales management).

Thus, the first originality proposed by the authors is to study this item, not as an isolated issue but really in a holistic way, to be consistent with practices of system engineering in order to assess globally its costs and profits by considering its impacts and relationships with the other items implemented in company.

Therefore, the book is developing material to attack the following general question: How to select a policy for a warranty assistance program and track its performance in order to improve the profit and the image of the company?

In response to this question, a second originality is defended in terms of a proposal of a warranty management framework as a conceptual and generic structure, which can be then instantiated to particular sector. Thus, the generic aspect assigned to the framework is enabling the book to be of interest to a wide audience of readers having different warranty needs with regard to products or services of different application domains.

The last main originality is that the book encloses a series of examples constituted by case studies related to the different chapters and that offer a practical overview. Indeed the intention of the authors is to synthetize case studies with practical applications that transmit in a simple way how the referred methodologies are implemented in the after-sales service field.

From these originalities, and as with all good reference works, this book provides a solid, formal, and, therefore, authoritative basis for the study, as well as the application in practice, of the after-sales service and warranty management. It is also an easy-to-ready book because it is going, by providing illustration, from fundamental issues to details on different methods and techniques that can be used to improve decision making in the different stages of the after-sales service management process. Therefore, the main strength of the book is, on the basis of a very clear text, to offer a way of thinking to asset provider, manager, engineer … in order to facilitate the requirement definitions with regard to after-sales service and transforming them into the realities of methodologies and techniques implementation well in phase with the profit and image of the company.

This way of thinking is structured on six parts in the book. The Part I details the context and the purpose of the work. In Part II, the fundamental issues and current research topics in warranty management and after-sales service are described. It is providing readers with a case study to illustrate not only a range of typical items for the topics addressed but also a global state of the art including models, techniques, methodologies, and tools. The Part III is defining the key contribution of the book in terms of warranty management framework proposal. This framework is divided into four steps considering the effectiveness, efficiency, assessment, and continuous improvement of a technique assistance program. The use of the framework is shown with regard to the case study given in Part II. Then, the authors are proposing to go deeper with the four different stages. It leads to describe relevant material because underlining practically how the different well-known methods and approaches such as Balance Score Card, ILS, LCCA, RAMS … are referred to each stage (Part IV). An extension of this framework is studied Part V by considering at least intellectual capital and maintainability index. The final part is devoted to conclusions and summarizing the obtained results.

Thus, the content of the book is well organized, attacking the right questions, and bringing significant answers enough formalized and generic to be usable both by academic and industrial readers. It leads to offer a real scientific and applicative added value for the maintenance and asset management communities. In that way, this book is going to sit on my bookshelf forever. I am confident the reader will find this book as useful and enlightening as I have. Thus, I wish the reader an enjoyable study of after-sales service and warranty management, and many productive applications of it.

Benoît lung
Université de Lorraine
Nancy
France

Preface

The aim of this book is to propose, to describe in details and to show the practical implementation of an advanced after-sales management framework devoted to warranty management. This framework is proposed for companies producing either standardized or customized products. Such a management tool will allow easy organizational improvement and will support innovative decision making processes for technical assistance in after-sales services.

The content of the book is divided into six parts:

- The Part I (Chap. 1) is an introduction to the book and describes the context and the purpose of the work. An introductory material to the process followed for the elaboration of the book is also provided.
- The Part II presents the fundamental issues and current research topics in warranty management and after-sales services. A literature review illustrates current state of the art including main international research contributions and best practices. Case studies are presented to illustrate these issues accordingly. This Part II is structured in the following chapters:

  - Chapter 2: This section presents a case study to illustrate a range of typical circumstances in the management of warranty claims, such as spare parts management, inter-departmental decision-making processes, cost related issues, etc.
  - Chapter 3: The state of the art in this area including models, techniques, methodologies, tools and other contributions developed by different authors will be here presented.

- The Part III contains the main contribution of the book: a proposal for a warranty management framework. Specifically, the following chapters are included:

  - Chapter 4: Existing models for assets maintenance management are commented and compared with each other, highlighting the difference between process-oriented versus declarative models. These models will be adapted to the after-sales management process showing the actions and stages in order to lead and manage the organization of a warranty assistance program.

- Chapter 5: Based on the above chapters, a framework for warranty management is proposed. This framework is divided into four steps or stages considering the effectiveness, efficiency, assessment and continuous improvement of a technical assistance program.
- Chapter 6: This chapter exemplifies the proposed framework for the initial case study in Chap. 2. This chapter also shows various ways to follow for the analysis of quality and maturity in terms of customer service management.

• The Part IV explores the different stages of the proposed framework more deeply. The idea is to specify different methods and techniques that can be used to improve decision making in the different stages of the after-sales service management process. This part includes the following chapters:

- Chapter 7: Deals with the warranty program effectiveness. Techniques like The Balance Score Card, Criticality Analysis and the Root-Cause Failure Analysis are applied to different case studies.
- Chapter 8: Focusses on the warranty program efficiency. The implementation of tools like Integrated Logistics Support or Cost-Risk-Benefit Analysis is considered within the after sales service.
- Chapter 9: Warranty assessment and control. The application of methodologies like: Life Cycle Cost Analysis (LCCA), and the Reliability, Availability, Maintainability and Safety (RAMS) analysis is suggested and exemplified.
- Chapter 10: Devoted to the continuous improvement in customer service management. This chapter considers the implementation of techniques such as Six Sigma, CRMs, and new ICTs in order to make data processing and communications easier.

• The Part V of the book presents extensions to the warranty management framework considered above. Different ideas and contributions are introduced to strengthen the structure and foundations of the framework. specifically the following chapters are added:

- Chapter 11: The value of Intellectual Capital in the customer service is here analysed, quantifying the worth generated for the company by the customer service department.
- Chapter 12: The maintainability index is defined and assessed by attributes related to staff, product design, as well as logistic support needs.
- Chapter 13: In the final part we explore how the proposed framework can be extended by applying System Dynamic Model and simulation to the after sales services. Also we tackle the problem of disassembly of a complex industrial asset planning using the Theory of Bayesian probability.
- Finally, the sixth and final part is devoted to the conclusions of the entire work, specifying the different parts and summarizing the results obtained (Chap. 14).

This book illustrates our personal experiences within this field and offers the readers a real perspective of the use of the tools defining a framework for warranty management of engineering industrial assets.

July 2013

Vicente González-Prida Díaz
Adolfo Crespo Márquez

Acknowledgments

The authors wish to thank specific people and institutions for providing their help during this last two years, making the publication of this book possible.

We would like to thank our colleagues from the research group "Intelligent Maintenance Systems" of the Department of Industrial Management of the University of Seville. They did provide their help and support to this work; they also co-authored some of the papers serving as basis for different Chapters of the book. We fully appreciate the amicable and friendly working atmosphere in our group over these years, where the area of asset management could importantly develop.

We would also like to express our warm recognition to the association INGEMAN for its valuable help, trusting us the development of different research projects in the field of reliability, maintenance and asset management in general.

Likewise, during this time international events, conferences and projects offered good opportunities for interaction with engineers and research fellows from universities abroad. We give special importance to our links with universities such as Lorraine (Prof. Iung) and Toulouse (Prof. Peres) in France, Politecnico de Milano (Prof. Garetti and Macchi) and University of Bergamo (Prof. Cavalieri) in Italy, University Federico Santa María (Profs. Stegmaier, Krjstianpoller and Viveros) of Valparíso in Chile, among others. We received precious benefits of their knowledge, which in turn have influenced the development of this book. At the same time, we would like to thank them for their friendship, enriching our personal life.

We want to give special thanks the companies Abengoa and General Dynamics—European Land Systems, these institutions were providing an excepcional professional experience to Vicente over many years. This experience was extremely valuable for the development of this book. Within these companies, many colleagues gave their support offering a practical view helping us to understand different aspects of supplier-business-client relationships.

The funding and recognition from the Spanish Ministery of Science and Education during the time this book was written (Research Project DPI: 2011-22806), and from our Regional Government "Junta de Andalucía" (Research Project TEP: 7303-2011), made many things related to this work possible.

The European Commission (FP7-PEOPLE-IRSES-2008) in addition to our Regional Government "Junta de Andalucía" (Programa de Incentivos a Agentes del Sistema Andaluz de Conocimiento) co-funded the project IMaPla, over the last four years. This project was endorsed by four institutions, three EU partners and a

Chilean partner, agreeing for a common exchange program built around the theme of Integrated Maintenance Planning.

Materials and knowledge gained during the development of these projects also serve as part of subjects that we are currently teaching like : "Production and Maintenance Engineering" and "Maintenance" in the School of Engineering of Seville; "Operations Management" in the University of Bergamo; "Advanced Models for Maintenance Management" in the School of Engineering of the Technical University "Federico Santa María de Chile" and many different Modules that we teach in collaboration with Asociación para el Desarrollo de la Ingeniería de Mantenimiento (INGEMAN) an other institutions in Spain and in other countries.

Last but not least, we wish to express our gratitude to our families for their unconditional trust, encouragement and unconditional support enabling this work to be accomplished.

To all of them, thanks.

Contents

Part I
Introduction

Chapter 1
Researching the After-sales Services and Warranty Management

1.1 Introduction

Day after day, enterprises must face a great variety of problems related to the management and reliability of the products they launch in the market. The setting where these companies carry out their activities is usually complex and globalized, and the one in which we must take into account high competitiveness of the market. In short, as we move in a changing world with competitors at any point of the globe and with different economic criteria, any business present many and diverse challenges to face, some of which cannot be found yet among the main decisions of executives nowadays. This research aims to observe these challenges, and specifically the after-sales management, and how it can improve reliability of industrial assets launched to the market increasing the satisfaction of the client and the financial objectives of the business. To this purpose, it is convenient to point out the reason behind the different keywords by defining each term [1].

- *Reference Framework*: "Frame" is defined as the pattern or conceptual model by which determined subject should be regulated or contrasted. If we add the term "of reference," it is implied as well that the frame is used as a base or support to compare, measure, or relate. That is, the concept of "reference framework" is used in this research as a conceptual structure to solve or guide complex issues about some of the areas of knowledge through a description or a generic approximation.
- *After-sales Management*: All those organizational and managerial activities that determine objectives or priorities, strategies or methods, responsibilities, etc. of the after-sales service, with the purpose of putting them into practice through planning of technical assistance, their control and supervision, allocation of resources etc., in a way that its economic aspects are included in the general accounting of the organization. The "after sales" concept is understood as the period of time during which the seller or manufacturer guarantees to the buyer assistance, maintenance, or repairs of what has been purchased.

V. González-Prida Díaz and A. Crespo Márquez, *After-sales Service of Engineering Industrial Assets*, DOI: 10.1007/978-3-319-03710-3_1,

- *Complex Industrial Asset*: An "asset" is defined as the whole concept of assets and rights with monetary value that are property of a business, institution, or individual, which at the same time are indicative of their accountability. The terms "industrial" and "complex" refer to all those material assets related to the industry, that have a certain level of technical complexity, and therefore require a determined technical control (for example, an automobile, an aeronautical component, an automaton, a reactor, etc.)
- *Warranty*: A temporary agreement of the manufacturer or retailer, by which it is obliged to repair with no charge any sold item in case of a breakdown.

Basically, this research is about products or services that are launched in the market and that present a certain level of technological complexity (*complex industrial assets*) and that after the sale require an assistance (*warranty*) that leads to a certain organization of resources that assures the buyer certain technical services (*after-sales management*). For which reason, some generic guidelines are proposed (*reference framework*) to facilitate and maintain team work, internal and external communication, business ethics, as well as human resources and corporative social responsibility.

1.2 Objectives of the Book

The main objective of this research is to develop a management framework that is applicable to organize the after-sales service, so that it remains aligned with the objectives of the company and it is not presented as a separate entity, nor as a mere cost center from a financial point of view. In other words, the idea is to consider all points of view related to the company and not just the technical or financial perspective. Sustainable development of the organization depends on its consideration as a holistic system where the component parts (highly specialized departments with their own management goals), collaborate to achieve collective objectives [2]. With such purpose (and even though the structure of the research will be developed in greater detail in a later/further section), this main objective can also be divided into the following secondary objectives:

1. Analyze the state-of-the-art regarding management and warranty of the after-sales service (Part II).
2. Identify the concepts and developed methodologies that come into effect for other areas of management such as maintenance, and to apply them to cases of warranty assistance (Part III).
3. Develop the implementation of these methodologies regarding the after-sales service (Part IV).
4. Justify the importance of the research with future extensions or possible lines of research in this area (Part V).

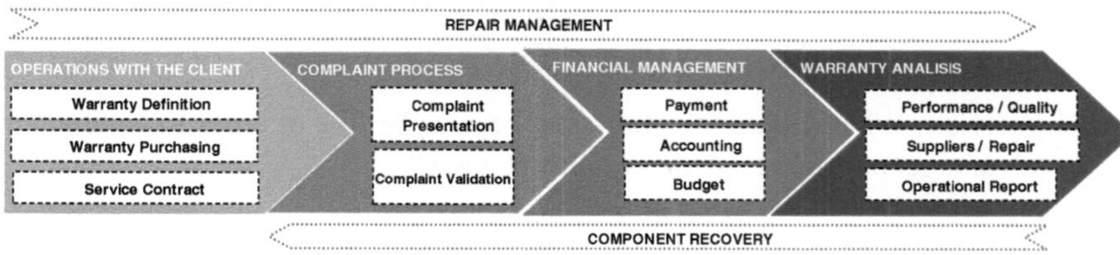

Fig. 1.1 General scheme of warranty management

Following this scheme of objectives, the research defines a framework as a support to the management of a warranty program (understanding program as a sequence of activities). This framework is divided into blocks that are also composed by actions centered mainly in the improvement of the reliability of the product as well as of the relation between manufacturer and user. The re-engineering in the management processes and the application of the right organization of the after-sales service allow (among other options) to influence on the design of the product and its fabrication, increasing its quality and reliability as it improves the flow of information (Fig. 1.1) relative to product defects and its causes. Due to the fact that the main objective of the research is vastly wide, in order to make the expected goals in the research more manageable, it is convenient to break down such objective into small questions of research. The answers to such questions will jointly contribute to the essential solution of the subject to investigate. The after sale, viewed from a wide perspective, can boost the business' economic benefits not just by the sale of secondary products (spare parts, maintenance contracts, and warranty extensions…), but also by identifying and analyzing problems to determine and eliminate the causes, avoiding constant problems of the users.

In order to facilitate a better understanding of the essential problems in this area, the first question of the research will be relative to the existing methodologies used in areas such as maintenance, improvement processes, and its capabilities:

- *1st Question*: How the different studies, models, and approximations… done so far have helped the identification, analysis, elimination, and prevention of problems when there has been a failure in an industrial asset?

With the previous knowledge, it is possible to suggest a management process that covers the already known and successfully implemented methodologies in other areas, so that the maturity in the organization improves in terms of the after-sales service. This leads us to the second question of the research:

- *2nd Question*: What systems are proposed to provide after-sales service to make the most of the advantages of the successfully applied methodologies in other areas?

One of the most important issues in after sales is decision-making: which policy should be adopted, or if the right follow-up procedure is being applied. For this a systematic procedure is needed to assure the decisions are based on facts and that a correct follow-up procedure is being properly applied. With this in mind, the third question of the research is:

- *3rd Question*: What should the after-sales service be like in order to be efficient and advantageous from a business point of view?

The book describes a cyclical procedure of improvement, selecting a determined management policy, and carrying out a follow-up check of its performance. It is required, in order to know which is the most appropriate policy, a form that indicates how to measure the impact in the decisions of the company, before and after the application of the selected policy. This leads us to the next question:

- *4th Question*: How should technical and financial performance in the after-sales service be assessed and controlled so as to allow improvement during the management of its procedure?

In order to measure the performance of technical assistance, access to the relevant information regarding the service is required. This is a complex problem owing to the after-sales impact of an industrial asset that can affect different areas of the business. On the other hand, the commercial information of the asset's behavior in different types of users is not always available or represents a wide variety of data [3, 4]. This leads us to the fifth research question:

- *5th Question*: What information should be relevant to make decisions regarding the after-sales service and what are the future researches that arise from this line?

Figure 1.2 shows how the suggested questions relate to each part of the research.

In few words, taking the previously mentioned into account, this book develops the following problem: How to select a policy for a warranty assistance program and track its performance, in order to improve the profit and the image of the company? The content of the indicated chapters is described in the next section where the structure of the book is summarized.

1.3 Rational of the Methodology Followed

This research has been structured in a series of *Parts* that are divided into chapters, with its content mainly theoretical and general. Some of them include cases that are concerned with mainly practical content, that attempt to specify with examples some techniques introduced previously in the same chapter. In other words, a classification of engineering tools will be presented throughout the chapters, developed mainly for the improvement of the assistance to industrial assets in the market through warranty management developed in cases, which will be shown in

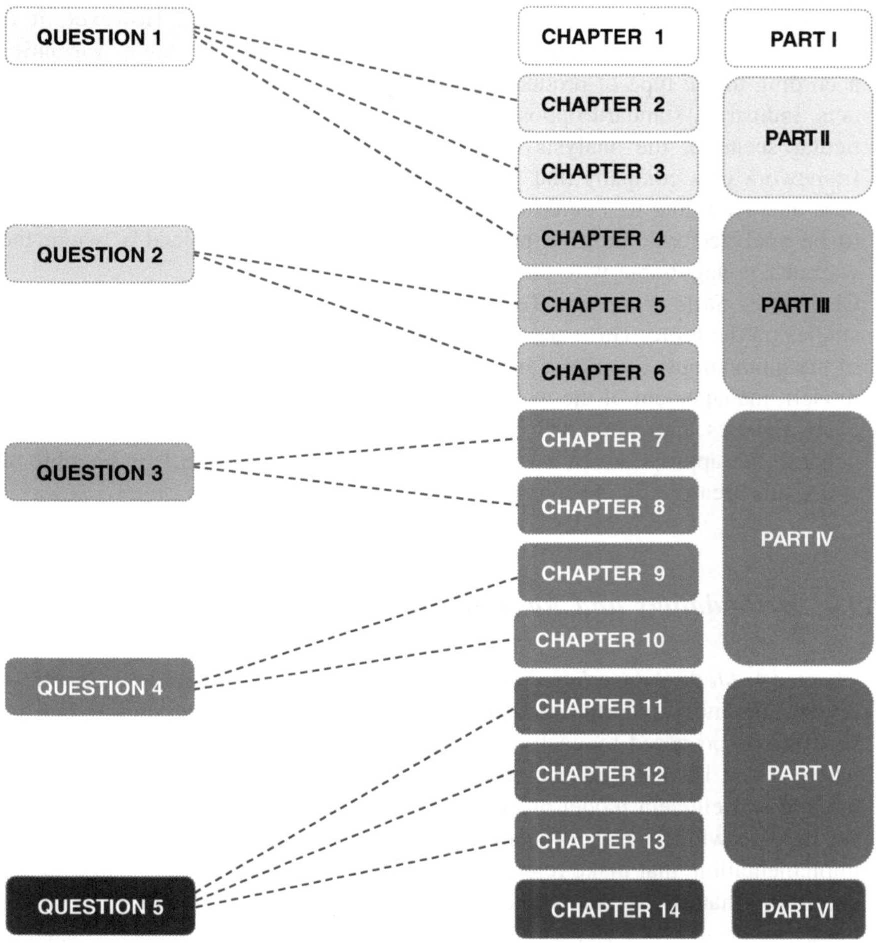

Fig. 1.2 Relation between the research questions and the chapters

practical applications. Next, there is a description on how the present document is structured:

1.3.1 Origins and Background (Part II)

- Chapter 2: *Initial scenario where the problem is proposed.* This chapter develops a current and generic example that proposes a realistic and concise vision of the chain of supply, the internal works of a typical manufacturer company and the clients and providers management. This context pretends to introduce the casuistry of the problems regarding warranty problems, all of this serving as a starting point of development of this research. However, it is important to highlight the existence of different types of suitable warranties according to the type of product (products for consumption, commercial products, industrial, standard opposite personalized products, etc.) [5]. In this particular scenario, the analysis of after-sales management is done within the framework of a company that provides its services during a specific period of time and according to a determined calendar of deliveries. The type of product to be analyzed will be of a personalized item (or "customized") where the warranty is negotiable with the client.

- Chapter 3: *State-of-the-art*. The first task was to analyze all those relevant studies on the topic. This state-of-art describes in a brief and general way some of the innovations, functions, and objectives that different authors propose for a modern management of the warranty. From this, analysis can be deduced from characteristics and properties of different scenarios and contexts and that, in general, are appropriate to a whole taxonomy of cases where it is possible to apply this area relative to management.

1.3.2 Methodology and Resolution Proposal (Part III)

- Chapter 4: *Methodology based on maintenance management models*. Chapter 4 reviews the most important maintenance management models, pointing out the innovations proposed by each author. From this, analysis derives some results where it can be distinguished from some of the desirable characteristics of a modern and efficient model of maintenance management. These characteristics are the following: that the system includes in itself a clear methodology for its implementation, that make reference to supporting tools that are required for its execution, that be preferably cyclic, etc.; considering these features and taking these conforming key elements of each model, finally a maintenance management model is proposed based on the operation of the norm ISO 9001:2008 [6].
- Chapter 5: *Reference framework*. This chapter corresponds to the central point of the research. It is here proposed a reference framework for the management of warranty assistance, using engineering techniques that are already developed for similar processes [7]. Within this generic framework, models are presented integrating models already developed and in application. They will be gathered in four sequential stages where the different techniques will play a crucial role. These four stages will contemplate effectiveness, efficiency, evaluation, as well as the constant enhancement of the management of after-sales service.
- Chapter 6: *Audit to the initial scenario*. In this occasion, the initial scenario presented in Chap. 2 will be compared to the criterion of the reference framework proposed in Chap. 5. This comparison will also allow us to introduce in this chapter a brief summary of the quality aspects for the enhancement of the

after-sales management together with some other practices done by businesses nowadays. All of this can facilitate the quantification and evaluation of the maturity level that the company reaches.

1.3.3 Development of the Proposed Stages (Part IV)

- Chapter 7: *On the effectiveness*. This chapter starts by defining the concept of The Balance Score Card, which intends to align local after-sales objectives with the global objectives of the business. The model also intends to avoid the creation of a compartmentalized company in which decisions are made in an independent manner. An organizational structure of this sort would negatively affect its own future development and consequently its image to the customer. Next, the chapter describes the *Analytic Hierarchy Process* as a tool that can facilitate decision-making related to some of the critical aspects of the management warranty permitting the alignment of after-sales actions with the business' objectives. The chapter ends with a study of the root failure analysis.
- Chapter 8: *On the efficiency*. In order to assist warranties with a minimum in loses, unnecessary expenses, or efforts, it is needed to take into account the efficiency in the after-sales service. This efficiency is here defined as the assistance of the same service but less costly. For this reason, the components that constitute the *Integrated Logistic Support*, its levels of repairs, and the frequency of duties are defined. The chapter continues with a brief *Cost, Risk and Benefits Analysis* applied to the after-sales service. These tools (originally developed for general project management), can be adapted in a very useful way to organize and control the technical assistance of the customer during the warranty period.
- Chapter 9: *On the assessment and control*. Equipment of any industrial asset must take into account a great deal of restrictions and conditions. The components or subsystems present ranges of potential failure modes that should be considered from the initial stage of the design of the entire system, according to its way of operation, environmental conditions, the time of the failure, etc. The *RAMS Analysis* offers an approximate assessment of the behavior of equipment that operate under severe conditions, such as those elements with complex systems of the aerospatiale sector, defense, self-propulsion, industry, etc. the terms of the acronym RAMS transmit the following concepts [8] that are usually expressed in terms of probability:
  - *Reliability* is the capacity of a system to carry out the functions that are expected from it, under established conditions and during a specified period of time.
  - *Availability* is the degree to which a system can be used during a determined period of time. It is usually expressed as a time ratio.

- *Maintainability* is the characteristic of the design and the installation of an element in which it must be re-established or substituted after a determined period of time, it is the maintenance done according to the prescribed procedures.
- *Safety* is the condition of being protected against failures, errors, damage, or any other incident that can be considered undesired.

  Besides the before mentioned, this chapter includes the Life Cycle cost of the physical asset. This is determined through the identification of the product's functions applied at each stage of its life, the calculation of these functions' costs and the application of appropriate costs throughout the life cycle.

- Chapter 10: *On continuous improvement*: The chapter defines the *Application of the ICTs* and the benefits that can provide to a physical asset in the market and that requires warranty assistance, defining qualitatively its influence within the proposed reference framework and its interaction with the rest of the stages, necessary to provide the customer the proper management in technical assistance. After that, the chapter describes how customer relations management can help decision-making regarding the after-sales service (strategic acquisition of replacements, classification of different types of customers, etc.), and how it is adjusted within a contractual and legal framework that regulates the relations with the client when the technical assistance warranty is applied. This legal point of view link management of after-sales issues to the current situation is presented in real markets. The chapter concludes with a definition of the Six Sigma Method, expanding and indicating the parts that consist of and carry out a review regarding how it interacts with other methodologies and tools.

1.3.4 Extensions to the Proposed Reference Framework (Part V)

- Chapter 11: *Technical assistance intellectual capital.* After-sales services are characterized as a highly complex field inside business and involve various disciplines: management, organization, human resources, company economy, safety, environmental management, and knowledge of production throughout the chain. In short, it is a key department by its contribution to look after the interests or to satisfy the needs of clients and enterprises profits. In order to improve maintenance, the intellectual capital is one of the best tools to achieve it. It has to appreciate technical aspects, personal flexibility, changes, adaptations, creativity, learning capacity, willingness to team work, ability of decision and easiness to work in multiple, and interrelated operations. Therefore, the knowledge of the maintenance staff is a strategic asset for the company, and it can be considered as a competitive advantage. These aspects must be evaluated in order to place correctly the contribution of maintenance within the company valuation and future plans.

- Chapter 12: *Maintainability in the after-sales management*. This chapter contains a procedure that obtains the maintainability indexes for industrial devices. This analysis can be very helpful when comparing or selecting a system from another, to achieve an improvement in the design of a device according to the demands of a technical assistance program, to improve the after-sales services to a device that is in a given time and in a determined industrial environment, or to prevent maintainability problems caused by temporary changes of location, among others. For this purpose, those characteristics are here defined, describing the different levels on which technical assistance can be divided. All of this derives in the evaluation of some indexes applicable to any type of industrial asset or user, in the phases of preparation as well as of its operation. So, for those measurements can be contrasted, it will be necessary to determine the moment in the life cycle of the device in which its index is evaluated and the conditions of use and environment made by the client at that precise time.
- Chapter 13: *Dynamic model and Bayesian nets in after-sales services*. Businesses that operate in mature markets have to acknowledge the importance of complementing the sale of industrial products with assistance services. That is to say, the contribution as an added value of tangible products along with intangible services, designed and combined in a way that are able to satisfy as a whole customers' specific needs [9]. The supply of spare parts, technical support, damage repairs, update installations, equipment reconditioning, performing inspections, constant maintenance, are some of the significant examples of offered services related to the product after the sale of an asset and that can be a stable source of income, being more resistant to economical cycles than the buying–selling equipment [10]. This chapter deals with the development of a dynamic model of systems that have been created to explore quantitatively the cause relation and the impact that represents the introduction of a new after-sales policy, with the aim of improving customer service [11]. As a second part of this chapter, a new approach on after-sales service is included, in this occasion at the end of the product's life cycle. This will serve to determining an optimal plan for the dismantling of a complex industrial system and its recycling, using for that the Bayesian nets theory.

1.3.5 The Case Studies

The book encloses a series of examples constituted by case studies related to the different chapters and that offers a practical overview. With the exception of the introductory and concluding chapters (Chaps. 1 and 14) and those related to the state-of-the-art and methodology Chaps. 3, 4 and 5, the rest of the chapters include an exercise or an example that would make easier for the reader the comprehension of the concepts developed in theoretical sections.

In general, the studies of specific cases are normally used, in engineering as well as in other fields of research, to support and help the understanding of different subjects and theoretical developments.

Generally, there is a great amount of information at our disposal when it comes to the description of the cases. Such information can trivialize the analysis, or complicate it even, to a level beyond unreasonable. That is why, the intention is to synthesize case studies with practical applications that transmit in a simple way how the referred methodologies are implemented in the after-sales service field, in order to reduce prices. How can they allow more efficient decision-making processes? How can they boost the business's image and its standing in the market? etc. An emphasis is placed on the range of study scenarios, as it is shown in Table 1.1. These several contexts are distinct from one another and analyze a different matter concerning the after-sales warranty management of an industrial asset under warranty.

1.4 Limitations and Evolution of the Research

With the task of reaching the indicated objectives in Sect. 1.2, the research has been affected by a series of limitations. Some have been on account of external issues (imposed by the economic and technological reality in which we found ourselves) and others of an internal one (imposed by the author in order not to extend the time in this research in an exaggerated fashion.). Basically, these limitations have consisted in:

- The research is about after-sales management in general. It is not specified to a particular sector (each sector will require this framework to be applied in a different way).
- Within warranty management, the research mainly highlights the warranty period, even though such service (depending on the terms of the service contract) could include other type of assistance such as maintenance, repairs, or gathering of supplies and repairs.
- In most parts of the book, the object is a complex industrial asset. This asset is considered to be the physical product that is up for sale. However, the object can also be a service that needs to be guaranteed, i.e., utilities.
- The different methodologies of the book are presented from a conceptual point of view. It is not the objective of this research to outline mathematical or computational models.
- The study limits itself to the selected tools of the available warranty management at the time of the development of this book.
- The research is based on warranty management [7]. The importance point here lies in the way of technical assistance is carried out. Instead of doing this in the asset's own place of manufacture, it is done on a sold asset. This is the reason why this research incorporates customer's voice in the casuistry of maintenance.

Table 1.1 List of case studies included in the chapters

Chapter	Methodology applied	Case Study	Question to solve
6	Quantification of maturity	Manufacturer of products which are launched into de market and which require assistance	What is the company's own perception of the after-sales service?
7	The balance scorecard	Farm machinery (haversters and headstocks) sold first and second hand	What after-sales service results more profitable?
7	Procedure of hierarchical analysis	Spare parts management for customized products manufacturing in a determined production moment	From where is the best place to get the spare?
8	Analysis of management risks of customer relations	Water distributing company that suspends the supply due to a failure in the network	What is the cost of maintaining or losing a customer after suffering a failure
9	Reliability, availability, maintainability, and safety analysis	Electromechanic item in a designing phase	Which would be the list of spare parts recommended for attending the warranty in a satisfactory way?
9	Life cycle cost analysis	Faulty engines are repaired with a repair warranty	How long would be the best warranty period in case of doing a repair?
10	Six sigma methodology	Automotive company that repeatedly receives the same time of reclamations in warranty	What is the procedure of a reclamation that affects design?
11	Quantification of the intellectual capital	Water distribution companies that attend similar areas of customers	Which company presents the best investment of their initial budget?
12	Maintainability assessment	Bridge crane requiring technical services	How easy or difficult can the assisted item be maintained?

- The notion of "reference framework" is used as a conceptual structure to solve and guide complex questions in a determined field through an approximated generic description. Therefore, it is not a "reference model" since it would represent the basis of a computational context for its implementation in a determined field.

In regard to what is stated above, it is necessary to add that this present work has taken a long period of time and its development consists of several years of labor. Throughout this time, the questions of the research have been evolving and arising from both its necessities and its own development. The value of chapters and case studies is intended to be supported through publications in congresses, scientific journals papers, or in book chapter, where the research does not present a strict temporary parallelism with different contributions.

Some of the most important contributions to the development, consolidation, and countersign of this book (published in different scientific journals, conferences, and book chapters, both in the international and national field) are included in the reference section of each chapter.

References

1. Real Academia de la Lengua Española. 2001. Diccionario de la lengua española, Edición 22. Espasa. ISBN 8423968138, 9788423968138.
2. Joan Costa. 1987. Imagen de empresa. Métodos de comunicación integral. Imagen global Ceac Ediciones, Barcelona.
3. Labib AW. 2004. A decision analysis model for maintenance policy selection using CMMS. Journal of Quality in Maintenance Engineering 10(3):191-2002.
4. Kans M. 2008. On the utilisation of information technology for the management of profitable maintenance. Doctoral Thesis, Växjö University.
5. Murthy DNP, Blischke WR. 2005. Warranty management and product manufacturing. Springer-Verlag London Limited, pp. 302 + xviii. ISBN 1852339330.
6. AENOR. 2008. Norma ISO 9001:2008 Sistemas De Gestión De La Calidad.
7. Crespo Márquez A. 2007. The maintenance management framework: models and methods for complex systems maintenance. Ed. Springer.
8. International Standard. 2004. ISO/DIS 14224/2004. Petroleum, petrochemical and natural gas industries. Collection and exchange of reliability and maintenance data for equipment.
9. Brandstotter M, Haberl M, Knoth R, Kopacek B, Kopacek P. 2003. IT on demand – towards an environmental conscious service system for Vienna (AT). Third International Symposium on Environmentally conscious design and inverse manufacturing – EcoDesign, pp. 799–802.
10. Wise R, Baumgartner P. 1999. Go Downstream - the New Profit Imperative in Manufacturing. Harvard Business Review, September-October.
11. Crespo Márquez A. 2010. Dynamic modelling for supply chain management – Dealing with front-end and back-end and integration issues. Springer-Verlag, London.

Part II
Current Issues in Warranty Management

Chapter 2
An Initial Case Study. Understanding Warranty Management Issues

Case studies have normally been used to support and help theoretical subjects in engineering and other research fields. When these cases are developed, it is usually found that such an amount of information can either trivialize the study or complicate it beyond a reasonable level. Therefore, the intention here is to synthesize a practical case, which transmits easily how a proper management of warranty assistance helps to reduce costs, enables us to make suitable decisions, and improves the image of the company in front of the client. As mentioned at the beginning of this research, there are different types of warranties according to the different products in the market (consumption products, commercial products, industrial, standard or personalized products, etc.) [1]:

- Standard Products: Free Replacement Warranty (FRW), Pro Rata Warranty (PRW), or a combination of both.
- Commercial Products (purchased in volume): Warranty applied to a fleet or group of items.
- Personalized Products: Reliability Improvement Warranty (RIW).
- Base Warranty and Extended Warranty.
- Etc.

The case examined here deals with a customized product and the warranty management with the client. This study starts summarizing the background related to warranty, highlighting the importance of a warranty cost management system. Once the problem is defined and along with the development of a particular case study, analysis of data and conclusions will be shown. Besides, a procedure is also proposed related to the way of working among different sections inside a generic company. This procedure will be examined succinctly using a workflow chart. Other practical cases in the field of warranty can be found in [2].

V. González-Prida Díaz and A. Crespo Márquez, *After-sales Service of Engineering Industrial Assets*, DOI: 10.1007/978-3-319-03710-3_2,

2.1 Background

In order to apply an effective warranty management, it is critical to gather proper data and exchange the different types of information between the modules into which a management system can be divided [3]. In our case, a warranty management will be proposed, as a system based on a several modules organization. In the literature review (this chapter), one can observe different interactions between warranty and other disciplines, and how they are dealt with by the different models and authors. In particular summarizing, three important interactions must be considered:

1. *Warranty and Maintenance*: In many cases, the warranty period is the time when the manufacturer still has a strong control over its product and its behavior. Additionally, the expected warranty costs normally depend, not only on warranty requirements, but also on the associated maintenance schedule of the product [4].
2. *Warranty and Outsourcing*: The warranty service or, in general, the after-sales department of a company, is usually one of the most susceptible to be outsourced due to its low risk and also to the fact that, among other features, outsourcing provides legal insurance for such assistance services [5].
3. *Warranty and Quality*: The improvement of the reliability and quality of the product not only has an advantageous and favorable impact in front of the client, but also this improvement highly reduces the expected warranty cost [6]. However, in the vehicle industry for example (see Fig. 2.1), it is probably a simplification to consider a low ratio of client's claims supposes high quality, that is because such situation has to do with those companies that reject customer claims redirecting them to the suppliers. In Fig. 2.1, the axis of ordinate indicates the percentage dedicated to warranties in respect of the income in vehicle sales.
4. *Warranty and Cost Analysis*: In reference to costs estimations (see Fig. 2.2), and apart from warranty issues, there are nowadays several methods to accurately estimate the final cost of a specific acquisition contract. In our case study, the method applied in a simplified way is denominated "Estimate at Completion" (EAC).

In few words, EAC is a management technique used in a project for the cost control progress. Here, the manager foresees the total cost of the project at completion, combining measurements related to the scope of supply, the delivery schedule, and the costs, using for that purpose a single integrated system. Warranty costs can be included in the global analysis of the project, providing the estimation costs of the same service generated at the end of contractual responsibility.

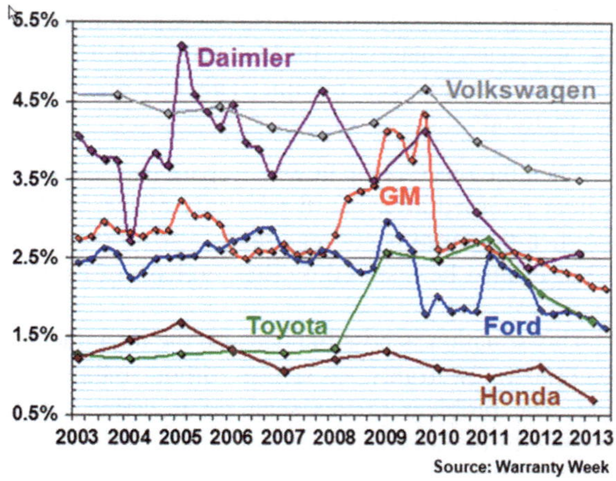

Fig. 2.1 Ratios of warranty claims from 2003 to 2013

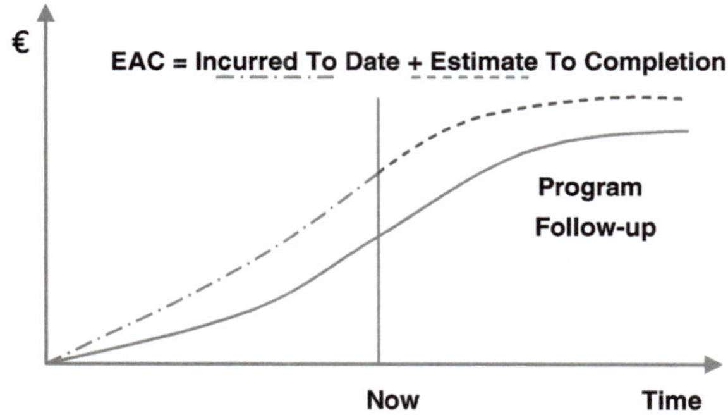

Fig. 2.2 EAC formula (adapted from [7])

2.2 Case Study Scenario

The company that is focused on in the case study is a large manufacturer in the metal industry that operates globally. The company designs, manufactures, and purchases a wide range of industrial vehicles (such as machinery used in forestry, hydraulic excavators, or track loaders) for industrial customers, as well as other related products such as spare parts. In addition to the purchase of standard vehicles, nowadays the customization of machines is also common. In our case, the company must supply the client with a specified amount of customized vehicles on a defined schedule. In the contract, warranty assistance is included for the vehicles of the fleet during a period of time that starts when each vehicle is delivered to the customer. To provide the after-sales service in a satisfactory way, the fulfillment of some conditions is required:

1. Teams formed by properly trained personnel.
2. Tools for maintenance/warranty tasks.
3. Materials and spare parts to carry out the repairs.

The first two conditions are considered fulfilled. Regarding the third condition, the necessary materials for warranty operations are obtained from the same warehouse as the assembly line. This way, there are two possibilities to recuperate the material:

- When the piece is repairable, a spare part is taken from the warehouse, which is later returned after the disassembly of the product being repaired.
- When the piece is not repairable, a spare part is also taken from the warehouse, but it must be restored by the purchase of another.

This situation is possible because the manufacturing stock allows the supply of material for warranty assistance without risking the necessities of the assembly line. The problem in this scenario is defined as follows: Due to the fact that manufacturing and warranty assistance share the same warehouse, there will be instances where the manufacturing is at a very advanced stage and simultaneously there will be many vehicles under warranty. From this moment onwards, every decision made prioritizes one of the two activities. Apart from the context described above, the study takes place during the lifetime distribution of deliverables. This means that, historical data regarding costs, faulty items, etc., are available for the research. In reference to the faulty items, it has been used a classification tree with several levels following a hierarchical structure based mainly on their functionality, and reaching a sufficient level of detail in terms of procurement aspects. In figures, the described scenario and the delivery schedule are shown in Table 2.1. Our case study will be developed considering also the following hypothesis:

- Every vehicle has the same reliability (they have the same failure probability).
- The warranty cost is constant with the time.
- The warranty time does not stop at any moment.

The EAC for warranty depends on company policy. Usually, the budget for warranty is determined as a percentage of the total project cost. In our case study, the total cost, manufacturing plus indirect costs for each vehicle, is supposed, amounts to ca. 375,000.00 € and the percentage for warranty attendance will be 2 % of the budget for total costs. That yields around 2,625,000.00 € for the attendance of warranties during the whole project.

2.3 Analysis, Development, and Results of the Case Study

In the literature review of the previous chapter, many case studies can be found which are related to the data analysis taken from warranty assistance (qualitative and quantitative date). This analysis is based on real data and offers a substantial amount of detail. Two references that offer a complete revision of this topic are [8] and [9].

Table 2.1 Data of the described scenario

Date	Accumulated amount of vehicles
March 2008	Rollout
April 2009	45 units
April 2010	100 units
April 2011	150 units
April 2012	200 units
April 2013	260 units
April 2014	315 units
April 2015	350 units

Total amount of customized vehicles to be delivered: 350 units
Warranty period for each vehicle: 2 years
Warranty expiration for last vehicle: March 2017
Time point of the case study (t_1): April 2011 (150 unit already delivered)

2.3.1 Costs Analysis of the Warranty Assistance

As mentioned, the study takes place when the company has already delivered a number of 150 vehicles. In this time, there are 105 vehicles under warranty. Some preliminary data are shown in Table 2.2. Together with this, there is also a sample regarding the number of vehicles under warranty according to the defined delivery schedule. Some figures here have been rounded off in order to simplify their use during the study.

Table 2.2 (as commented) is only a sample extracted from the complete delivery schedule. From this complete schedule, it is possible to note that the warranty expiration date of the first vehicles is obviously on March 2010 and, also, that the most critical moment (t_2) will be in September 2013. The graphic in Fig. 2.3 can help to illustrate it.

In April 2014 (t_2), the already delivered fleet—315 units—will have a maximum number of vehicles simultaneously under warranty—115 units—(see continuous graphic line). In this moment, we can observe how close the due date of the deliveries is (April 2015). Consequently, much closer (and critical) is the manufacturing of the last vehicles. In t_2, our teams of maintenance/warranty technicians will have to attend a high number of vehicles which will demand a huge number of spare parts. At the same time, the operators of the assembly line will request pieces for the production of the last vehicles. The shared warehouse will have in storage enough pieces for manufacturing, but not for more, so the supply of any spare parts demanded by the after-sales personnel must be carried out taking into consideration the importance of the material, the time to repair the disassembled piece, and/or the time to restore it by purchase. Every piece in the classification tree (see Fig. 2.4) belonging to the lowest level (level where materials can be procured) will have a weight (or critical) which changes with time. Every piece will be considered much more critical as the end of the manufacturing process gets closer. Therefore, and taking also into account a costs analysis, it is

Table 2.2 Extract of the delivery schedule

Date	Accumulated amount of vehicles	Monthly delivery (units)	Vehicles in warranty (units)
March 2008	Rollout	5	5
April 2009	45 unit	3	45
April 2010	100 unit	4	91
April 2011	150 unit	4	105
April 2012	200 unit	4	100
April 2013	260 unit	5	110
April 2014	315 unit	4	115
April 2015	350 unit	2	90
April 2016	350 unit	0	35

Number of delivered vehicles t_1: $V_1 = 150$ units
Number of warranty claims in t_1: $R_1 = 1{,}200$ warranty claims
Warranty incurred cost in t_1: $C_1 = 1{,}000{,}000.00$ €
EAC for Warranty: $\text{EAC}_w = 2{,}625{,}000.00$ €
Number of vehicles to be delivered: $V(t) =$ (according to delivery schedule)

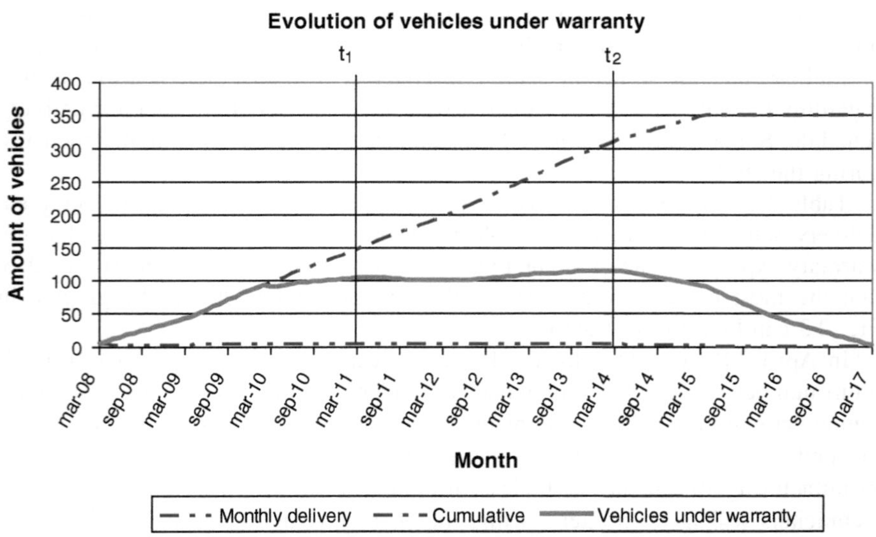

Fig. 2.3 Warranty evolution graphic, in terms of deliveries

necessary to bear the investment of a minimum strategical stock in mind in order not to leave warranty claims unattended.

Considering the data indicated above, it is possible to carry out a simple costs analysis obtaining some average values. Calculation of some values:

- Warranty cost per complaints:
  $C_C = C_1/R_1 = 1{,}000{,}000.00/1{,}200 = 833.33$ €

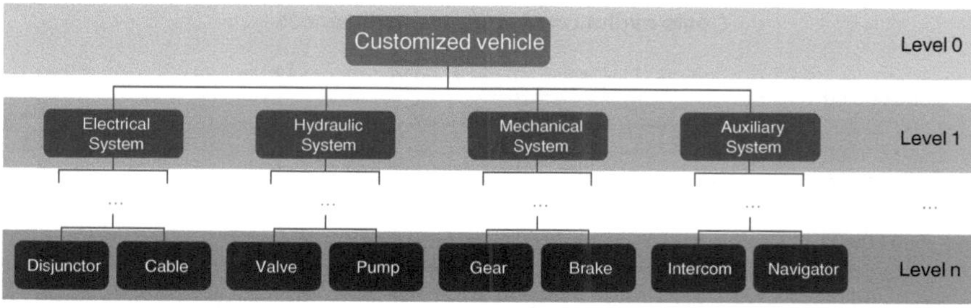

Fig. 2.4 Classification tree of components

- Warranty cost per vehicle:
  $C_V = C_1/V_1 = 1{,}000{,}000.00/150 = 6{,}666.67$ €
- Complaints per vehicle:
  $R_V = R_1/V_1 = 1{,}200/150 = 8$ complaints

With these values, and in order to illustrate them more clearly, a graphic (Fig. 2.5) is included in the warranty evolution in terms of costs. That is because the total incurred warranty cost of every vehicle has been considered in a conservative way. That means they have been treated as already incurred costs when each vehicle is delivered to the customer. Therefore, the accumulate warranty cost does not increase after the delivery of the last vehicle. In further studies, it will be possible to consider as well several destinations of the vehicles, where a maximum number of local vehicles can happen in different moments of the defined lifetime and costs must include the movement of warranty teams to different locations.

Comparing the above results with the foreseen costs indicated in the EAC, a graphic is obtained as the one set out in Fig. 2.5. The EAC is formed by a first part already known, which refers to the Incurred to Date (ITD), plus a second foreseen part (continuous gray line), which refers to the Estimate to Completion (ETC). Apart from the EAC line, the warranty cost line obtained from the cost average in t_1 is also here implemented (discontinuous black line). As a result from this graphic comparison, one can see that the cost at the end (ca. 2,335,000.00 €) is slightly lower than the budget considered at the beginning of the project (2,625,000.00 €). This means, there is a budgetary buffer of ca. 290,000.00 €, which can be used for the investment of a strategic stock of spare parts. This amount would correspond to the budget for attending the warranty of around 43 vehicles, or equivalent as if the warranty assistance should be taking the advantage of ca. 15 months before the end of the manufacturing process. Other interesting average values that can be obtained from this exercise are for example the estimated total amount of warranty claims, which shall be around 2,800 reclamations. Anyway, and as the main conclusion of this analysis, the procurement of these strategic spare parts should avoid the use of the stock shared with the assembly line, offering this way an appropriate service to the client. That is due to the possibility of assisting warranties independently of the manufacturing department and consequently, not affecting the final goal of the project.

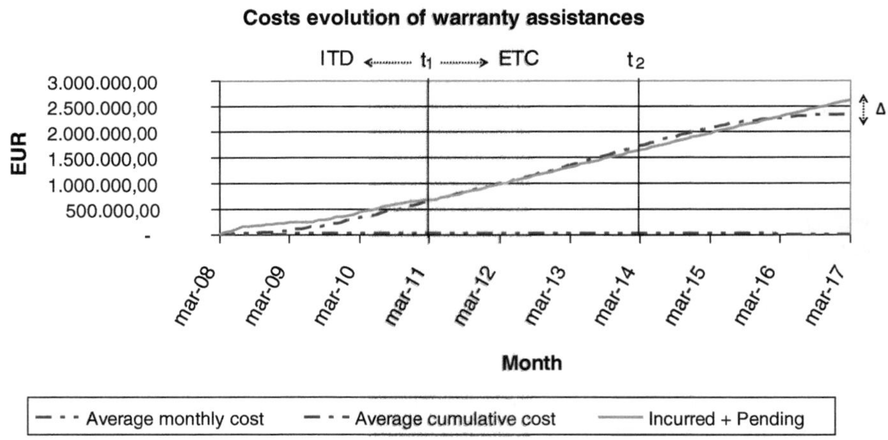

Fig. 2.5 EAC versus average cost line

2.3.2 Quantitative Analysis of the Claims

It has been possible to compile data for a large variety of items with customer complaints. These items are classified according to their functionality and also divided into components that can be procured (see Fig. 2.4: Classification tree of components). Fig. 2.6 exposes a sample of the gathered data as an example for this case study.

This kind of analysis usually helps not only the quality department, but also the manufacturing process, in order to focus on the components which have more incidents during the warranty period. By improving the manufacturing process or taking special care during the assembly of components, it is possible to reduce complaints regarding a specific item. Due to the large amount of components in complex systems such as an industrial customized vehicle, it is suggested to select the items to make all the gathered information easier to manipulate. The set of criteria to select a group of items is not exclusively related to the number of failure occurrences. The cost of such components is also important, as well as the delivery time to procure them, etc. In general, it is important to know how critical each component is for the company and for the fulfillment of the production line. All these features will be conditions to bear in mind when the time to make a decision comes. In other words, these features will be turned into factors which will give a specific weight to each component. This weight will finally help the manager to make the proper decision. By taking this into account, and referring back to the previous figures and data included in the graphic, it is possible to transform the data in terms of relative frequency. This relative frequency refers to the number (n_i) of times that an event (i) takes place (in our case, failures) and divided per the total number of events (Σn_i). Considering therefore statistical concepts (and together with other factors) is possible further on to weigh, as mentioned, the value

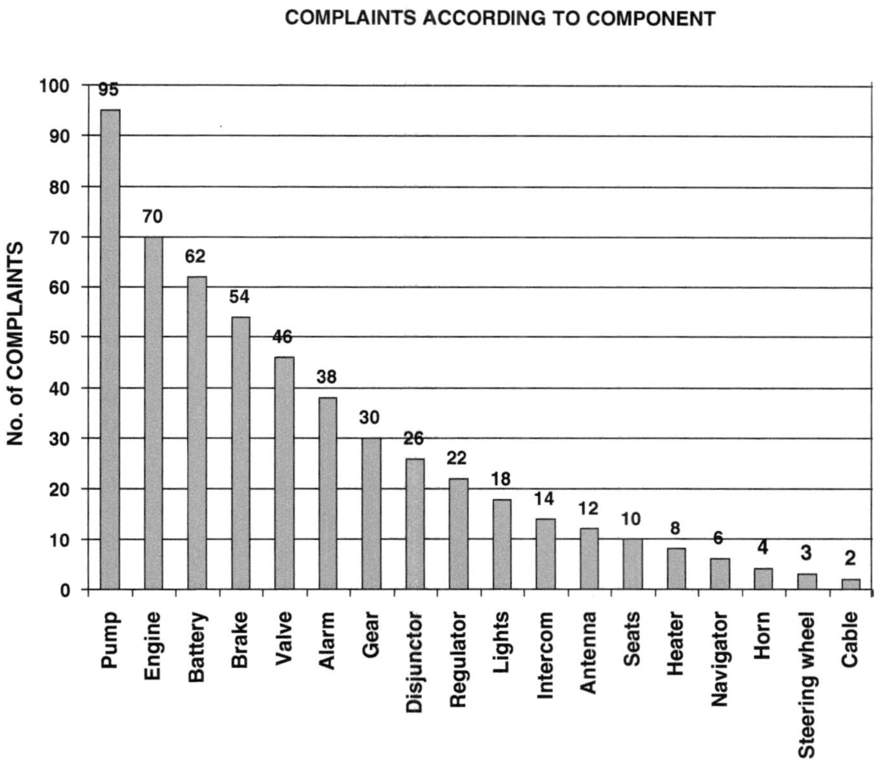

Fig. 2.6 Number of complaints per component

of each component in order to prioritize between the supplies for warranty assistance or to keep the available piece for the manufacturing process (Table 2.3).

A very interesting reference regarding analysis of data statistics is [10], where data analysis is used for the improvement of TQM. The rest of the components are basically not considered because:

- They have been affected by a very small number of failures.
- They have been delivered fast enough and mostly in time.
- There is extra stock in warehouse due to the purchasing of a minimum amount of pieces, greater than the real necessity.
- Or they are not, definitively, under the interest of the project managers' point of view, due to other reasons.

Summarizing, with the tasks previously explained in order to obtain a set of chosen components (those acknowledged as critical), what we are really composing is a list of strategical spare parts. This means that, in case the company approves the use of the budgetary buffer that supports of the warranty service, the purchasing process can be quickly launched. All these actions will finally lead the company to positive returns:

Table 2.3 Relative frequencies

Component	Number of claims	$f_i = n_i / \sum n_i$
Pump	95	0.1827
Engine	70	0.1346
Battery	62	0.1192
Brakes	54	0.1038
Valve	46	0.0885
Alarm	38	0.0731
Gear	30	0.0577
Disjunctor	26	0.05
Regulator	22	0.0423
Lights	18	0.0346
Intercom	14	0.0269
Antenna	12	0.0231
Seats	10	0.0192
Heater	8	0.0154
Navigator	6	0.0115
Horn	4	0.0077
Steering wheel	3	0.0058
Cable	2	0.0038

- By reducing the probability of paying penalties due to global delay in the project delivery.
- By improving the confidence of the client due to the completion of contractual terms, such as the warranty assistance.

It is necessary to remark that everyone of the failures referred so far are incidences considered to be under warranty. For further research in this field, incidences not considered under warranty are also included. The analysis of such events must take into account the reasons why these situations occur (bad training of the user?, poor information for maintenance?, clients accustomed to other family product with different behavior?, etc.). Anyway, in each case and even when the failure is not attributed to the manufacturer, the company must take interest in the possible causes.

2.3.3 Spare Parts Management for Warranty Assistance

Regarding logistics of warranty services, Ref. [11] discusses diverse questions in this context. The change in the utilization of pieces from assembly line to warranty assistance has a negative effect in terms of cost on the whole project. The extra costs associated with the spare parts are of course due to the different prices between the acquisition of a piece at the beginning of the project for the whole fleet, and the acquisition of a piece punctually during the lifetime of the project and for a specific incident. Therefore, management must provide compensation between the difference of values in order not to keep such increments in the total costs of manufacturing, but to incur in the total warranty costs.

Consequently, the percentage incremented in the final price of purchase can also be a factor to bear in mind when estimating the weight for each component. In order to ascertain a correct warranty support service, the proposed action is basically to acquire various reserves that would allow repairs without delays in the vehicle manufacturing and simultaneously during the process and to supply spares to the warranty service from the assembly line in a reasonable way. According to the aforementioned considerations, and together with the data collected, the experience of warranty/maintenance technicians, the knowledge of the engineering department, and of course using the already developed techniques in maintenance (Fig. 2.7), it is possible not only to elaborate a spare parts purchase plan for warranty, but also to improve the business decision-making process as well as to contribute with improvement actions for engineering and manufacturing. This purchase plan refers to an applicable list of essential parts to assure the satisfactory assistance of a large number of reclamations. At the end of the warranty period, the remaining spare parts can be negotiated with the client for their use in further maintenance tasks. This fact forces strict control of all these material parts so that they remain available for supply to the customer at the conclusion of the project. At the same time, this action will represent an opportunity to recuperate, at a determined point in the future, one part of the incurred cost. In general, decision-making will be the result of a process focused on a final choice among several alternatives. In our case, in order to lead the company to a fast and adequate decision-making process, every department should clearly identify what needs to be done and what is the scope of their responsibility. For the company in this case, we have adapted the idea of a warranty management system divided into modules, proposing furthermore, certain interactions among different departments within the company, which share the information, make suitable decisions according to their responsibilities, and coordinate activities to a common and profitable goal for the whole company. In order to illustrate such interactions, activities, etc., a workflow has been used (Figs. 2.8, 2.9) following a business process modeling notation (BPMN) methodology as a graphical representation for this specific business process, making it more easily understandable. The considered departments here (including the client) are as follows:

- Logistics department (LD)
- Manufacturing department (MD)
- Administration (MB)
- After-sales department (AD)
- Quality department (QD)
- Purchasing department (PD)
- Engineering department (ED)
- Customer (C).

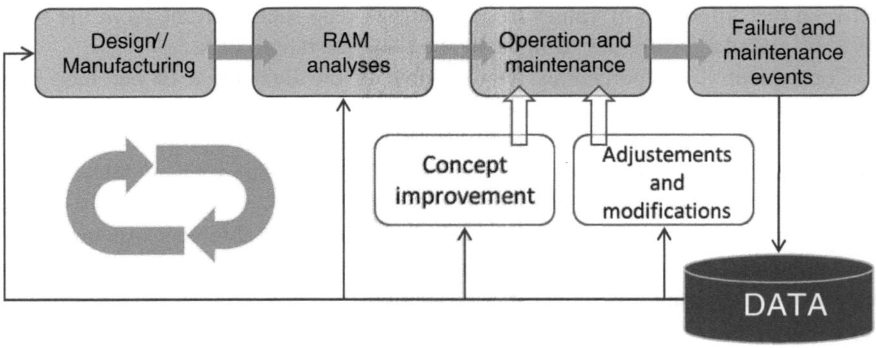

Fig. 2.7 Typical feedback from collected reliability and maintenance data

The process starts when the customer detects a failure in a vehicle and consequently informs the company. Communication can be addressed to different department of the company, but the most appropriate way is to channel them in only one communicator as, for example, the service or after-sales department. Anyway, the after-sales department can also detect failures in the course of its maintenance activities (see Fig. 2.8). In any case, once the information reaches the after-sales department, the facilitated information is analyzed. If the incidence is considered not to be object of repair under warranty (for example, when the cause of the failure has been a wrong or bad utilization), the after-sales department informs the management board who finally decides if, in spite of this, the incidence is repaired as warranty. If the incidence is discarded as warranty repair, the management board should also inform the customer. The customer can of course disagree with such a consideration. Therefore, a list of interventions (those not considered firstly as warranty) must be negotiated between the parts. If the incidence is considered under warranty conditions, the after-sales department must carry out a diagnosis of the incidence, detecting the problem, analyzing its solution, and determining the resources (staff and materials) and the necessary time for repair. In reference to the material, the warranty technicians must identify between the repairable and the non-reparable/consumable materials. In reference to the materials, warranty technicians must identify those repairable from those that are not or material for sale. The necessities in general are communicated to the company management who addresses the actions to the corresponding department (logistics, manufacturing, and/or purchasing department), in order to finally facilitate the material to the after-sales department. This is the moment when the company administration must make the most important decisions in terms of costs and manufacturing prevision. Once the after-sales department has the material (either by a loan from warehouse, by a loan by cannibalization, or acquisition by purchasing), they inform the management board (and afterward the client) (see Fig. 2.9) the subsequent action plans. The damaged material is sent to the company where the quality department (together in some cases with the engineering department) analyzes the failure. If the repair has been by replacement and the

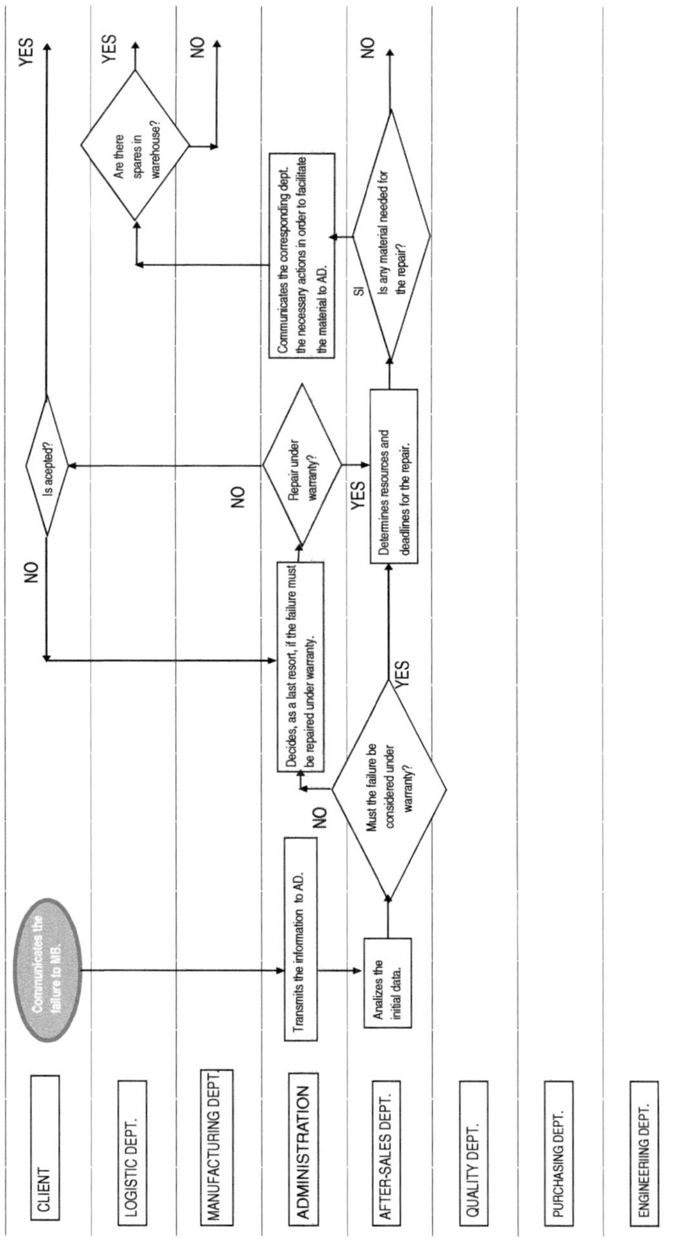

Fig. 2.8 Workflow of the warranty management process (part 1 of 2)

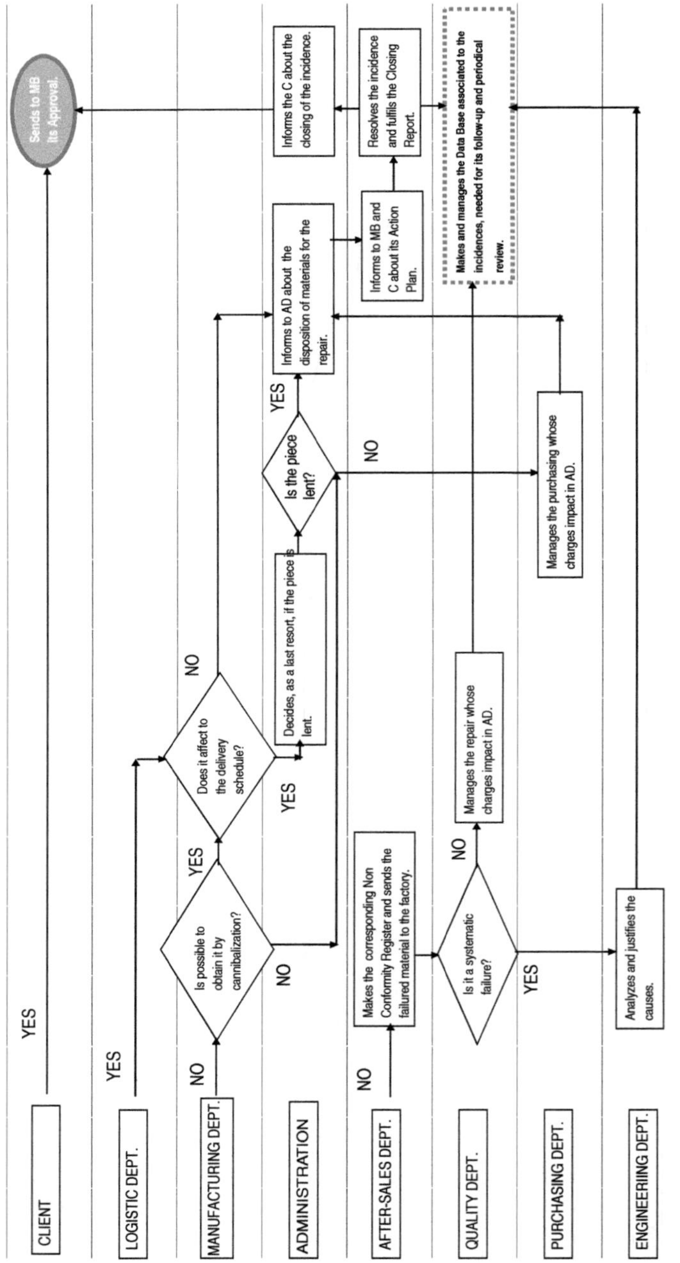

Fig. 2.9 Workflow of the warranty management process (part 2 of 2)

material is identified as repairable, the quality department manages the repair, taking into account the appropriate certification.

The material, once repaired and certificated, will be stored again in a warehouse for its usage in the assembly line. In this process, every piece of data about the incidence, damaged material, repair, etc., gathered by the after-sales, quality, and engineering departments are introduced in a database which is followed up and reviewed by the quality department. Once the incidence is resolved, the after-sales department communicates the closure of the assistance to the company administration, who transmits this to the client. It is important to receive a document from the customer with the approval of the performed tasks and the acceptance of the service closure. The database associated with these incidences and necessary for their follow-up and should include, not only those incidences considered to be under warranty, but also the data about preventive and corrective maintenance performed on every vehicle, in order to enable the analysis of, for example, repetitive or systematic failures among others studies. The described workflow refers to a particular management of a specific case study. This workflow does not intend to be generic and therefore is susceptible to improvement. It is proposed in Chap. 5 a general management with the aim of, later in Chap. 6, criticizing or evaluating the way management works. Likewise, in continuation to this chapter, it is suggested to consult Appendices 1 and 3, where matters such as quantification of maturity and analytical hierarchy processes are dealt with based on what it has been developed in this chapter.

References

1. Murthy DNP, Blischke WR. 2005. Warranty management and product manufacturing. Springer-Verlag London Limited, pp 302 + xviii. ISBN 1852339330.
2. Blischke WR, Murthy DNP. 2002. Case Studies in Reliability and Maintenance -, Wiley, New York, pp 661 + xxiii.
3. Lyons K, Murthy DNP. 2001. Warranty and manufacturing, integrated optimal modelling. Production Planning, Inventory, Quality and Maintenance. Rahim MA, Ben-Daya M (Eds.). Kluwer Academic Publishers, New York, pp. 287–322.
4. Boyan Dimitrov, Stefanka Chukova, Zohel Khalil. 2004. Warranty Costs: An Age-Dependent Failure/Repair Model. Wiley InterScience, Wiley Periodicals, Inc.
5. Gómez J, Crespo Márquez A, Moreu P, Parra C, González-Prida V. 2009. Outsourcing maintenance in services providers. Safety, Reliability and Risk Analysis: Theory, Methods and Applications – Martorell et al. (eds), pp. 829-837. Taylor & Francis Group, London, ISBN 978-0-415-48513-5.
6. Stefanka Chukova, Yu Hayakawa. 2004. Warranty cost analysis: non-renewing warranty with repair time. John Wiley & Sons, Ltd. Appl. Stochastic Models Bus Ind, 20:59–71.
7. D. Christensen. 1993. Determining an accurate Estimate At Completion. National Contract Management Journal 25:17-25.
8. Lawless JF. 1998. Statistical analysis of product warranty data. International Statistical Review, 66:41-60.
9. Meeker WQ, Escobar LA. 1998, Statistical Methods for Reliability Data. Wiley, New York.

10. Murthy DNP, Solem O, Roren T. 2004. Product Warranty logistics: Issues and challenges. Euro. Jr. Oper. Res., 156:110-126.
11. International Standard. 2004. ISO/DIS 14224/2004. Petroleum, petrochemical and natural gas industries. Collection and exchange of reliability and maintenance data for equipment.

Chapter 3
State of the Art

Warranty is usually defined as the assurance policy applied to all customers where the purchased goods or services are (or shall be) as presented and, if not, they will be replaced or repaired. This assurance is applied during a period of time after a product has been sold, the management of such policy combines technical, administrative, and managerial actions during the warranty [1]. The purpose is to maintain or restore the item to a state in which it can perform the required function, which are considered necessary to provide a given service. Therefore, a structured approach to warranty preparation must take into account that offering warranty involves risk of high warranty servicing cost [2].

Warranty management is considered as all those activities of organization that determine the after-sales objectives: priorities, strategies, and responsibilities, and implements them by control, supervision, and a proper management of the customer complaints. As priorities are understood those targets assigned and accepted by the management of the after-sales department. These objectives may include availability, cost reduction, custom satisfaction, etc. In the same way, strategies are understood those management methods in order to achieve warranty objectives. It is important to indicate that warranty can be observed basically from the customer or client perspective, or from the seller or supplier. The most generic case would be of a business that is in the middle of a supplier chain (see Fig. 3.1), offering a customer after-sales service (managed by the after-sales department) and demanding a warranty service from the suppliers (managed by the operation or purchase department). In many instances, it is found in practice that both managements are interrelated to each other. This means that there are cases where the failure communicated by the customer can be repaired by the supplier within the warranty period. The ideal case would be that of acquired products from suppliers and those sold to customers after a manufacturing process. Those products do not present defects or failures. This ideal situation is represented in Fig. 3.2.

As mentioned, the structured warranty preparation must take into account that by offering warranty, there is an involved risk in the warranty servicing cost [2]. This case is logically viewed from the perspective of a company that provides warranty assistance to its customers. On another respect, there are cases where there is a warranty from the supplier. Businesses to which the purchase of new products

V. González-Prida Díaz and A. Crespo Márquez, *After-sales Service of Engineering Industrial Assets*, DOI: 10.1007/978-3-319-03710-3_3,

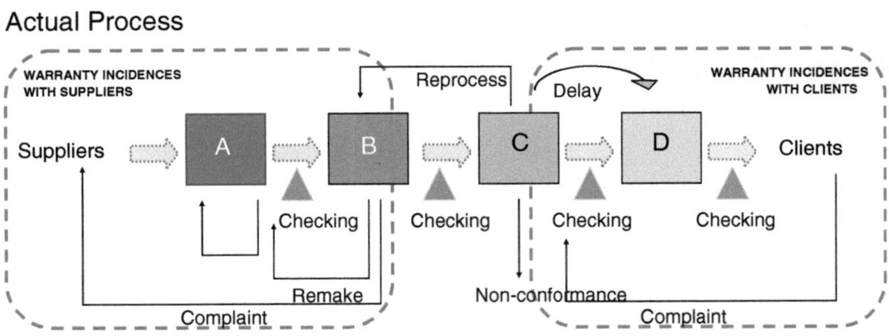

Fig. 3.1 Warranty of a business in a supplier chain

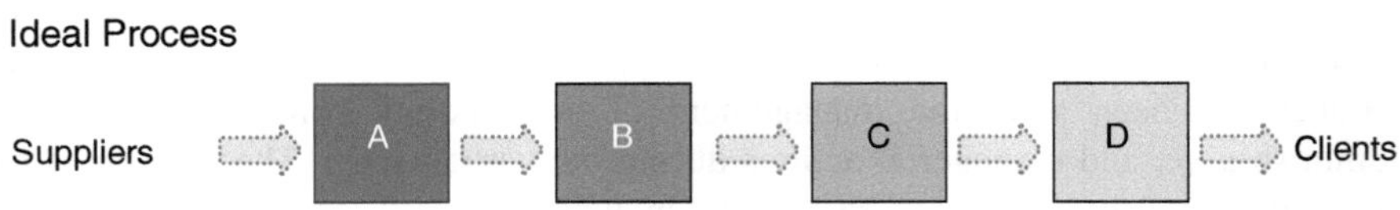

Fig. 3.2 Ideal case without any defects

represents an important element in the total overall operations budget, an efficient management of warranty programs, have a significant impact in the total operations cost [3]. For both of these reasons, it is possible to find methodologies that include the economic aspects of the organization of warranty management (in both directions). For that purpose, different authors here propose models, action frameworks, or systems that allow those businesses in specific cases to manage warranty costs in the best way possible. Although in any case, the authors will also present both strong and weak points. In general, studies of warranty are usually observed through a manufacturer optic that provides after-sales services to its clients. This is in short the context in which this research is developed. The research starts by highlighting the importance of a warranty management system and by describing the methodologies followed to reach the state of the art and that has simply consisted of an intense search and the gathering of models and studies that appear in the current literature, classifying and comparing them. This global search of systems of warranty management will be presented in a chronological order, and later making comments on them in terms of the different criterion. The result of this first part of the research has been integrated in the general conclusions at the end of the document.

3.1 The Importance of a Warranty Management System

Many manufacturing companies spend great amounts of money of their budget on their product and service warranties. However in most of cases, this decision does not receive any more attention. Today, effective warranty management processes

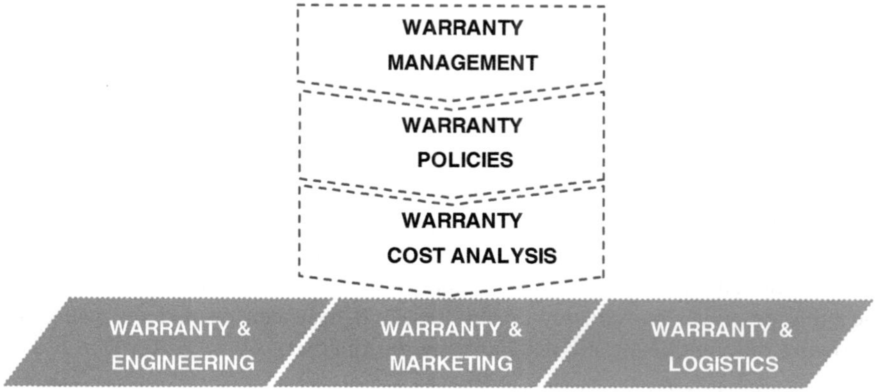

Fig. 3.3 Warranty interactions (adapted from [3])

are not yet decided by most executives. In practice, companies often pay attention to their warranty management processes only when high levels of liability are involved with a claim or recall. As a consequence, they are missing opportunities for significant cost savings, as well as for increasing the business value through better-quality products and higher levels of customer satisfaction.

Warranty management deals with decision-making at strategic and operational levels. This requires taking into account the interactions (see Fig. 3.3) between engineering, marketing, and post-sale support elements of the manufacturing firms [3]. Therefore, warranty management needs to be treated today not as a cost center but as an asset that can create a higher value of the business performances. An effective warranty management requires the integration of leading practices and to find critical points in the decision process. These critical points will depend on the already existing service management of the organization and its warranty management capabilities. These are different actions that will help improve business performance and value creation. Some examples are as follows:

- Reducing the time for claims handling;
- Reducing costs by introducing more automated processes;
- Reducing customer loss due to poor product quality;
- Improved supplier selection and collaboration, cost sharing and quality programs.

The result is lower warranty costs, better products, and higher levels of customer retention. It is known that warranty management requires attention not only to warranty processes, but also to better strategies, more effective tools and integrated systems, as well as better organizational structures.

3.2 Literature Review

Up to this point, the provision of a global vision is demanded including, as much as possible, articles, essays, and important works dated from the initial commencements of this subject up to the latest known sources. For this reason, it is necessary to point out those articles and revision works that had previously dealt with this field, such as [4] and [5], where approximately 1,500 classified references in different categories with very detailed documents can be found. Besides the aforementioned and in order to find studies related to the warranty cost modeling, the bibliographical search used in this research has been carried out using the following different electronic databases on the Internet:

- ABI/INFORM Global—ProQuest
- Blackwell Synergy
- Business Source Premier—EBSCOhost
- Compendex (Engineering Village)—Elsevier Ingeniería de la Información
- Current Contents Connect—ISI
- ISI Web of Knowledge—ISI
- NTIS—Ovidio (SilverPlatter)
- Scirus
- Scopus—Elsevier
- Springer Link
- Wiley Interscience.

Up to now, the management of warranty has received very little attention, at least as consumer products are concerned [3]. Nevertheless, searching concepts as "warranty," "warranty management model," it is possible to find many related studies treated from different points of view. Some of these studies are regarding the following issues in Table 3.1, listed chronologically.

This summary chart tries to list certain key articles, relevant for the focus of this research, providing a global vision of warranty management. Many other articles could have been considered due to the assigned key words in different magazines and journals. However, they had to be discarded because of their different scope and final objective. Basically, the criteria to the above-mentioned articles were as follows:

1. The article proposed a warranty cost management model.
2. The article was published in scientific journals.
3. The article presented not only a review or an application, but also a model proposal.
4. The model in the article had preferably a graphical/mathematical representation or a software application.

Table 3.1 Summary chart of models and authors in chronological order

	Year	Author	Scenario	Reference
Case 1	1976	Henry N. Amato, Evan L. Anderson, and David W. Harvey	Modeling of general warranty in the future	[6]
Case 2	1989	D.N.P. Murthy	Warranty costing with an usage-dependent model	[7]
Case 3	1993	Amitava Mitraq and Jayprakash G. Patankar	Warranty cost estimation and production with an integrated multicriteria model	[8]
Case 4	1999	Xuemei Zhang and Hoang Pham	Model taken into account warranty cost, error, removal times, and risk costs	[9]
Case 5	2001	Chih-Hsiung Wang and Shey-Huei Sheu	Warranty effects on an imperfect model about economic manufacturing quality	[10]
Case 6	2001	Shau-Shiang, Kulkarni, A. Mitra, and Jayprakash G. Patankar	Warranty policy with time-dependent costs for renewable minimal repair	[11]
Case 7	2003	Sandeep Mondal, Surajit Pal, and D. K. Manna	An application of cost estimation considering a renewing warranty policy	[12]
Case 8	2004	Stefanka Chukova and Yu Hayakawaz	Non-renewing warranty with repair time regarding the warranty cost analysis	[13]
Case 9	2004	Boyan Dimitrov, Stefanka Chukova, Zohel Khalil	An age-dependent failure/repair model about warranty costs	[14]
Case 10	2005	Yu Hayakawa, Stefanka Chukova	Quasi-renewal inter-repair times related to warranty cost analysis	[15]
Case 11	2006	Hongzhou Wang	Warranty cost models considering imperfect repair and preventive maintenance	[16]
Case 12	2006	Shaomin Wu and Min Xie	Analysis for non-repairable services products regarding warranty cost	[17]
Case 13	2007	D. R. P. Williams	Warranty cost model related to software reliability	[18]

In short, the mentioned articles represent a simple way the current state of the art related to warranty cost management. Nevertheless, besides these contributions, there are also other papers, books, and essays found in the bibliographical search, which deal also with the same subject and that bring also other points of view of the object of study. Investigators of diverse disciplines have analyzed different aspects of warranty that are also relevant for an efficient management. More details in this respect can be found in [19].

3.3 Models and Support Tools for Warranty Cost Management

A key aspect of the strategic warranty management is that decisions with regard to warranty must begin at a very early stage in the product life cycle and not as an after thought just prior to the launch stage [3]. Other usual problems during the application of warranty services are for instance the following ones:

- Information systems support often limited.
- Long cycles of time for claims review.
- Excessive number of invalid claims.
- Not to distinguish warranty responsibility.
- Warranty data not used to improve product quality.
- Additional claims due to slow corrective actions in manufacturing, engineering, or product design.
- Others.

All these issues are negative circumstances that a proper warranty management model should avoid. And effective management of product reliability must take into account the link between warranty and reliability [20] (see Fig. 3.4). Interesting contributions as the one above mentioned can be found in the literature regarding this topic. The authors of these contributions usually try to identify the processes, the actions, stages, tools, methods, or necessary support techniques to manage properly the warranty costs. Basically, a well-established model on warranty management will help achieve a successful goal in the performance of the company warranty service. Hence, by reengineering of management processes and the application of a correct warranty cost model, it is possible to:

- Improve sales of extended warranties and additional related products (maintenance contract, spare part supply, technical assistance).
- Improve the quality by improving the information flow about product defects and their sources.
- Improve relationships with the customer.
- Reduce expenses related to warranty claims and processing.
- Better management and control over the warranty costs.
- Reduce invalid-related expenses and other warranty costs.

Fig. 3.4 Link between warranty and reliability (adapted from [20])

In all previous cases, a difference in warranty organization should be considered when the product is still pending manufacture, and when the product is already in the market. In both cases, for an effective warranty management, it is critical to collect the proper data and to exchange the different types of information between the modules (see Fig. 3.5) in which the management system can be divided [21].

3.4 Main Contributions of the Selected Models and Authors

As it has been mentioned before, the selected models and authors can be summarized in the following chronological chart (see Table 3.1). Up to now, these contributions have been commented in a very succinct fashion. However, the study is going to be focused on particular interactions between warranty and other disciplines, and how they are dealt with by the different models and authors. Other important aspects for future researches will be the view on warranty from different points of view, such as the manufacturer, the customer, and society perspective.

3.4.1 Warranty and Maintenance

Firms which offer warranties frequently require that certain guidelines be observed in the use of maintenance of the product [6]. A guideline for determining warranty issues versus maintenance issues is that information which can assist to define if failure should be considered to be warranty, maintenance, or neither. Warranty usually includes (but it is not limited to):

- Defects in materials or workmanship, and
- Replacement of parts with valid manufacturing failures during warranty coverage.

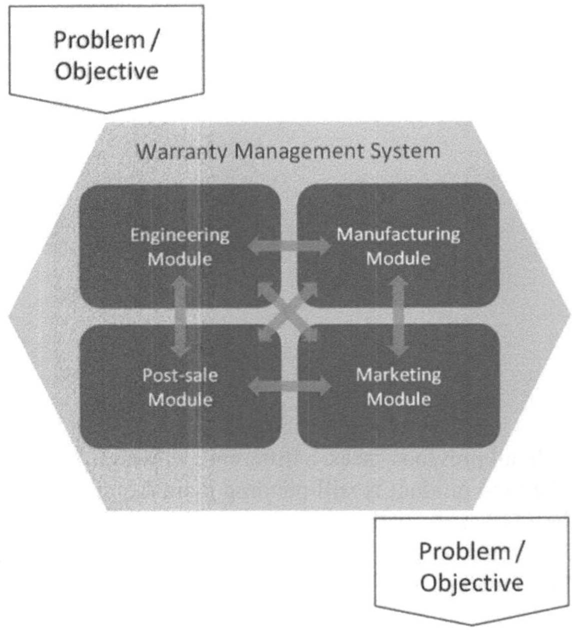

Fig. 3.5 Warranty management system in four modules (adapted from [21])

On the other hand, the definition of maintenance [22] includes all those technical, administrative, and management activities to preserve or restore equipment to a state in which it can perform the function it is required to. Some management activities could be considered under warranty issues. In any case, maintenance requires us to identify the objectives of maintenance. This, as it will be shown further, is in order to complete its functions according to the characteristics of those service provider companies [23]. In the literature review, there are models which consider warranty and maintenance together [14]. In any case, maintenance usually includes (but it is not limited to):

- Scheduled maintenance and inspections.
- Adjustments, lubrication, replacements, etc., after determined time of use.

While typical components of direct costs are direct material and direct labor, maintenance together with insurance, taxes, and depreciation is usually considered indirect costs [8]. In order to integrate maintenance service with other information systems, as for instance could be information about warranty, there are also computerized management systems which provide efficient methods of integration and support all maintenance, combining management and anticipated warranty services [24]. In spite of all this, there are issues not covered by warranty or maintenance (but that are not limited to):

- Missing parts.
- Fuel/battery servicing (if needed).

- Upgrading of software (if needed).
- Environmentally induced corrosion.
- Parts changed for customer convenience.
- Storage.
- Any issue related to accidents, negligence, etc.

Regarding the effects of warranty cost on an imperfect model that can be applied to economic manufacturing quality, if a manufacturing process is out of control, it stays there unless maintenance inspection actions are applied [10]. Preventive maintenance ensures that a process is in control and in the as-good-as-new state at the beginning of the next production run. This, as it will be shown further, affects to the warranty in the sense that a process under control will produce articles with an expected quality, reducing consequently the cost of its warranty. Therefore, it is essential a good maintenance of the productive process. Analyzing non-repairable services products in reference to warranty cost, there is a situation in which the authors analyze replacement cost according to different warranty policies [17]. That is, the author considers renewing or non-renewing free replacement warranty policies; applying preventive maintenance or only replacement upon failure is carried out. There is also a scenario about software reliability and its warranty cost [18]. In this specific case, it is considered that the life cycle of a piece of software shall include a series of production activities, namely design, coding, testing, and operation/maintenance phases. In addition to these cases, there are other authors mentioned in the literature review who deal with warranty and maintenance. For instance, in the scenario where failure or repair is age dependent [14], the evaluation of the warranty cost depends on the failure process, the degree of repairs, and the prescribed maintenance of the item. The expected warranty costs associated with the product sold under warranty costs depend on the warranty contract and the maintenance schedule associated with the product [14]. Most technical items have some prescribed maintenance schedule such that at certain moment of time, preventive inspection must be applied. In this case, the model proposed by the authors is appropriate in many practical situations for calculating costs associated with service and maintenance policies of repairable products. While in most models for warranties and for preventive and corrective maintenance, the repairs assumed to be instantaneous in the case of non-renewing warranty with repair time [15], the main contribution is allowing nonzero repair time for the warranty claims and assigning costs, which are dependent on the length of the repair. Additionally, the evaluation of the warranty cost or any other parameter of interest in modeling warranties depends on the failure and repair processes as well as on the assigned preventive warranty maintenance for the items [15]. The type of repair which takes place depends on the warranty reserves, related costs, assigned warranty maintenance, reliability and safety requirements of the product regarding the warranty cost analysis. When the scenario is taking into account imperfect repair and preventive maintenance [16], warranty policies with preventive maintenance could be a good candidate since they provide extra protection to consumers at a relatively low cost. Many maintenance and warranty models using renewal theory were actually based on this assumption, i.e., "as good as new" after each maintenance. "As good as new" represents one extreme type of repair result and usually a system may not be good as new, but updated to an even later/improved version after maintenance; i.e., repair is imperfect. Optimal maintenance actions for one component depend on the states of the other components. Obviously, the warranty cost policy proposed in this case incorporates opportunistic maintenance and characterizes the classes of possible maintenance actions during the entire warranty period. In summary, in this section, it will be possible to observe two types of analysis toward warranty and maintenance: One focused on maintenance and its impact in the warranty (free anticipated maintenance provided by the manufacturer) in respect to maintenance given after warranty; and second type focused on the role of maintenance (mainly preventive maintenance) in the warranty service. Other relevant references to this respect are the following [25, 26] and [27].

3.4.2 Warranty and Outsourcing

Outsourcing has become a strong complement to the resources of a company that wants business development and expansion. The topics of warranty and subcontracting are currently in active research, where subcontracting appears to be linked to the theory of agency. An important reference to this respect [28]. As definition of outsourcing everyone understands that it is when a company purchases a product or process from an outside supplier rather than producing it in house. In other words, outsourcing is like subcontracting and, consequently, it has advantages and disadvantages (see Fig. 3.6). This fact has caused to numerous organizations the re-evaluation of procedures when engaging outsourcing activities. The outsourcing of technical tasks does not guarantee successful completion nor ensure that the best interests of the company will be accomplished. Nevertheless, within the advantages, it is important to mention that (among others features) outsourcing provides legal insurance for services [23].

In addition to this and regarding the warranty outsourcing (as in maintenance), it is suggested to compose levels of externalization progressively in time, increasing internal knowledge and control about activities before recruiting a new one, that is, to carry out a partial outsourcing. Another advantage by using an independent third-party warranty servicing company is that the main organization can focus on new and existing projects rather than spending a disproportionate amount of time and effort on warranty works. In any case, in order to implement strategies (as a possible externalization), one needs an effective warranty management system that deals with aspects such as data collection, model building and analysis, and tools for effective decision-making [20]. A computer-based management system to assist the decision-making process must contain the following three subsystems:

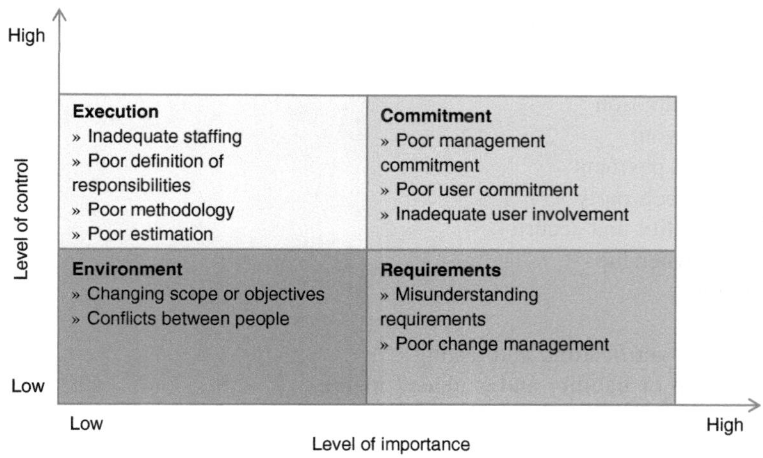

Fig. 3.6 Risk factors of outsourcing

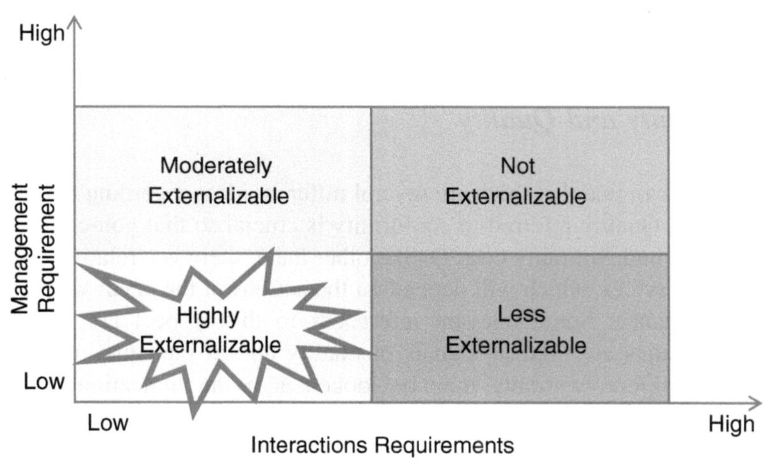

Fig. 3.7 Decision of outsourcing

1. Databases;
2. Prototype models;
3. Packages for modeling, analysis, and optimization.

When times are good, companies are maybe not too concerned taking decisions and tend to spend money at deficiencies. However, during difficult economic times, companies should evaluate every cost. In order to compare costs between in-house warranty service and outsourcing to a third party, a calculation is needed of what is being spent on personnel time and expenses for trade scheduling time, traveling expenses, taxes and insurance, non-warranty items repaired or replaces among others (see Fig. 3.7).

A contract model [29] between the outsourcer and the outsourced companies can be as follows:

1. Contract definition
2. Contract layout
3. Pricing and payment
4. Duties of each party
5. Confidentiality and security
6. Product ownership
7. Warranties
8. Penalties
9. Validation, verification, and testing
10. Limitation of liability and excluded matters
11. Termination provisions.

Nowadays, companies are expanding into markets outside their usual region. By standardizing their warranty and customer management process, aggressive companies can make use of outsourcing just to extend to major markets worldwide.

3.4.3 Warranty and Quality

Nowadays, we can find that there are several different ideas regarding the meaning of quality [30]. Quality referred to conformity is crucial so that non-conformative items present a high warranty cost. On the other hand, there is a relative quality in the technical services, which will depend on the actions of the agent which carries out such assistance. Some relevant references to this respect [19, 31–34]. In general, warranties are used as signals of quality and as elements of marketing strategies. Therefore, warranty may be looked at as an indication of quality. Looking at those companies that are leading the way in customer satisfaction, making and selling quality products, it is possible to see that they are the ones making aggressive changes to post-sales services, starting with their warranty programs. Often, manufacturers use warranty as a means to provide information about the quality of the products and it represents an important tool to influence the market [7]. At the same time, warranty is imposed on the manufacturer through legal processes so that customers' interests are defended and protected. When organizations decide to increase their competitive level, they often decide to enhance the service quality as a differentiator (see Fig. 3.8). Although there are nowadays more suppliers to choose in any field, customers are also more informed about product and service quality. They expect an attentive service. Organizations can only provide attentive, trustworthy, and resourceful service if they have an internal customer culture. If employees are discouraged, they will not be focused on delivering high levels of service quality. It is a well-know fact that consumers

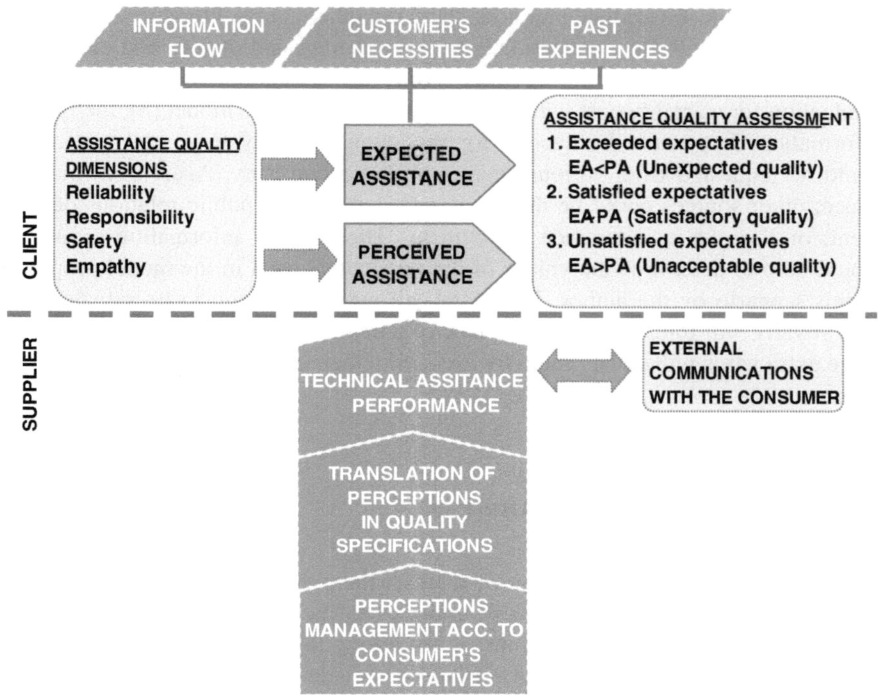

Fig. 3.8 Perception of technical assistance quality

have more knowledge of quality issues. Otherwise, reasons closely related to quality and the uses of warranties by the sellers are as follows:

- Many consumers have neither the time nor the inclination to deal with products' failures or repairs.
- Due to the increasing complexity of the products, consumers are often unable to judge quality before buying a product.

As it has been mentioned before, warranties are used as signals of quality and as elements of marketing strategies. Some other authors [14] focus warranty and quality in the way that warranty provides indirect information about the quality of the products and represents an important tool to influence the market. The use of warranty is widespread and serves many purposes, including protection for manufacturers, sellers, insurance, buyers, and users.

In the same way as above mentioned, warranties also serve as indicators to inform customers of the product quality and reliability [17] that could give the company a competitive advantage. Regarding the situation about software reliability and its warranty cost [18], software reliability is one of the most important characteristics of software quality, being its measurement during the software life cycle essential for producing and maintaining quality/reliable software systems.

Apart from the just mentioned authors, there are others in the literature review that have also dealt with the connection between warranty and quality and even gone into this aspect in depth. For instance, from the case about the estimation and production for warranty cost with an integrated multicriteria model [8], the type of information that would be needed for modeling warranty in a real application could be requested from various sources within a company. Apart from maintenance, other sources could be the marketing department, public relations department, or the quality assurance department. The type of information requested would help to determine the values of the parameters used in the model proposed in this scenario by the author. In general, all customers want to be satisfied and their loyalty will remain if there is a lack of a better alternative. Giving customers some extra value will satisfy them by exceeding their expectations and insure their return.

3.4.4 Warranty and Cost Analysis

Just as it happens with the notion of quality, there are also various conceptions regarding warranty costs [35]. These notions can include, for example, the warranty cost of a sold product, warranty costs during the products' life cycle or the warranty costs for a time period (for those sales made in a determined time). Once again, there are costs for the manufacture and the consumers and that are widely discussed and detailed in [31]. Regarding more recent works related to warranties, it is recommended to refer to [36]. In general and in respect of warranty cost modeling for a future period of time [6], the author of the same reference mentions that warranty increases manufacturer responsibility before the consumer, besides the obligation to meet certain specific quality standards or performance that is expected of the product. In a situation like the one outlined in [9], the model takes into consideration warranty costs, time of replacement, and risk costs. In this context, the quality of the system usually depends on the quality of the testing methodologies used and the duration of their application.

This situation is applied to software, but can be also considered in other scenarios. In this particular case, when the test takes more time, the more reliable the software can be expected to be; nevertheless, the total cost of the software development will also increase. In the scenario when it is considered a warranty policy with time-dependent costs for non-renewable minimal repair [11], most of products provide some sort of assurance to the consumer regarding the quality of the product sold, and this assurance is in the form of a warranty. With the philosophy of improving quality adopted by manufacturers, that impacts on the product in the development on warranty costs. Contrary and in Ref. [13] to the warranty cost analysis in the case about non-renewing warranty with repair (Case 8), authors consider here that the improvement of the reliability and quality of the product will highly reduce the expected warranty cost. It is easy to see that when the penalty cost is high, low warranty cost is needed; the best strategy is to decrease the average repair time by improving the quality and reliability of the product. Customer assistance quality costs (C_c) can be expressed as follows:

$$C_c = C_f + C_d + C_p + C_o \quad (1)$$

where the terms of (1) correspond to:

- C_f: Failure Costs (customer claims, warranty charges, responsibility insurance, legal trial, or losses due to repetitive assistance, reprocesses, labor, and materials);
- C_d: Detection Costs (process controls, inspections, revisions and supervisions, or customer remarks); and finally
- C_p: Prevention Costs (quality planning, training programs, quality audits, data analysis and acquisition, preventive maintenance, or supplier assessment).
- C_o: Other costs.

As previously mentioned in other parts of this research, because of the difficulty of providing future costs, often companies consider a percentage of the total value of the product a business decision to be able to fulfill warranty. The challenge here lies in the estimation of such warranties in order to be feasible as much as possible to adjust reserved budget to attend to customer claims. This estimation, as we will see further on, is evidently related to product reliability that is why the calculation of this estimation will be fundamental to obtain a good provision of warranty costs. This estimation of warranty costs together with the anticipated sales of a determined product make it possible for an organization to determine and manage its customer service properly without unnecessary excess or deficiencies in order to be better prepared to face future claims. Regarding the effects of warranty cost on an imperfect model that can be applied to economic manufacturing quality [10], it is there indicated that traditionally the lot size for manufacturing a product was economically determined based on the assumption that a production process always produces items of acceptable quality. Product quality, however, is not always perfect, and it is usually a function of the production process. Therefore, it is not desirable to isolate the problem of economic manufacturing quality (EMQ) from the state of a production process. When the production process is in the "in-control" state, items produced are of high or perfect quality [10]. On the other hand, items produced contain defects or are of substandard quality when the process is out of control. It is assumed that at the beginning of the production cycle, the production process is "in-control," producing items of acceptable quality. In a situation about imperfect repair together with preventive maintenance [16], warranty cost could be a significant portion of the overall product cost depending on the type of product, its inherent quality/reliability, and a multitude of other variables including the warranty period. To gain some advantages in highly competitive markets, manufacturers must continuously improve product quality or upgrade their products creatively. In general terms, warranty management involves models in order to evaluate the objective in terms of decision variables, to assist the decision-maker in the decision-making process [3] (see Fig. 3.9).

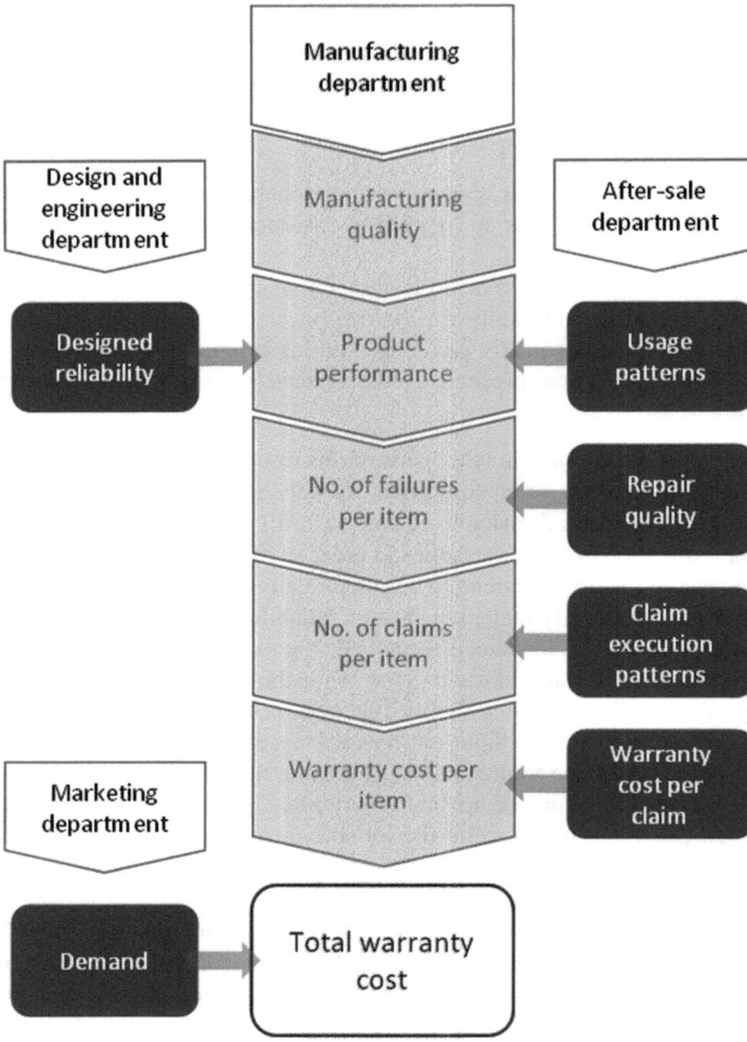

Fig. 3.9 Integrated model for total warranty costs (adapted from [3])

In the scenario of non-renewing warranty with repair time [15], manufacturers must have effective warranty management due to the fact that on the one hand customers expect a warranty program, and on the other hand, and in order to be successful in the marketplace, the manufacturer needs it. Furthermore, and related to the scenario on software reliability and its warranty cost [18], management technologies employed during the life cycle are essential for producing and maintaining quality/reliable systems. There should be efficient management during the testing and maintenance phases of the life cycle since any delay may cause a

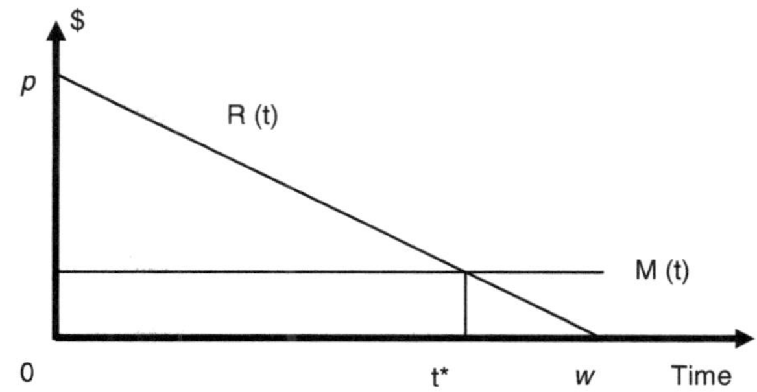

Fig. 3.10 Consumer claims costs and benefits (from [6])

company significant financial loss. Other contributions are for instance when a consumer experiment a product failure during the period of warranty and makes a claim again the seller [6]. The implication of claim costs for the buyer has to be into consideration. In this scenario about the modeling of general warranty costs for a period in the future, there are situations when the benefits of the warranty at time t are less than the consumer's personal costs of making a claim. That means, let us suppose that the length of the firm's expressed liability may be constrained by warranty. However, the actual duration of the warranty period may be shortened by a point at which cost and income are equal and there is neither profit nor loss in consumer claim costs and benefits. That happens at a point in time prior to w (that is t^* in Fig. 3.10). If the benefits available in time t are given by the linear function $R(t) = p(1 - t/w)$ and the consumer's claim costs are given by $M(t)$, the consumer would reach an intersection point at time t^*. Rational consumers would not redress warranty claims beyond t^*, even when they have a valid claim, because their claim costs exceed potential claim benefits. This analysis has important implications for the firm because it will tend to reduce the quantity of claims which will be made against the firm and, hence, the costs of the warranty program. In the scenario related to the estimation and production for warranty cost with an integrated multicriteria model [8], it is shown how top management should be interested in the integration of functional aspects of the company as marketing, production, and finance in order to be collectively incorporated in models with a systems focus. This management is assumed to be able to prioritize the different goals in a descending order of importance. Additionally, it is expressed that top management would be much more at the disposal to set aside larger amounts of warranty reserve if sales were high.

Regarding the warranty policy with time-dependent costs for non-renewable minimal repair [11], the manufacturer can estimate the amount that may be spent on minimal repair with an estimation of sales during a period. Such information helps management in selecting an adequate warranty period that is consistent with the goals and objectives of the company. The magnitude of the warranty expenses should be established serving this as a planning tool for management.

References

1. Murthy DNP, Blischke WR. 2005. Warranty management and product manufacturing. Springer-Verlag London Limited. Pp 302 + xviii. ISBN 1852339330
2. Hadel JJ, Lakey MJ. 1993. A structured approach to warranty preparation and risk assessment. Proceedings of the Annual Reliability and Maintainability Symposium. Atlanta.Pp. 298–304.
3. Murthy DNP, Djamaludin I. 2002. New product warranty: A literature review. Int. J. Production Economics 79:231–260
4. Djamaludin I, Murthy DNP, Blischke WR. 1996. Bibliography on Warranties, Chapter 33 of Product Warranty Handbook. Marcel Dekker (eds), 33:839–916
5. Murthy DNP, Djamaludin I. 2002. Product warranty - A review. International Journal of Production Economics. 79:231–260
6. Amato HN, Anderson EL, Harvey. DW. 1976. A General Model of Future Period Warranty Costs. The Accounting Review. Vol. Ll, No. 4, October.
7. Murthy DNP. 1989. A usage dependent model for warranty costing. European Journal of Operational Research 57:89–99.
8. Mitra A, Patankar JG. 1993. An Integrated Multicriteria Model for Warranty Cost Estimation and Production. IEEE Transactions on Engineering Management. 40(3).
9. Pham H, Zhang X. 1999. A Software Cost Model with Warranty and Risk Costs. IEEE Transactions on Computers. 48(1).
10. Wang CH, Sheu SH. 2001. The effects of the warranty cost on the imperfect EMQ model with general discrete shift distribution. Production Planning & Control. 12(6):621–628.
11. Shau-Shiang J, Kulkarni VG, Mitra A, Patankar jg. 2001. A Nonrenewable Minimal-Repair Warranty Policy with Time-Dependent Costs. IEEE Transactions on Reliability. 50(4).
12. Mondal S, Surajit P, Manna DK. 2003. Cost Estimation Under Renewing Warranty Policy - An Application. Taylor & Francis Online.
13. Chukova S, Hayakawa Y. 2004. Warranty cost analysis: non-renewing warranty with repair time. John Wiley & Sons, Ltd. Appl. Stochastic Models Bus. Ind. 20:59–71.
14. Dimitrov B, Chukova S, Khalil Z. 2004. Warranty Costs: An Age-Dependent Failure/Repair Model. Wiley InterScience, Wiley Periodicals, Inc.
15. Chukova S, Hayakawa Y. 2005. Warranty cost analysis: quasi-renewal inter-repair times. International Journal of Quality & Reliability Management. Emerald Group Publishing Limited. 22(7):687–698.
16. Wang H. 2006. Warranty Cost Models Considering Imperfect Repair and Preventive Maintenance. Bell Labs Technical Journal Lucent Technologies Inc. Published by Wiley Periodicals. 11(3):147–159
17. Shaomin W, Min X. 2008. Warranty cost analysis for nonrepairable services products. International Journal of Systems Science. Taylor and Francis Group. 39(3):279–288.
18. Williams DRP. 2007. Study of the Warranty Cost Model for Software Reliability with an Imperfect Debugging Phenomenon. Turk J Elec Engin. 15(3).
19. Blischke WR, Murthy DNP. 1996. Product Warranty Handbook. Marcel Dekker ed. New York. Pp 925 + xxiii.
20. Murthy DNP. 2006. Product warranty and reliability. Springer Science + Business Media, Inc. Ann Oper Res. 143: 133–146.
21. Lyons K, Murthy DNP. 2001. Warranty and manufacturing, integrated optimal modelling. Production Planning, Inventory, Quality and Maintenance, Kluwer Academic Publishers, New York. Pp. 287–322.
22. AENOR. European Standard. UNE-EN 13306. 2010. Maintenance terminology. Madrid. ISBN: 2910009588394.
23. Gómez J, Crespo Márquez A, Moreu P, Parra C, González-Prida V. 2009. Outsourcing maintenance in services providers. Taylor & Francis Group, London, ISBN 978-0-415-48513-5. Pp. 829–837.

24. Da Silva CM et al. 2008. Proactive reliability maintenance: a case study concerning maintenance service costs. Journal of Quality in Maintenance Engineering. 14(4):343–355.
25. Murthy DNP, Jack N. 2002. Warranty and Maintenance, in Reliability Engineering Handbook. Springer-Verlag. London. Pp 305–316
26. Murthy DNP, Jack N. 2009. Warranty and Maintenance in Handbook on Maintenance Engineering and Management. Springer Verlag. London. Pp. 461–478.
27. Murthy DNP, Jack N. 2007. Warranty Servicing, in Encyclopaedia of Statistics in Quality and Reliability. Wiley, New York. Pp. 2091–97.
28. Murthy DNP, Jack N. 2008. Outsourcing of Maintenance, in Complex System Maintenance Handbook. Springer Verlag. London. Pp. 373–394.
29. European Standard. 2006. EN 13269. Guideline on preparation of maintenance contracts.
30. International Standard. 2002. ISO/TS 16949. Quality management systems - Particular requirements for the application of ISO 9001:2008 for automotive production and relevant service part organizations.
31. Blischke WR, Murthy DNP. 1994. Warranty Cost Analysis, Marcel Dekker, New York.
32. Murthy DNP. Djamaludin I, Wilson RJ. 1997. Lot-sizing and inspection for items sold with warranty, in Optimization in Quality Control. Kluwer Academic Publishers. New York. Pp 317–340.
33. Lyons K, Murthy DNP. 2001. Warranty and Manufacturing, in Integrated Optimal Modelling in PIQM. Production Planning, Inventory, Quality and Maintenance. Kluwer Academic Publishers. New York. Pp. 289–324
34. Murthy DNP, Djamaludin I. 2001. Warranty and Quality, in Integrated Optimal Modelling in PIQM. Production Planning, Inventory, Quality and Maintenance. Kluwer Academic Publishers, New York. Pp. 325–360
35. Issacson DN, Reid S, Brennan JR. 1991. Warranty cost-risk analysis. Proceedings of Annual Reliability and Maintainability Symposium, Orlando, FL. Pp. 332–339.
36. Murthy DNP, Baik J, Wilson T, Bulmer M. 2006. Two-dimensional Failure Modelling, in Springer Handbook of Engineering Statistics. Springer-Verlag., London. Pp. 97–111.

Part III
A Reference Framework for Warranty Management

Chapter 4
Learning from Maintenance Management Models

4.1 Introduction

The European Standard UNE 13306 (2001) [1] defines maintenance management as the management activities that determined maintenance objectives, strategies, and responsibilities, and that are carried out through planning, maintenance control supervision, and improving organizational methods including the economic aspects of them. In this respect, different authors have proposed models, frameworks, or systems that look for better methods of maintenance management. Using the most advanced techniques and proposing innovative concepts; each model that has been proposed has strengths and weaknesses. This chapter chronologically presents a series of developed models for the management of maintenance, of which their applications can be seen when dealing with the organization of warranty management. Analysis results from different models have generated a management proposal based on the ISO 9001:2008 [2].

4.2 Maintenance Management Models

As previously set out in the Chap. 3, different contributions were observed in the reviewed literature regarding warranty management; in this case, an equivalent task is carried out related to maintenance management, which will permit us to identify the process, the actions, stages, tools, methods, or support techniques necessary to properly manage maintenance. Based on these methodologies, that will serve as base to develop later a management framework applied to after-sales management. The contributions regarding models of maintenance management appear in chronological order in Fig. 4.1. In the similar way as the Chap. 3, criteria to select the models were as follows:

1. The proposed management model has to be global and not focused on just one phase of management or maintenance tool.
2. The model must not be an automated software or CMMS type.

V. González-Prida Díaz and A. Crespo Márquez, *After-sales Service of Engineering Industrial Assets*, DOI: 10.1007/978-3-319-03710-3_4,

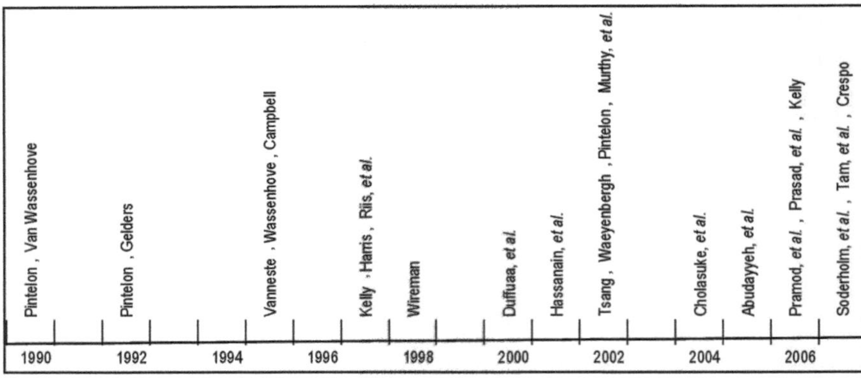

Fig. 4.1 Chronology of maintenance management models

3. The model must be published in a scientific journal or review.
4. The model proposed must be new and not a revision or application of an existing one.
5. The model preferentially must have a graphic representation.

Each of the models proposals have their qualities and limitations, that is why it is useful to recognize and analyze them so as to be able to take the best of each one and later apply them according to the particular merits of the practice. The revision allows us to visualize an important aspect of the study that has to do with the chronological order of the contributions of different authors that are considered most relevant to the improvement of maintenance management models. Table 4.1 summarizes the proposed innovations for each model in chronological order. Here, it must be noted the work of gathering models made by the Ref. [15], which justifies in greater detail the different innovations that those models suggest. It is necessary to mention that aforementioned innovations correspond to those appeared for the first time within a maintenance management model, which does not mean they are innovative concepts inside or outside the area of maintenance.

Briefly discussed, results presented in Table 4.1 can distinguish that throughout the time, maintenance management models have been acquiring new elements such as:

1. Process focussed.
2. Innovative proposal in technical aspects, for example oriented to maintenance re-engineering.
3. The development of models in a standard language of information exchange with the view to be used in GMAO applications.
4. The quantity and quality of the maintenance information required (operational and financial data) have led to the successive incorporation of quantitative techniques and software tools, mainly to deal with matters of maintenance

Table 4.1 Innovations on maintenance management models

Year	Innovations	Author(s)
1990	1. Suggest a complete system of maintenance indicators	Pintelon and Van Wassenhove [3]
1992	2. Expose the necessity of a link between maintenance and other organizational functions	Pintelon and Gelders [4]
	3. Highlight the importance of the use of quantitative techniques for management	
	4. Propose organization by levels to execute maintenance functions	
	5. Discern the usage of expert systems	
	6. Mention TPM and RCM	
1995	7. Propose an analysis of efficiency and effectiveness of maintenance	Vanneste and Wassenhove [5]
	8. Emphasize the importance of leadership in maintenance management	
	9. Introduces the concept of Reengineering of maintenance	Campbell [6]
1997	10. Suggest the model base don the concept of situational management theory	Riis et al. [7]
2000	11. Propose the usage of a great variety of tools and Japanese concepts for the statistical control of maintenance processes, using a module called "feedback control	Duffua et al. [8]
2001	12. Orientate their model to automated use, expressing it in the language IDEF Ø (standard language of model)	Hassanain et al. [9]
2002	13. Discern usage of e-maintenance	Tsang [10]
	14. Propose a guide to analyze the convenience of subcontracting as an element of entrance to maintenance system	
	15. Incorporate tacit knowledge as well as explicit and integrates it in an automated database.	Waeyenbergh and Pintelon [11]
	16. Values specially the management of knowledge within a maintenance model	
2006	17. Suggest the unión of tools QFD and TPM within a maintenance management model	Pramod et al. [12]
2007	18. Propose that maintenance must focus on the fulfillment of requirements of all interested parties	Soderholm et al. [13]
	19. Contributes a model with a methodology of application clearly expressed, orientated to the improvement of operational reliability and the life cycle cost of industrial assets	Crespo Márquez [14]

management efficiency. Later efforts have been developed to introduce new methodologies and instruments to improve the efficiency of maintenance.

5. The software tools have made it possible to continuously improve and evaluate the function of maintenance. This is due to the quality and quantity of indicators and developed methods to study the efficiency and effectiveness of their operations.
6. There is a recognized necessity that the area of maintenance be effectively linked to productive assets and to other areas of organization. The revolution of

e-maintenance [16] leads us to a new model of potential management, but also to new demands in the business capabilities.
7. The management model begins to include the evaluation of the products' life cycle and also the maintenance function evaluation. The management tends to integrate the asset strategy with that of maintenance.

In summary, it is possible to identify the tendency of models developed in the last few years which consists of creating systems focused on the process and quality, of a closed cycle and that include supporting techniques, methods and tools to improve the decision-making and to achieve the greatest organizational efficiency.

4.3 Process-Oriented Models Versus Declarative Models

In Sect. 4.2, there has been described some of the most important management models published in recent years. Table 4.2 proposes an initial division classification of the models in two types: declarative models and process-oriented models.

Declarative management models mentioned the components in the management system, even if they do not refer explicitly to the intercommunication link between them. In this type of models, there is no clear distinction in the flow of information between the components, and therefore, some aspects related to the functional, interrelational, or synchronization cannot be clearly perceived. However, some of these models are very complete, including a great variety of aspects related to maintenance. On the other hand, process-oriented models normally offer a clear transfer of information between the components of the model. In others, a closed circuit sequence is also clearly represented. In many cases, it seems like these models are more easily applied in a organization than declarative management models [17]. It is undeniable that each type of model has its pros and cons that is why it is convenient to study and analyze them to better identify which is more convenient when applied in a specific situation and condition.

4.4 Maintenance Management Models Comparative Analysis

In the view of being more precise when comparing models, it is opportune to create a verification list that in someway reflects the distinct elements that appear as a general tendency in different management models throughout the years. A first group of verification elements can be inspired by the norm ISO 9001:2008 (see Table 4.3).

Table 4.2 Classification of declarative and process-oriented models [15]

Declarative models	Process-oriented models	
Pintelon and Van Wassenhove [3]	Vanneste and Wassenhove [5]	Waeyenbergh and Pintelon [11]
	Riis [7]	Pramod et al. [12]
Pintelon and Gelders [4]	Duffuaa et al. [8]	Soderholm et al. [13]
	Hassanain et al. [9]	Crespo Márquez [14]
	Tsang [10]	

Table 4.3 Verification list for the analysis of management models

Typical elements of a quality system	Supporting techniques to maintenance management
Quality management system	1. Study techniques for economical and financial aspects of maintenance
1. Process-based system input/output	2. CMMS
2. Sequence and interaction of processes	3. Specific techniques for human resources management
3. Description of elements of each process	4. Application of operation research of or administrative sciences
4. Generation of documents and/or registers	5. Life cycle analysis
Responsibility management	6. TPM
5. Link with policies or strategic objectives of the organization	7. RCM
6. Definition, objectives	8. Maintenance policies: FBM, UBM, CBM, OBM, DOM
7. Commitment to high management	9. Simulation
8. Clear definition of responsibilities and authorities	10. Inventory models
9. Adequate communications	11. Reliability theory
Product manufacture	12. Expert systems
10. Proposed tools for its execution	
Resources management	
11. Resources management (people, material, infrastructure)	
12. Purchase control	
Measurement, analysis and improvement	
13. Audits, internal studies on customer satisfaction, information analysis, corrective and preventive actions, etc.	
14. Focus on continual improvement	

This norm [2] is the international reference of a quality management system, so it serves as a generic guide to the operation of any process that needs to fulfill with determined requirements such is the case of maintenance. It can be observed that the number of elements that appear in Table 4.3 is less that the number of quality elements that originally appeared in the norm. This is due to the fact that some components have been grouped and simplified in function to their importance for maintenance, considered as an integral part of the organizations' general system of quality. As it has been mentioned previously, some models limit themselves to the enumeration of elements that a maintenance management system should contain, only taking into account the system's own dynamics. However, there are relatively few models that define a clear method to implement and operate a maintenance management system. In fact, the following can be commented upon:

1. *About management system*:

- Within the models, the declarative models do not have an input/output focused on a process and do not consider a clear methodology for their application. In general, these models do not mention the advance quantitative techniques in maintenance.
- In all of the models, documents and registers are generated as entries for the decision-making process.

2. *About management responsibility*:

- All of the models define the objectives of the function of maintenance; however, not all of them link these objectives with the strategic objectives of the business.
- The majority of the models do not make a clear reference to principles of responsibility, authority, and good communication.

3. *About maintenance support*:

- Approximately half of the models incorporate the use of support techniques of operative research and management methodologies.
- The techniques TPM and RCM are the most mentioned maintenance methodologies.
- In general, TPM and RCM tend to appear together in the management models.
- CMMS is mentioned as an indispensable tool in the majority of the models. The most recent models include other functions of ICTs in the improvement of maintenance management (e-maintenance and expert systems). These tools also permit us new necessities of organization.
- The majority of the models are not linked to the functioning systems of technical support tools.
- Very few models include the use of simulation, reliability theory, and/or expert systems.

4. *About resources management*:

- Almost a third part of the models does not mention inventory management and purchase control techniques.
- The majority of models refer to economic aspects and financial analysis of maintenance function.
- Curiously in previous models, one can see a greater emphasis in aspects related to management of human resources.
- In the majority of recent models, the study in human resources is carried out from the perspective of security and hygiene. It is supposed that modern organizations have already adequately covered issues-related salaries,

personal satisfaction, training, and other basic aspects of traditional human resources.

5. *About measurement, analysis and improvement*:

- All models take into account different stages for maintenance (evaluation, analysis, and improvement), and little more than half of them (even with a rising tendency) mention literally the concept of "continuous improvement."

6. *About the methodology and functioning of the model*:

- A very important attribute of some of the models is the inclusion of an application of methodology, which stimulates continuous improvement.

4.5 Maintenance Management Model Based on the Standard ISO 9001:2008

From the previous analysis onwards, it is possible to identify a series of concepts (tools, elements, activities) that are repeated in some of the models previously mentioned. With the aim of ordering them, such concepts were classified into distinct categories: planning, support processes, maintenance, control, and improvement execution. These categories have been framed in the model suggested by the norm ISO 9001:2008 for quality management with the purpose of conferring a sense to the grouping of terms. In this manner, a model can include in itself the tools that can be used to fulfill each of its stages. The norm ISO 9001:2008 is taken as an international reference for any quality management system, making it (as it has been mentioned in previous sections) a generic guide to the operation of processes in which requirement fulfillment must be demonstrated, such is the case of maintenance function. The result is a maintenance management model that intends to take the best from the studied contributions and that is based on the operation of the norm ISO 9001:2008 (see Fig. 4.2).

This model begins and ends with the requirements and the satisfactions of the interested parties for proper maintenance management [15]. Here, it takes the "interested parties" focus that is proposed by [13]. The model is also designed to be effectively used in all levels of organization (reminding us of the proposals of [4] who manage their model in three levels of activity within the business). Note that planning involves the company's senior management. The middle levels are those that perform support processes and control at maintenance level and generates data for the continuous improvement of the system and the replanning. Likewise, the way in which this model is structured makes possible the existence of a link between function of maintenance and the rest of the organizational functions.

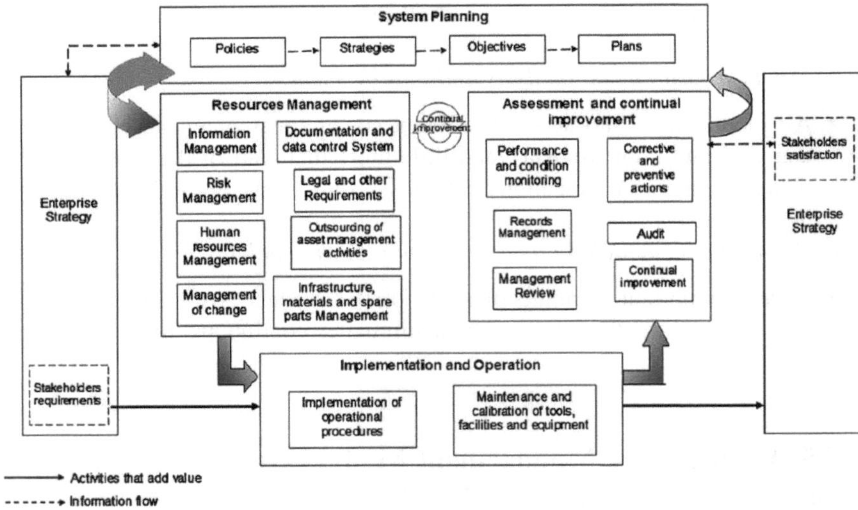

Fig. 4.2 Maintenance management model based on ISO 9001:2008

4.6 Maintenance Management and ICTs

4.6.1 Concept of Electronic Technology

The e-technology is a new concept where the use of ICTs and new technologies are integrated. In order to develop, implement, and support them [18], they require intranet, extranet and Internet, as well as electronic devices as: sensors, wireless communication devices, Personal Digital Assistant (PDA), among others. All these technologies increase the possibilities of: using data from many origins and different kinds [19], processing huge data volumes, performing advanced analysis as a support for the decision-making and putting into practice cooperative or collaborative activities. The e-technologies infrastructure is constituted by all those resources (applications, services, devices) needed to carry out the operations identified and defined by the organization, but in a remote way (MCT or Mobile Collaborative Technologies). Thus, the e-technology encompasses the use of a net with one or several servers, work stations, applications, databases, sensors, PDA devices… elements in general terms which provide a control, follow-up and diagnosis of the equipment conditions in real time. The use of MCTs [20] allows optimal data integration and a high flexibility in the adaptation to the different operational environments, increasing at the same time their access. The functioning of this infrastructure is grounded to guarantee the quality level of the foreseen services for the process performance in terms of scalability and availability. More in particular, the goal is to guarantee an acceptable answer time for the company and the client. To achieve this, it is used Key Performance Indicators

(KPIs) in order to asset the global process performance and to compare them with the foreseen ones. The current advances in the mobile collaborative technologies (MCT) can change radically the management of industrial assets. The combination using wireless sensors, nets and mobile handy devices, solves the access problem to the information existing in many organizations. Some of the MCTs main features are as follows [20]:

- Synchronized communication,
- Spatial distance handling,
- Portability improvement,
- Geospatial location,
- Maximal accessibility,
- Identification simplicity,
- Instant information and
- Process agility.

The influence of all these capabilities is even more important when the tasks are characterized by the immediacy and the high mobility of the involved staff [21]. Currently, the fast convergence of technologies is causing that many of the main features in a particular mobile device are more and more shared by most devices as PDAs, tablets, smartphones, or ultra-portable laptops (UPL). As commented before, the mobile technology has a positive and very important impact thanks to the elimination of space–time barriers, which allow the development of intelligent user interfaces that share information contextualized to each kind of operation or function. A large amount of innovative products already belongs to the e-technologies family. Some of them are mentioned here below [18]:

- New intelligent sensors MEMS, (autonomous microsensors, memory cells, etc.), wireless sensors, sensor nets. Sensors are the key element to carry out basic processes in reference to electronic technologies as they permit us to know in real time the operational conditions of a unit or device.
- Label RFID. It identifies operator and component of the unit (geolocation). It allows us to store data about the unit or device performance and functioning and makes it possible to outline actions and operations performed on it (for example as in maintenance).
- Global positional system (GPS), in addition to the labels RFID, it is used for the calculation of location of an operator or the necessary tools.
- Wireless technologies provide considerable savings in net costs and a high degree of flexibility that it is not possible with cable systems. The main wireless technologies include nets, such as IEEE 802.11, 802.15.4 ZigBee, 802.15.1; Bluetooth; wireless local area network, such as WiFi, WiMax, GSM-UMTS (of long distance).
- Innovative communication systems (virtual reality) that support the interaction people/device or people/people, for example, allowing information reception directly in text form or in fixed or animated image form.

- Tools for the intelligent diagnosis and prognosis provide support to the decision-making process.
- PDA, smartphones, graphic tablets, laptops, etc. (all of them provided with WiFi, Bluetooth, RFID reader, Windows Mobile), for the development of a high number of functions that enables the right execution of tasks performed by the operator, with a minimum time requirements.
- Specific standards of integration and interoperability among the ICT components that constitute electronic technologies (for example MIMOSA o Machinery Information Management Open Systems Alliance, IEC62264, OPC foundation for Open Connectivity).
- Web services (for control, diagnosis, prognosis, and planning) based on a group of technical norms and standards (Internet-based technologies) used to exchange data between applications in heterogeneous environments: Simple Object Access Protocol (SOAP) for message exchange, Web Service Description Language (WSDL), etc.
- Web-CMMS (e-CMMS) is a computerized maintenance management system (CMMS). It is capable of monitoring and managing the preventive maintenance activities of the company, offering also new functions such as application service provider/providing (ASP) that allows us to connect with mobile technologies to recover or to send data.

The use and integration of all the above described technologies (among others) in a single infrastructure changes the denomination of the processes adding the prefix "e-", which is generally accepted. Therefore, new concepts appear as for instance the one referred to maintenance management (e-maintenance) or for the warranty management (e-warranty), generating at the same time an infrastructure or platform (e-platform) for maintenance or warranty.

4.6.2 Definition of E-Maintenance

Nowadays, the necessities of information and the ICT development are causing the transformation of the whole business process. The ICT use is generating changes in the management of the different processes as, for instance: e-business, e-operation, e-manufacturing, e-production, or e-maintenance. In general terms [16, 22], e-maintenance can be defined as a maintenance management concept by which the assets are controlled and managed via Internet. Nevertheless, e-maintenance can be considered as a maintenance strategy, that is an electronic management method, using in real time that data obtained from the equipment by digital technologies; as a maintenance plan; as a kind of maintenance (based on CBM, proactive and predictive maintenance), or as a maintenance support (resources and services). E-maintenance constitutes an interdisciplinary approach which consists of supervision, diagnosis, prognosis, and control [23]. The objective is to reduce the amount of "in situ" inspections and to eliminate unnecessary maintenance activities. E-maintenance integrates the already applied principles on maintenance, adding a Web service as well as collaborative principles [16]. The collaboration is not limited to share and exchange information, but knowledge and intelligence among the different units, departments, expert operators, and companies in a collaborative environment for the development of better decision-making on maintenance, along the asset life cycle. The e-maintenance platform appears when the conventional and hierarchical structure is replaced by an intelligent structure [23] (Intelligent Manufacturing Systems, IMS). It consists of providing "intelligence" to the component through electronic technology. For that purpose, it is required to have support nets for information exchange, real-time communications between the devices, etc. Through a prefixed alarm system and based on the monitoring of equipment conditions, it is possible to optimize the priority of the maintenance activities. The costs based on criticality (CBC) combines information about the technical status or equipment health, cost of the maintenance activities, cost by production losses and risk factors. We can summarize these improvements in the following [19]:

1. *Remote assistance*: The warranty service technicians can remotely access to the product which presents a failure or where a complaint has been made, allowing them the remotely start-up, control, configuration, condition monitoring, data collection, and analysis of the product. One of the best advantages is to connect products distributed in different and distant geographic locations, to a warranty service center that allows a remote decision-making.
2. *Cooperative warranty service*: The e-warranty would allow implementing an information infrastructure connecting clients (users) with suppliers (manufacturers), or connecting engineers and technicians with the products using Internet. The resulting platform would allow then a strong cooperation between different areas of the company (production, maintenance, purchasing, etc.) and different companies (suppliers, customers, equipment manufacturers, etc.). This way, errors in the process are minimized, communication are optimized (enables bidirectional flow), feedback cycles are generated; in short, quality is enhanced.
3. *Online assistance*: The remote follow-up of sold units during the warranty period, together with the possibility of programming status alerts, allows the manufacturer or supplier to respond quickly to any situation and even before the user has been detected failure, preparing accordingly any intervention in an optimal way. For example, PDA plays a key role in remote assistance, since it permits maintenance staff to get direct information from devices and equipment in any place.
4. *Predictive warranty service*: The e-warranty would allow a failures prediction in those units offered for sale taking into account the product current state and the intended use on it, together with the life predictions on component (supplied by the manufacturer) that allows it to achieve a reduction in the life cycle cost of those units. As well as the adjustment of operation needed for the right maintenance.

5. *Failure diagnosis*: The e-warranty allows technicians to carry out online diagnosis to those occurred failures, as well as to suggest solutions to the end user. Consequently, it reduces the communication time between supplier and customer, improving the quality of the shared information and, therefore, the resolution time. All these factors contribute to increase the availability of production equipment reducing time of repair (MTTR), as well as minimizing necessary resources (costs).

4.7 The Warranty Management Process

4.7.1 Methodology

As we already know, management is usually defined as a process for leading and directing all or part of an organization. This means, management corresponds to all those activities that determine targets, strategies, and responsibilities within an organization, putting them into practice by means of planning, control, supervision, and considering economical aspects of the company. In our case, the warranty management process will suppose consequently the course of actions, series of stages, or steps to follow, in order to lead and direct the organization of a warranty program. Management here will be focused on a new product recently launched into the market, coexisting production, and postproduction activities.

More details about the different focuses of strategical management can be found in Refs. [24, 25]. As it is shown in Fig. 4.3, a warranty program management, viewed from a strategic perspective, begins with a strategy linked to technical and commercial planning from the very start of the product development process [26, 27]. The steps in this scheme happen in many cases simultaneously; hence, an after-sales support (for example) is frequently offered while the production line is still open. That means, that the information obtained from this after-sales service can be very helpful for the production lines and, of course, very useful to improve the design and development of the product. For such context is exposed here a proposal for reference framework. In order to characterize the processes related to warranty management, it is possible to implement different tools of representation. In a very simplified way, our reference framework will be surrounded by boundary conditions such as the ones indicated in Fig. 4.4. This black box-like representation (level 0) will be analyzed and studied in depth (level 1 and level 2) using standard integrated definition methods (IDEF) [28, 29]. This standard is easy to apply and comprehend and enables us to represent each part of our framework in boxes with inputs, outputs, resource uses, and signal controls [30, 31] that have been implemented in this document in a very simple way.

In addition to this brief definition about the management of a warranty program and its inference on the product life cycle, there are currently different international standards which pretend to approach the whole business activity, dividing this in a chain of processes. Nevertheless, the management process

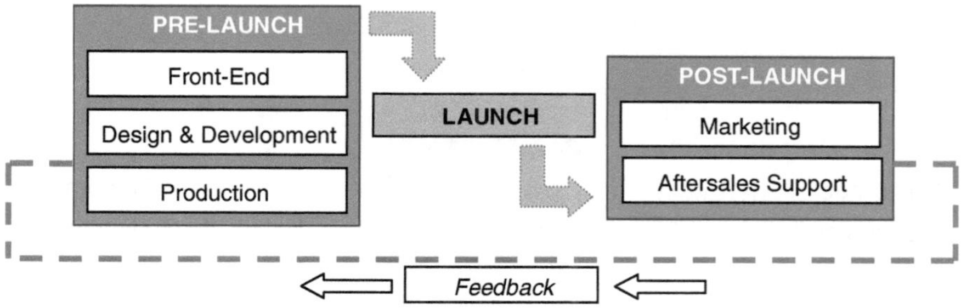

Fig. 4.3 Life cycle from a warranty perspective (adapted from [24])

Fig. 4.4 Representation of boundary conditions of the framework (*Level 0*)

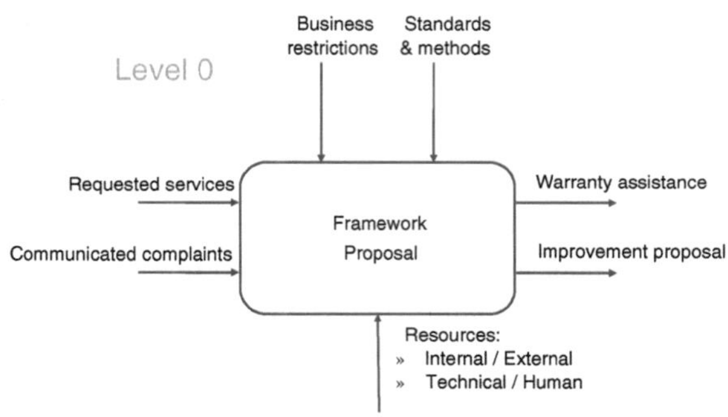

defined in such standards and quality management systems usually fulfills the following features [32]:

- Customer focus.
- Involvement of people.
- System approach to management.
- Factual approach to decision-making.
- Leadership.
- Process approach.
- Continuous improvement.
- Mutually beneficial supplier relationship.

Particularly, the EN-ISO 9000 [33] defines a quality management system, which is designed to attend the needs of all interested parties: customers, suppliers, employees, and owners.

4.7.2 Warranty Management Process

Using the model of a process-based quality management system in a simply way and adapted to the warranty management, it is possible to obtain an approach shown in Fig. 4.5. This improvement process is a variable of the Deming cycle

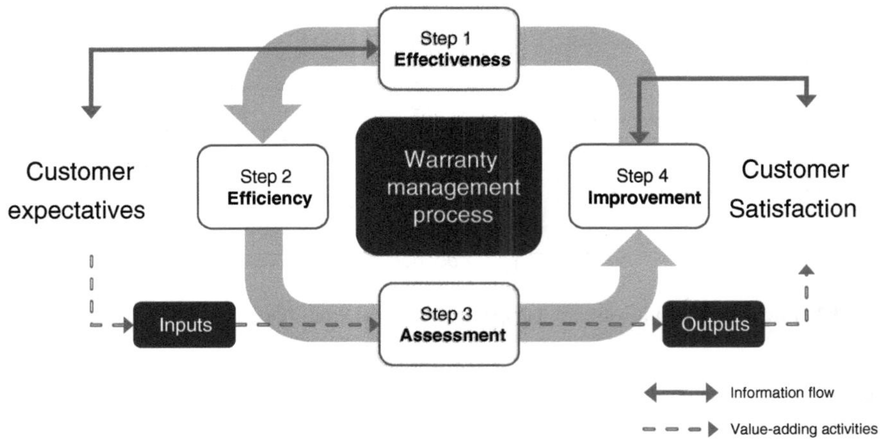

Fig. 4.5 Approximation of a warranty management process, according to ISO

also known as plan–do–check–act (PDCA). It is important to highlight the fact that warranty data will play a critical role in this process. This data collection comes from customer's claims (that is, data that can be collected when a unit is receiving warranty services), as well as other complementary data.

Further researches can approach problems related to data requirements, taking into consideration for example their overall impact on the analysis and the related consequences. In short, this adjustment to the warranty management process is possible due to the flexibility of the standards, since they can be applied to every type of work organization regardless of type and size of the business. On the other hand, following is the justification to the selection of the four steps in the warranty management process (effectiveness, efficiency, assessment, and improvement), in which our proposed framework will be based on.

The first step is related to the warranty program effectiveness. Here, it is defined the warranty objectives, priorities, and strategies, as well as those key indicators for the proper performance of the warranty program. In this moment, it is important the soon intervention on those already detected weak points with high impact on the customer satisfaction. Effectiveness shows how well the company (or particularly, the after-sales department) meets its goals or fulfills the customer/ company needs. It is often discussed in terms of the quality of the service provided, viewed from the customer's perspective. This will allow us to arrive at a position where we will be able to reduce warranty costs associated ultimately with customer dissatisfaction. In the case of warranty, effectiveness represents the customer satisfaction related to the capacity and condition of the company assets or, in other words, the reduction in the after-sales company cost because product service is available for the user when it is needed. Therefore, effectiveness concentrates itself on the correctness of the warranty process and if such process produces the required result.

As a second step, it is considered the warranty program efficiency. That is to attend warranties with minimum waste, expense, or unnecessary effort. Efficiency is therefore understood as the provision of the same or better service under the same cost, giving a measurement of how well the warranty assistance is being performed. The warranty support efficiency can be considered as the ratio between the planned or expected resources necessary to fulfill the required warranty task and the resources actually used. By the properly performance of this step, it is possible to obtain not only a fast and relevant enhancement in terms of reliability, but it is also possible to assure a very effective definition of the warranty plan activities from the after-sales department.

The third managerial block is devoted to the warranty program assessment. In order to assess the warranty service, warranty program assessment is needed to use suitable measurements which should be defined during the strategy phase of the warranty program. The measurements in general must enable mainly the comparison of reliability data as well as the inclusion of the life cycle cost assessment. Once warranty assistance is done, or based on a periodical review to the product performance, is when the warranty program assessment must be carried out. The company should therefore establish and use a standard and repeatable method for collecting and analyzing data and interpreting results, which may be based on corporate or industry factors. The results should be used to support and justify enhancements.

Finally, the fourth step is related to the warranty program improvement. This will be made based on accepted management performance indicators. As everyone can imagine, with the already mentioned steps, there is a huge amount of data related to warranty with a wide variety of information. To assure the continuous improvement cycle is therefore necessary a computerized information system for the warranty program, which can enable the cyclical process of managing data and analyzing results [34]. In this step is included the concept of e-warranty, which can increase the possibilities to use information from multiple origin and from different types, to process larger volumes of data and to make more advanced reasoning and decision-making, as well as to implement cooperative activities. E-warranty is therefore the use of information technology in the context of after-sales service, for example, in the data storage and its analysis in relation to customer's claims. According to the described process and divided into four steps, it is possible to represent in a very simple and easy way the different connections for data exchange (see Fig. 4.6) that can be useful for a company when it comes to properly organize a warranty program as well as to manage all information in regard to the field.

In Chap. 5, there will be a deeper analysis on each phase (level 2) trying to offer a complete representation of all stages. Practical application of the proposed framework will become evident in later chapters. Nevertheless, it is important to highlight the role of the client in the warranty management process [35]. Likewise, due to a complexity in the structure here proposed, it seems logical that the final contextual element for the application of this framework will be those industrial customized products that present a high level of technology and sophistication (for example aircrafts, defense systems, naval machinery, agricultural industry, or construction machines).

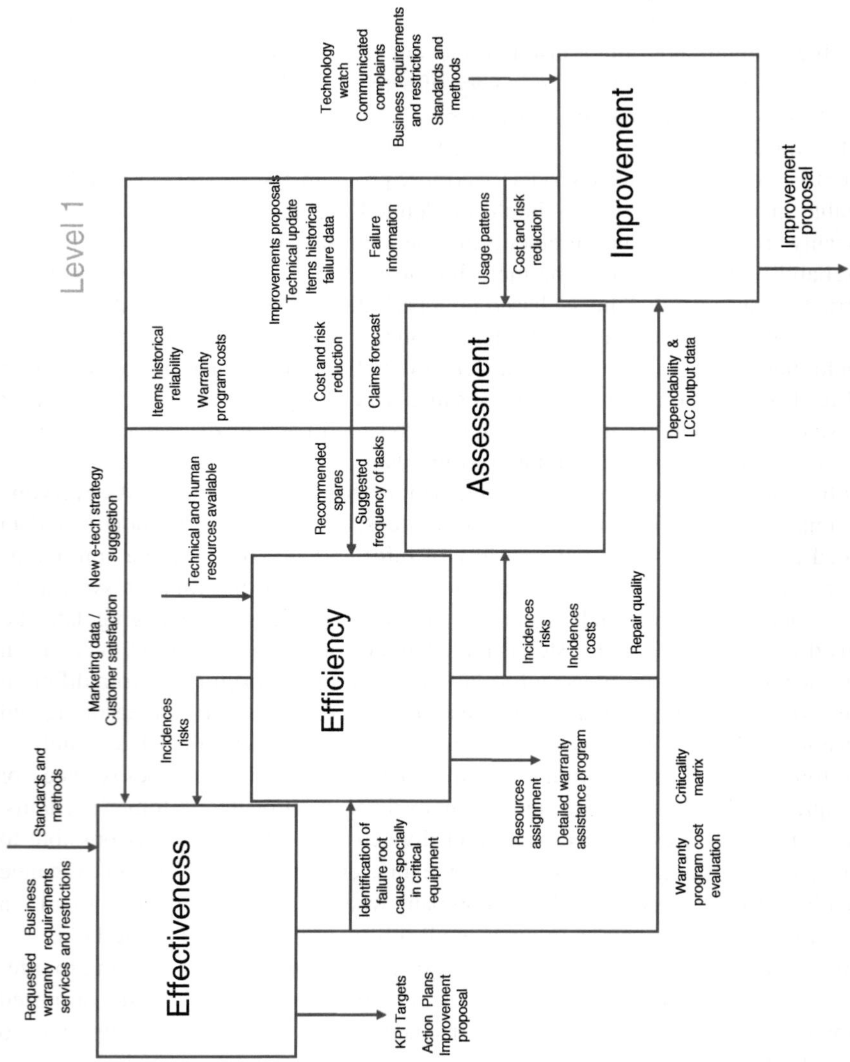

Fig. 4.6 Relationship and exchange of information between steps (*Level 1*)

References

1. AENOR. 2010. Norma UNE-EN 13306. Terminología De Mantenimiento.
2. AENOR. 2008. Norma ISO 9001. Sistemas De Gestión De La Calidad.
3. Pintelon L, Van Wassenhove LA. 1990. Maintenance Management Tool. Omega. 18(1): 59–70.
4. Pintelon LM, Gelders LF. 1992. Maintenance Management Decision Making. European Journal of Operational Research. 58(3): 301–317.
5. Vanneste SG, Van Wassenhove LN. 1995. An Integrated and Structured Approach to Improve Maintenance. European Journal of Operational Research. 82(2): 241–257.
6. Campbell JD. 2001. Organización y Liderazgo Del Mantenimiento. TGP Hoshin. Madrid.
7. Riis J, Luxhoj J, Thorsteinsson U. 1997. A Situational Maintenance Model. International Journal of Quality and Reliability Management. 14(4):349–366.
8. Duffuaa S, Raouf A, Dixon Campbell J. 2000. Sistemas De Mantenimiento. Planificación y Control. Ed. Limusa. México.
9. Hassanain MA, Froese TM, Vanier DJ. 2001. Development of a Maintenance Management Model Based on IAI Standards. Artificial Intelligence in Engineering. 15: 177–193.
10. Tsang A. 2002. Strategic Dimensions of Maintenance Management. Journal of Quality in Maintenance Engineering. 8(1): 7–39.
11. Waeyenbergh G, Pintelon L. 2002. A Framework for Maintenance Concept Development. International Journal of Production Economics. 77: 299–313.
12. Pramod VR, Devadasan SR, Muthu S, Jagathyraj VP, Dhakshina MG. 2006. Integrating TPM and QFD for Improving Quality in Maintenance Engineering. Journal of Quality in Maintenance Engineering. 12(2):1355–2511.
13. Söderholm P, Holmgren M, Klefsjö B. 2007. A Process View of Maintenance and its Stakeholders. Journal of Quality in Maintenance Engineering. 13(1):19–32.
14. Crespo Márquez A. 2007. The Maintenance Management Framework. Models and Methods for Complex Systems Maintenance. Springer. United Kingdom.
15. López MA, Crespo Márquez A. 2011. Modelling a Maintenance Management Framework Based on PAS 55 Standard. Quality and Reliability Engineering International. John Wiley & Sons, Ltd. 27(6):805–820.
16. Crespo Márquez A, Benoit I. 2006. Editorial Special issue on e-maintenance. Computers in Industry. 57:473–475.
17. Mishra PR, Anand D, Kodali R. 2006. Development of a Framework for World-Class Maintenance Systems. Journal of Advanced Manufacturing Systems. 5(2):141–165.
18. Iung B, Levrat E, Crespo Márquez A, Erbe H. 2009. Conceptual framework for e-maintenance: Illustration by e-Maintenance technologies and platforms. Annual Reviews in Control 33:220–229.
19. Muller A, Crespo Márquez A, Iung B. 2008. On the concept of e-maintenance: Review and current research. Reliability Engineering and System Safety. 93:1165–1187.
20. Emmanouilidis C, Liyanage JP, Jantunen E. 2009. Mobile solutions for engineering asset and maintenance management. Journal of Quality in Maintenance Engineering. 15(1):92–105.
21. Campbell DA, Sarker S, Valacich JS. 2006. Collaboration using mobile technologies: when is it essential? Proceedings of the International Conference on Mobile Business (ICMB'06), IEEE, Copenhagen.
22. Han T, Yang BS. 2006. Development of an e-maintenance system integrating advanced techniques. Computers in Industry. 57:569–580.
23. Ucar M, Robin GQ. 2005. E-maintenance in support of e-automated manufacturing systems. Journal of the Chinese Institute of Industrial Engineers. 22(1):1–10.
24. Murthy DNP, Jack N. 2009. Warranty and Maintenance in Handbook on Maintenance Engineering and Management. Springer Verlag. London. Pp. 461–478.
25. Murthy DNP, Blischke WR. 2000. Strategic warranty management - a life cycle approach. IEEE Trans. on Engineering Management. 47:40–54.

26. Murthy DNP, Blischke WR. 2005. Warranty management and product manufacturing. Springer-Verlag London Limited. Pp. 302 + xviii. ISBN 1852339330.
27. González-Prida V, Crespo Márquez A. 2010. Book Review: Reliability Engineering. Warranty Management and Product Manufacture (By D.N.P. Murthy and W.R. Blischke). Production Planning & Control: The Management of Operations. 21(7):720–721.
28. Feldmann CG. 1998. The Practical Guide to Business Process Reengineering Using IDEF0. Dorset House Publishing.
29. Giaglis GM. 2001. A Taxonomy of Business Process Modelling and Information Systems Modelling Techniques. The International Journal of Flexible Manufacturing Systems: 13:209–228.
30. IEEE Std. 1998. IEEE Standard for Functional Modelling Language-Syntax and Semantics for IDEF0. New York. IEEE Std 1320.1.
31. Kim SH, Jang KJ. 2002. Designing performance analysis and IDEF0 for enterprise modelling in BPR. Int. J. Production Economics 76:121–133.
32. Mihok L. 2004. Quality management systems—New approach by ISO 9000/2000. Metalurgija 3:193–198. ISSN 0543-5846.
33. European Standard. 2008. EN-ISO 9000. Quality Management System. European Committee for Standardization, Brussels.
34. Koç M, Ni J, Lee J, Bandyopadhyay P. 2003. Introduction of e-manufacturing. 31st North American Manufacturing Research Conference (NAMRC), Hamilton, Canada.
35. Murthy DNP, Blischke WR, Kakouros S, Kuettner D. 2007. Changing the way we think about warranty management. Warranty Week.

Chapter 5
Reference Framework

In every field, any management process can be usually split into two basic parts: a moment for "thinking" and a moment for "doing" (Fig. 5.1). That means there is a time to define a strategy and another time to implement such strategy [1]. Related to the management of a warranty program, the first part will require the definition of warranty targets, which will be derived directly from the business plan. These objectives will be not only to decrease costs, but also (and very important) to purchase a reliable product, offering an appropriate after-sales service to the user.

In general, the definition of the warranty program strategy will determine the success of such warranty service conditioning of course the implementation of plans, schedules, controls, and improvements. The second part of the warranty management processes (the implementation of the selected strategy) will face different difficulties. These difficulties can be for instance the capability to assure proper skill levels in the staff, to carry out correctly the work planning and the schedule fulfillment, or to assure the use of suitable tools, among others. Basically in this study, a generic framework is proposed for the management of warranty assistance with the intention of improving the quality of the product avoiding weak points during the manufacturing and, consequently, reducing costs from the after-sales service. In order to reach these purposes, the experience and the data gathered during the warranty assistance must be used as a feedback to the engineering/manufacturing process. Along four steps, this framework will include, as mentioned, moments for "thinking" and moments for "doing," just in order to perform finally an adequate service to the customer in a satisfactory way, applying a continuous improvement in the warranty management process and, consequently, a continuous improvement in the whole productive system.

V. González-Prida Díaz and A. Crespo Márquez, *After-sales Service of Engineering Industrial Assets*, DOI: 10.1007/978-3-319-03710-3_5,

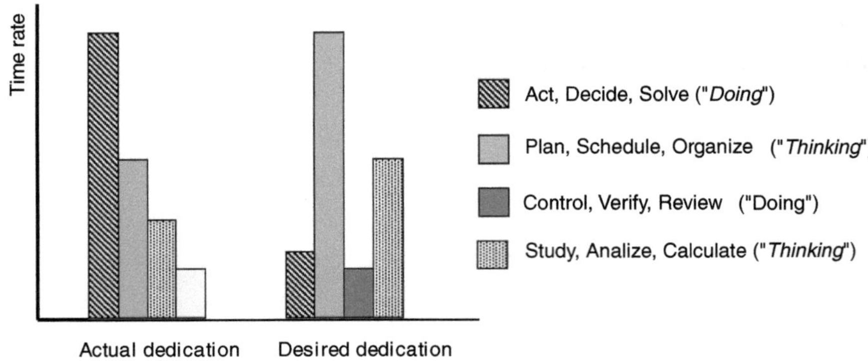

Fig. 5.1 "Doing" and "thinking" in warranty management (adapted from [2])

5.1 Management Framework of a Warranty Assistance Program

This section will briefly introduce blocks and methods that can be used to enhance the decision-making process in the after-sales service context. The proposed reference framework for a warranty management program supposes here an essential supporting structure and a necessary basic system for the management of such program and that influences engineering and manufacturing phases. It is important to insist on the fact that there are many other relevant tools and techniques and not just the ones mentioned in this document. As the reader will be able to see, some of the techniques here mentioned are related to the reliability of the product, but the intention is not limited just there. Besides product reliability, other aspects must be also taken into account like overall operations such as production and services. For this task, it has been necessary to elaborate joint strategies for warranty improvement in products already in the market and strategies for new products not yet developed. Due to the fact that this field is very wide and it is considered a challenge to attempt to integrate into it, this document will try to underline in a succinct way a great part of the material to that respect. Nevertheless, and for further information on this subject, other works in this broad field are for example the following sources: [3–6].

5.1.1 First Step: Warranty Program Effectiveness

With the purpose of avoiding contradictions in the warranty strategy program and the company's general management, the study proposes the implementation of a balance scoreboard (BSC). Basically, this methodology considers goals and targets through four points of view as it is shown in Fig. 5.2.

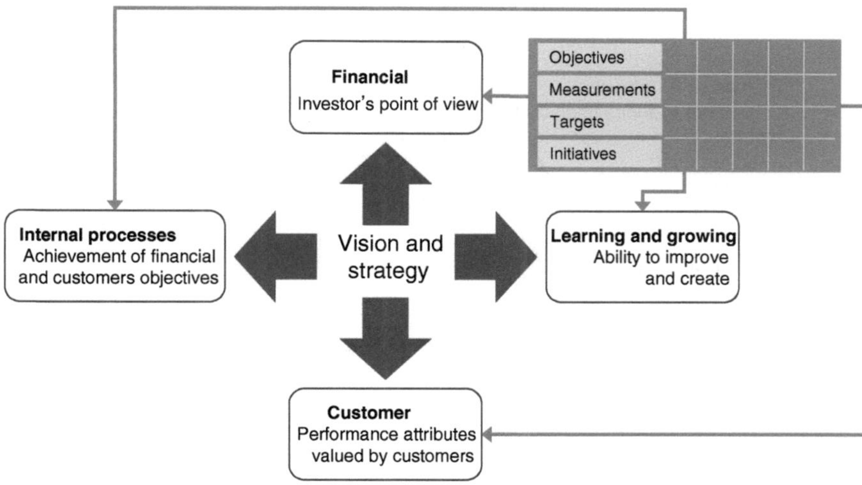

Fig. 5.2 Perspectives to reach goals by a balanced scoreboard

BSC applied to the management of a warranty program can involve the following steps:

(a) Strategical formulation for the warranty program (such as the outsourcing [7] of warranty assistance or the development of warranty teams with multiskilled technicians).
(b) Translation of warranty strategic into middle-term reachable objectives (such as the identification of measures designed to pull people toward an overall vision).
(c) Development of action plans (in order to reach the targets identified in step b, and taking into account any necessary changes in the organization's support infrastructure).
(d) Periodical review of performance and strategy (just to quantify the progress in order to formulate new strategic objectives, action plans, or the revision of the whole scoreboard).

Once the warranty program's objectives and strategies are defined, it is necessary to determine the product criticality. The criticality analysis is understood here as how crucial is the complaint of a client due to a failure and its consequences to the business. Decisions and actions from the warranty program can involve the possibility of a certain deviation from business objectives (in terms of profit losses, redirection of resources, possible delays, use of assembly pieces as spares, etc.). Therefore, the application of a technique which helps systematically to decide which assets should have priority related to the management of the warranty program is required, in accordance of course with the existing program strategy. One of these techniques can be based for example on risk–cost assessment. The criticality analysis based on risk–cost combines the probability of an

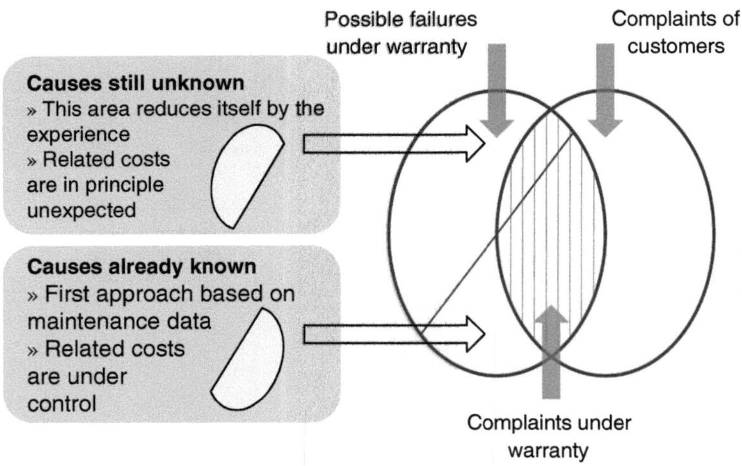

Fig. 5.3 First classification of causes

occurring event, with the impact that this event would cause. As a result, we obtain a criticality matrix that offers a prioritized view of probable events related to warranty issues, which allow us to align after-sales actions with the business targets.

By this way, it is possible to foresee where to apply available resources to mitigate risk reducing warranty cost and customer dissatisfaction. Finally, when the strategy to follow has been defined and the probable events have been prioritized, it is the moment to focus on those customers' complaints related to repetitive or chronicle failures that suppose high-priority events. If it is possible to find and to eliminate the causes of those failures by a soon intervention, it will provide a fast and important initial advantage of the warranty program strategy. Anyway, there are different methods already developed to perform this analysis of weak points. For instance, the failure root cause analysis considers actions in order to discover the reason for the appearance of a specific failure and how to correct its causes. A first classification of the causes could be of course if the failure must be attended under warranty or not (see Fig. 5.3). This analysis will help to readapt the considerations taken initially on the balance scorecard, improving by this way the warranty program effectiveness. A detailed root cause failure analysis may consist of the following:

- Determining a root cause of failure;
- Proposing, testing, and validating hypotheses;
- Recommending preventive actions;
- Implementing improvements;
- Forming a team of experts;
- Gathering evidence;
- Analyzing the results and determining failure causes.

Table 5.1 Inputs/outputs in the effectiveness of a warranty program

	Inputs	Outputs
Balance scorecard	• Incidence risks	• Selection of KPI • Warranty costs evaluation • Targets KPI • Action plans
	• Historical reliability in items • Warranty program costs • New technologies strategies • Customer • Marketing data • Improvement proposals • Commercial requirements • Required warranty services	• Improvement
Criticality analysis	• Selection of KPI • Historical failure data • Failure information	• Matrix of criticality
Failure root cause analysis	• Matrix of criticality • Historical failure data • Failure information • Standards and methods	• Identification of the failure root in critical equipment

After all that has been stated here, the effectiveness of a warranty program could require specific information (inputs) and also provide determined results (outputs) that can be summarized in Table 5.1.

These data constitute a sample that intends to illustrate the way in which the proposed methodologies can be implemented in a generic company that commercializes industrial products. Inputs and outputs will depend on, naturally, the particular organization as well as on the boundary conditions where the company develops its commercial and industrial activity.

5.1.2 Second Step: The Efficiency of the Program

To attend warranties with minimum waste, expense, or unnecessary effort is necessary to design an adequate plan for the warranty program. The plan for a certain product will require identifying its functions, the way these functions may fail, and then establishing a set of applicable and effective tasks, based on considerations of product safety and service economy. A method that can help to elaborate such plan at the very beginning of the warranty program can be extracted from management techniques of the maintenance field. Particularly, it will be useful for the application of an adapted FMECA or RCM, as for instance using

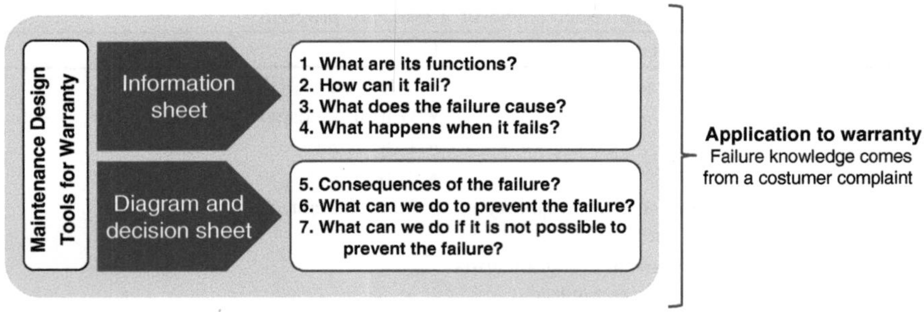

Fig. 5.4 Application of maintenance techniques to warranty casuistry

FMECA for generic maintenance plan design (see Fig. 5.4). An initial maintenance plan, applied to the warranty time horizon, can suppose a good first approach for the warranty capacity planning, the spare parts' provisioning, the warranty task schedule, skill level of the technicians, etc. The planning and scheduling improvement applied to a warranty program can of course enhance the effectiveness and efficiency of program policies. Such improvement will depend on the time horizon of the analysis. About costs and failure modeling, there is a huge range of diverse approaches, analytical and empirical, related to other fields. In order to carry out a Warranty Policy Risk-Cost-Benefit Analysis, such activity will depend of course on the objective that the own model builder keeps in mind. Previously in Chap. 3, in Fig. 3.9, an integrated model for total warranty costs is presented.

In any case (as here with the management of a warranty program), the problem is influenced by the information available. In addition to this, the complexity of the issue is usually high and considerations must be taken under certain assumptions in order, either to simplify the analytical resolution or to reduce sometimes the computational needs. In the same way as it happens in an effectiveness program, efficiency can also require specific information, such as to provide determined results that can be shown as an example for a generic company in Table 5.2.

Inputs and outputs will also depend on the particular organization as well as on the boundary conditions where the company develops its commercial and industrial activity.

5.1.3 Third Step: The Evaluation of the Program

Once designed, planned and scheduled, activities of the warranty program using the above described techniques are carried out, and the performance of such tasks has to be evaluated and assessed. As a complement of the above-mentioned techniques, a study related to reliability, availability, maintainability, and safety (RAMS analysis) takes its importance due to the huge amount of restrictions and conditions that any

Table 5.2 Inputs/outputs in the efficiency of a warranty program

	Inputs	Outputs
RCM adapted to warranty	• Failure root identification in critical equipment • Recommended replacements • Proposed task frequency • Historical failure data • Failure information • Technological update • Improvement proposals	• Resource estimation • Repair quality • Warranty assistance plan
Warranty policy of cost risk benefit analysis	• Resource estimation • Repair quality • Reclamation forecast • Cost and risk reduction • Availability of technical and human resources	• Incidence risk • Incidence cost • Resource assignment • Detailed assistance program

complex product, currently on the market, holds in itself. Its components or subsystems present ranges of potential modes of failure that must be considered from the initial state of design. Some standards (as the ISO 14224/2000) [8] deal with this matter. In any case, the RAMS analysis gives an approached evaluation of the product behavior when it will be in use. As a result of such analysis, one can obtain a deep knowledge about the product capacity to perform properly the functions expected from it, under established conditions and during a period of specific time. It is also possible to know in which degree a system or component can be used during a certain interval of time or must be restored or substituted. In addition to this, this analysis takes into account the condition of the product to be protected against failures, mistakes, accidents, damage, or any other incidents considered as not desirable. These concepts are usually expressed in terms of probability, and for their measurement, it is taken into account the mechanisms and relationships between failure cause and effect. Due to the fact that, as mentioned, an after-sales support is frequently offered while the production line is still open (see Fig. 4.3), the purpose here is to improve the product engineering and manufacturing with the feedback of warranty program data. With the data gathered till now, a reliability assessment can highly influence changes on the product design as it is shown in Fig. 5.5. With an adequate reliability and availability assessment, it is possible to demonstrate in the first stages of the product, how requirements expressed in initial technical specifications can be incompatible or even impossible to accomplish for determined product [9]. If the product is already launched, this analysis can help to take quickly the necessary measures to correct and/or improve the product, foreseeing also probable claims from the users due to the real lack of reliability on the product, in comparison with the previous reliability, sold a priori.

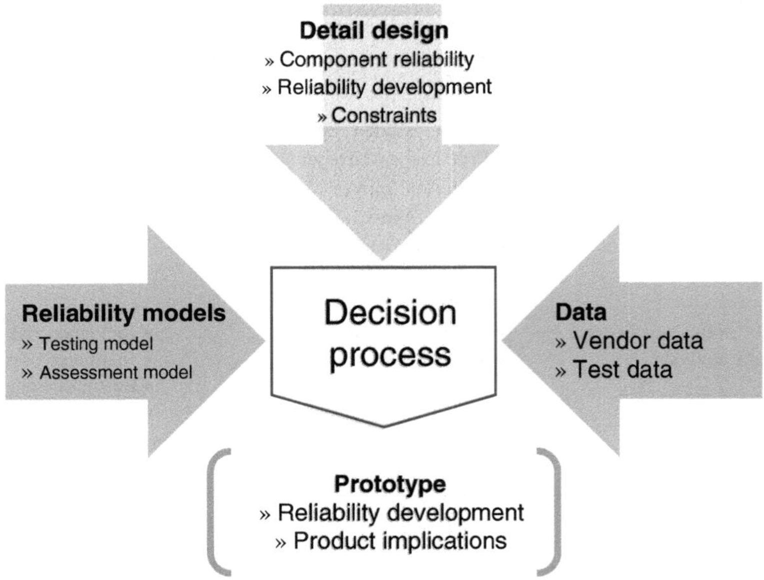

Fig. 5.5 Reliability assessment and the product development (adapted from [10])

To complete this third step about warranty program assessment is important to include a life cycle cost analysis, which calculates the cost of the product for its entire life. The typical analysis includes costs for planning, research and development, production, operation, warranty, and disposal [11]. From the consumer's point of view, the life cycle cost will suppose the acquisition costs, purchase price, costs of operation and maintenance, etc., that means, in general, the total cost of the item ownership. In any case, the life cycle cost regarding warranty issues is highly influenced on the values for reliability and failure rate, cost of spares, repair times, and component costs. Normally, a low budget for product engineering leads to high warranty costs in the future. Those customer complaints related to important or costly failures should be soon attended and the failures fully analyzed to identify not only further tasks to proceed with the repair, but also preventive actions which can avoid or at least to decrease future claims due to similar reasons. Therefore, this consideration involves performing a root cause failure analysis already included in step 1. An overall review of all the warranty complaints can show repetitive failures and trends related to vendor/buyer problems, quality issues, manufacturing conditions, product design, etc.

Table 5.3 provides a sample of the information that can be required or obtained in the assessment and control step of a warranty program for a generic company.

Table 5.3 Inputs/outputs in the assessment of a warranty program

	Inputs	Outputs
Reliability, availability, maintainability, and safety	• Incidence risk • Patterns of use	• Items historical reliability • Recommended replacements • Proposed task frequency • Data about "dependability"
Life cycle cost analysis	• Incidence cost • Data about "dependability" • Recommended replacements • Proposed task frequency • Patterns of use • Costs and risk reduction	• Warranty program cost • Data about "dependability" and CCV • Duration and potential extension

5.1.4 Fourth Step: Program Improvement

As a result of the previous steps of this management process, a continuous improvement in the warranty program management is possible thanks to the application of emerging techniques and technologies in areas considered of higher impact. The implementation of new technologies to warranty brings with itself the introduction of the concept "e-warranty." This concept is proposed as a component of the e-manufacturing, which takes an advantage from the emerging information and communication technologies, in order to implement a cooperative and distributed multi-user environment. Therefore, e-warranty can be defined as a warranty program support which includes the resources, services, and management necessary to enable proactive decision. This support includes not only e-technologies, but also e-warranty activities such as e-monitoring, e-diagnosis, and e-prognosis. Together with the application of new technologies to warranty management, the involvement of after-sales technicians in the improvement warranty process is fundamental to successfully achieve targets and goals that require a high level of quality in the product and efficiency of the technical assistance service. Consequently, higher levels of knowledge experience and training will be required.

It is clear that one of the most important parts inside the management of a warranty program is the relationship with the client. The techniques that manage these relationships are usually considered part of an enterprise resource planning (ERP) system, and sometimes it is treated as a complementary part. The customer relationship management (CRM) is used not only to define a strategy of business centered in the client, but also to include a group of applications useful to deal with data related to clients, complaints, and, in general, the commercial activity of a company. A CRM module can be adapted for the warranty management, which should allow the company

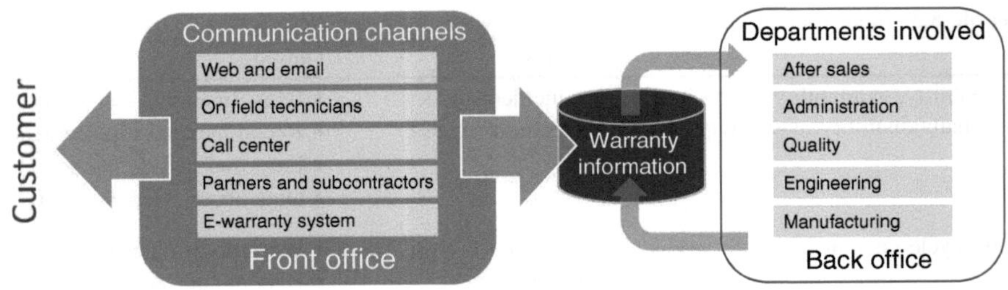

Fig. 5.6 Customer relationship management of the proposed framework

- To identify products and services that clients require to be attended.
- To optimize assistance times and information channels.
- To identify client's groups in order to develop warranty strategies.
- To be aware about the current after-sales necessities of the clients.
- To increase sales of the company together with the client's satisfaction.
- Among others.

The CRM (just as part of an ERP or as a complementary system) includes operational areas and tasks that deal directly with the client ("front office") and other ones more analytical where different parts of the company are closely involved ("back office") representing the business intelligence, as it is shown in Fig. 5.6. The main part of any CRM strategy will be the database, which can provide very interesting information about complaints, repetitive failures, hidden defects, quantitative and qualitative analysis, statistical results, etc.

To lubricate all this machinery, the setup of programs is necessary to optimize the organizational efficiency, in this case focused on the warranty program management. For that purposes, it is possible and interesting to adapt, for instance, concepts from the total productive maintenance (TPM) philosophy. The TPM tries to achieve a continuous improvement applied in that case to a maintenance practice. This objective resides also in other methodologies such as total quality management (TQM) or even the ISO 9000 (ISO 9000, 2000) [12], which has been used here as a base for the framework. In our case, and with the intention to go beyond, many organizations are recently implementing a new management philosophy that can be also very useful for the warranty management. The Six Sigma is a methodology that integrates the human factor with improvement tools, mainly statistical, to engage the complex mechanisms inside a company (Fig. 5.7). As it is known, Sigma in statistic means a dispersion measurement (variability). This variability is considered as the main enemy of quality; therefore, Six Sigma proposes the following steps:

- To select indicators or variables, which are critical for quality.
- To measure these variables in order to compile real data of the current situation.
- To detect the variation source of those interesting variables and their allowed tolerance.

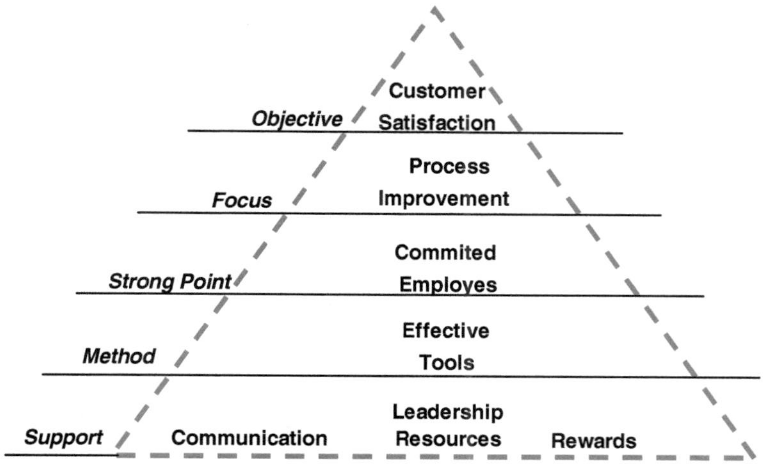

Fig. 5.7 Integration of total quality principles in Six Sigma

Fig. 5.8 Fundamental pillars of the Six Sigma methodology

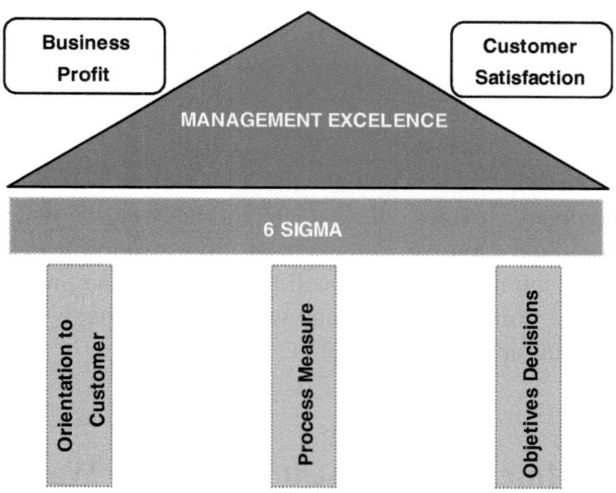

- To implement solutions.
- To apply statistical controls that allows the long-/middle-term improvement.

The application of the Six Sigma philosophy (Fig. 5.8), supposes the necessity of real data and the use of a huge amount of variables.

Therefore, this methodology combined with new e-technologies can bring very good results in the field of the warranty program management. For the implementation of the Six Sigma, it is required a team dedicated full-time to such matters. Regarding warranty issues, this team should be composed by middle managers, independents of after-sales, engineering, manufacturing, or quality department, but establishing a management team who watch over the warranty matters dealing with the complaints from customers and/or to providers.

Table 5.4 Inputs/outputs in the enhancement of a warranty program

	Inputs	Outputs
Communication and information technologies	• Assessment of warranty management program • Data of "dependability" and CCV • Technology observation	• New strategies, e-technology • Historical failure data • Reclamation forecast • Patterns of use • Historical data • Patterns of actions and claims
Customer relationship management	• Matrix of criticality • Reclamation forecast • Historical data • New technologies, e-technology • Business requirements • Communicated claims	• Customer satisfaction • Marketing data • Failure information • Claim execution patterns
Six Sigma	• Assessment of warranty program costs • Repair quality • Incidence costs • Data "dependability" and CCV • Claim execution patterns • Standards and methods	• Improvement proposal • Technological update • Costs and risks reduction

In the similar way to the previous stages, Table 5.4 offers a sample for a generic company. It shows all that specific information that is required or obtained in the continual improvement stage of the warranty. It is important to point out that all these inputs and outputs will depend of course on the organization in question and of the boundary conditions where the company develops its industrial and commercial activity.

5.2 Integral Representation of the Warranty Management Framework

The proposed warranty management framework [13] achieves the highest level of maturity as it is a cyclical process (see Fig. 5.9), whose decisions are taken according to the overall targets of the business, adopting the feedback taken from quantitative and qualitative data for the enhancement of the productive system and applying known technologies, already developed in other areas (see also Chap. 6).

Just as in Figs. 4.4 and 4.6 of the previous chapter, we try to characterize the process related to warranty management using representation tools such as integrated definition (IDEF) methods [14, 15]. These representation methods are easy to use and understand, and we are going to represent each proposed stage of the warranty management framework as a chart with inputs, outputs, and used

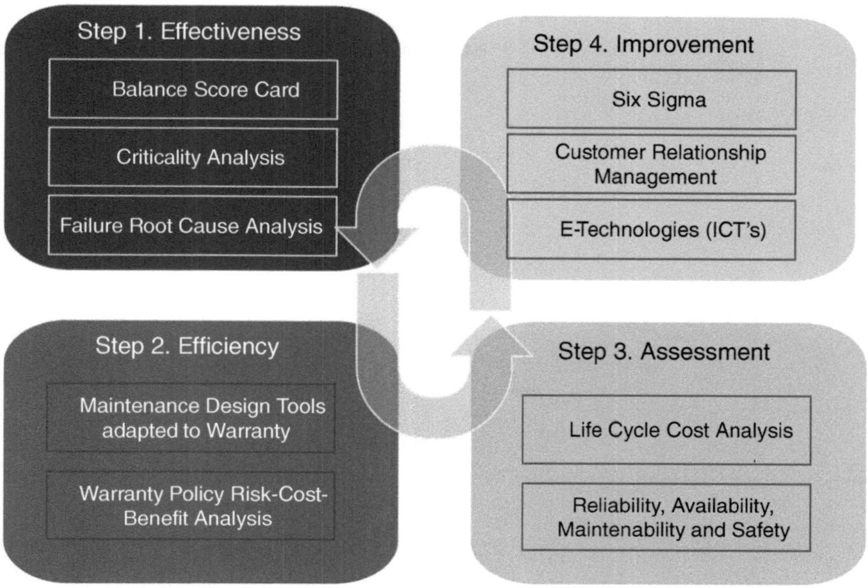

Fig. 5.9 Framework proposal for a warranty management program

resources [16, 17]. The after-sales department, as any other, has to operate according to a business plan and to contribute to the scope of company objectives complying requirements [18]. When the strategy is defined, the following step is to implement it considering the restrictions of the organization. For that goal, it is necessary to set all those functions required to perform the strategies and classifying them in processes and activities. The overall warranty management process can be divided into blocks, as the ones proposed for maintenance by Refs. [2, 19, 20].

Consequently, the objective will be to define a process diagram where the different stages of the proposed framework can be distinguished. The main processes and activities will be similar to the case of maintenance. Table 5.5 indicates the basic and common activities in the case of warranty and the proposed framework, indicating their relation to every phase of the reference they are led to. However, technical performance for each activity can be different from one company to the other. These blocks represent different processes for particular warranty management scenarios, which need different resources, and contribute in a different way to the service quality, according to the type of knowledge they are based on and to their control possibilities.

The process to coordinate all the interactions among warranty functions corresponds to the "strategy organization." This process also leads the warranty management interactions with the rest of the company organizational functions, enhancing the knowledge management.

Table 5.5 Inputs/outputs between stages of the proposed framework

INPUTS OF THE ACTIVITIES:							
Criticality Analysis	Failure Root Cause Analysis	Maintenance Design Tools adapted to Warranty	Warranty Policy Risk-Cost-Benefit Analysis	Reliability, Availability, Maintenability and Safety	Life Cycle Cost Analysis	E-Warranty	Customer Relationship Management
» KPI selection						» Warranty program costs evaluation	
	» Criticality matrix						» Criticality matrix
		» Identification of failure root cause specially in critical equipments					
			» Estimation of resources » Repair quality				
				» Incidences risk	» Incidences cost		
		» Recommended spares » Suggested task frequency			» Depend. data » Recom. spares » Suggested tasks frequency		
						» Dependability & LCC output data	
» Items historical failure data	» Items historical failure data	» Items historical failure data	» Claims forecast	» Usage patterns	» Usage patterns		» Claims forecast » historical data » New e-tech strategy
» Failure information	» Failure information	» Failure information					
		» Technical update » 6s improvemts proposals	» Cost and risk reduction		» Cost and risk reduction		
	» Standards and methods		» Technical and human resources available			» Technology watch	» Business Requirements » Communicated complains

From an operating point of view and according to the responsibilities involved, three possible interactions are here presented among the warranty processes:

1. A first possible interaction is when a complaint is communicated by a customer which must be answered by warranty assistance.
2. A second possible interaction in case of an incident not detected because a complaint from a customer, but through (for instance) a proactive (e-warranty) or a detection during another intervention.
3. Finally, activities from the after-sales department are also requested when there is any modification in parts of the product already in the market. When it happens, this is usually due to mistakes in product design or failures in a product batch. Many of these types of activities are sometimes approved to be accomplished during preventive maintenance actions.

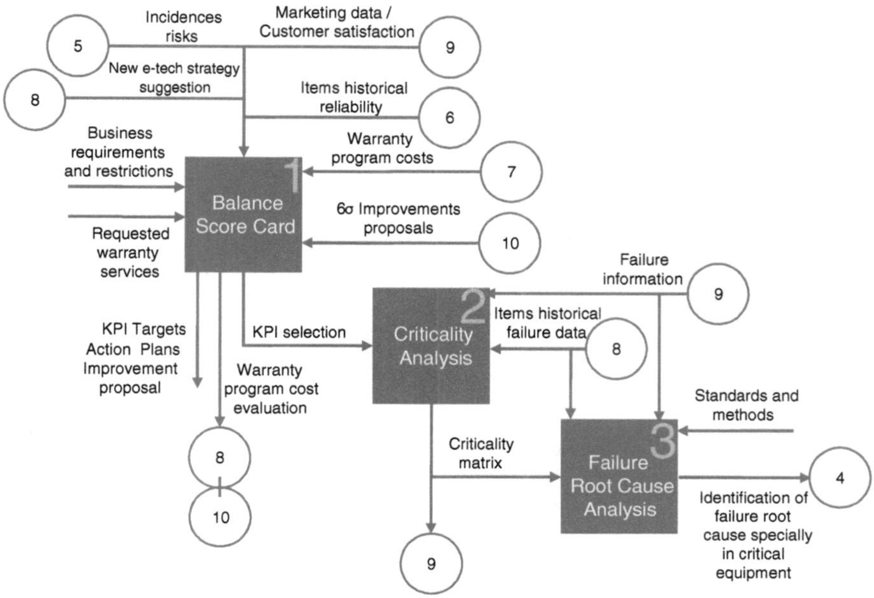

Fig. 5.10 Integrated representation of level 2A: effectiveness

In a few words, Table 5.5 summarizes all inputs and outputs that have been discussed throughout the present chapter, also adding the way the different methodologies and proposals interact. As it has been previously stated, the performance of the organization varies from one to the other and will depend on the particular scenarios where it is applicable. To illustrate in greater detail the interactions between stages, it is possible to represent next level (level 2), obtaining the processes shown in Figs. 5.10, 5.11, 5.12 and 5.13.

As it has already been introduced with level 0 and level 1, the representations shown in Figs. 5.10, 5.11, 5.12 and 5.13 are a simplified way of illustrating how the different proposed stages interact with one another, in a way that if we took one of them in an isolated way, we would observe the boundary conditions in which it is immersed, as well as the flows of information in which it participates, either as receptor or as generator. This form of illustrating the proposed reference framework is based on the standard IDEF methods [21, 22] that, also as previously mentioned, are easy to implement and understand, allowing us to represent each stage of our reference framework in boxes with a series of inputs and outputs [16, 23], applying here in a particular manner to serve our purposes of study.

That is to say, the intention here has not been to follow inflexibly the criteria of the IDEF standard, but to take advantage of its illustrative potential to reveal in a distinct way the same information shown in Table 5.5. In these four figures specifically, the indicated numbering in circles makes reference to the phase of the proposed framework from where it emerges or where information or data are

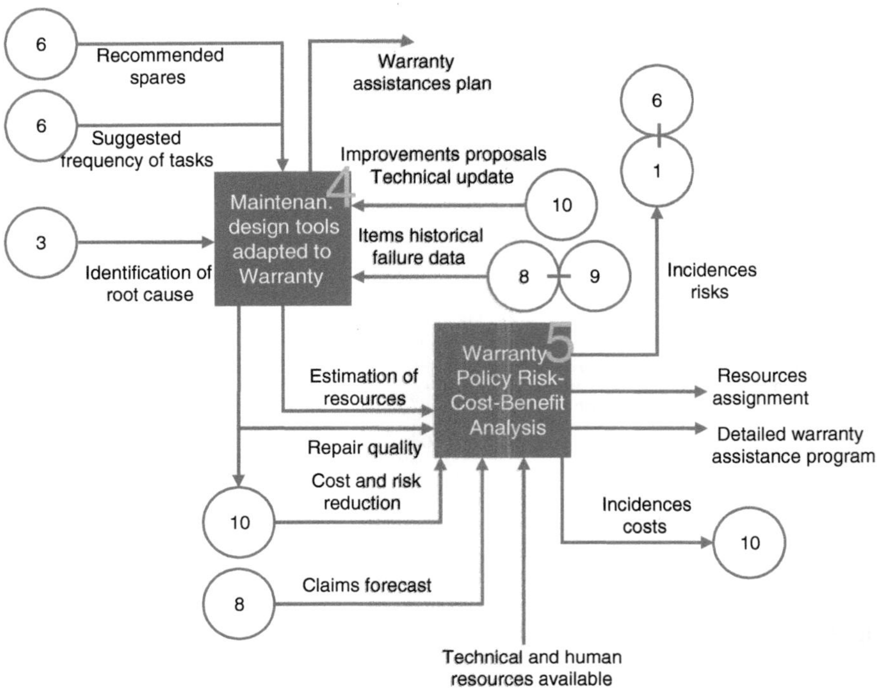

Fig. 5.11 Intergrated representation of level 2B: efficiency

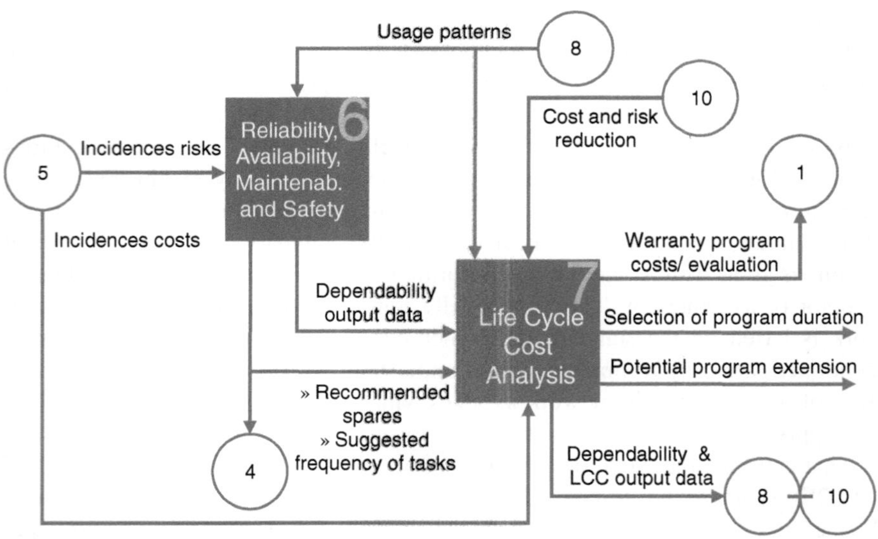

Fig. 5.12 Integrated representation of level 2C: assessment and control

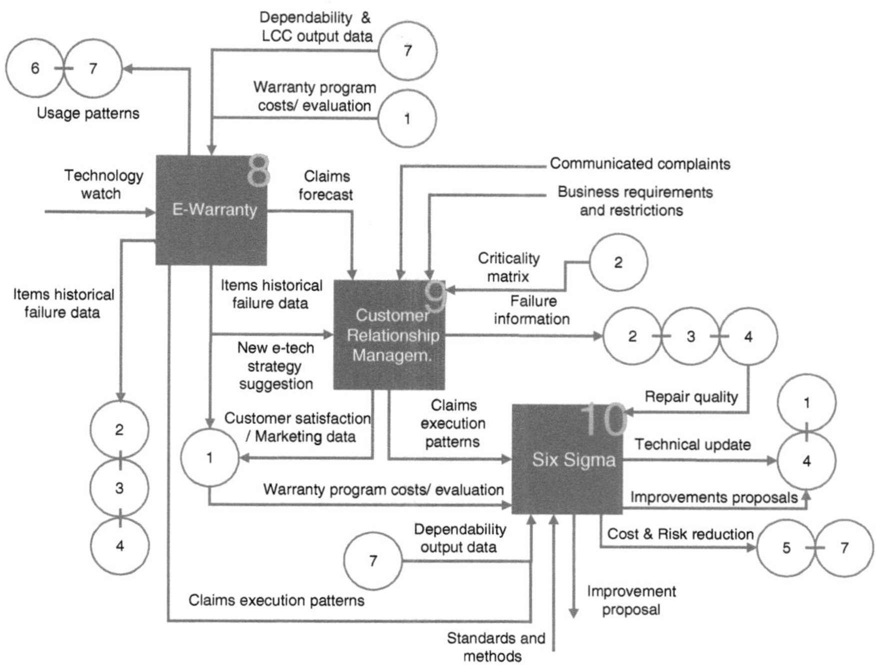

Fig. 5.13 Integrated representation of level 2D: continual improvement

directed that are indicated on the flow lines. Due to that at the beginning stage this study was founded on the intention of applying tools developed in the civil industry to the defense sector, it is important to highlight here (even though it is not within the limits of this research) that currently there are information software packages that facilitate organization of a high quantity of interactions and information, principally for complex industrial assets from the military fields such as combat vehicles, armored boats, and military transport airplanes.

5.3 Framework Implementation

Once the proposed framework has been described, this section will briefly describe the way in which it would be applied in a real case. In general terms, the previously described framework for warranty management requires the following conditions or characteristics for a successful application:

1. *General support of the organization*: The implementation of this framework requires the agreement, support, and comprehension of the whole company.
2. *Complex industrial assets*: This framework will be useful when the product is highly complex and technologically sophisticated.

Table 5.6 Modules of supporting information systems

Logistic support analysis	Provisioning	Maintenance management	Technical publications	E-learning
Management of logistic support analysis (LSA)	Integration of data from LSAR	Configuration management	Technical content management	Content management
Reliability, availability, and maintainability	Creation of the illustrated catalog of replacements	Operational use of failure follow-up	Knowledge database of SGML/XML	Reuse of content in technical publications
Maintenance task analysis	Interaction with the supply chain	Inventory and material management	Applicability management	Model of content in XML
Generation of task modules	Support through international standards	Parameters and data analysis	Complete publication capability	Multimedia support
Integrated edition of XML		Exchange capabilities and management of parts	Interactive electronic technical publications	Support to pedagogical structured objects
Standards for logistic data of products		Logistic based on performance	Specifications for technical publication	

3. *Supporting information systems*: It is mandatory that all the applied information systems should be appropriate to coordinate and organize large amounts of data as well as different involved areas. Specifically, the application of information systems can be organized into three main modules: logistical support analysis (LSA), provisioning, maintenance and supporting services management, as well as two additional modules: technical documentation and e-learning (Table 5.6).
4. *Training staff and managers*: The people in charge of making decisions in the field of application of this reference framework have to be trained in all supporting techniques and methodologies, also to be needed to be provided with a proper vision of the warranty management framework.
5. *Training of technical staff*: Complementing the last paragraph, personnel performing warranty interventions have to assure an accurate data collection (symptoms, ways of failure, causes, etc.).
6. *Assurance and commitment in the control*: It is important to guide and properly manage the warranty program at a tactical, operative, and strategic level within the applied and proposed reference framework.

As an example of a successful application of the warranty management framework, some businesses use as a supporting tool the "Eagle Tool Kit," developed by the company [24]. "Eagle" is an information system designed in an efficient fashion to help creating, administrating, and delivering supporting logistical data, being currently used by government agencies, contractors, service suppliers, and companies worldwide to manage integrated logistical support (ILS) of the equipment and assets in operation, following the corresponding military norm [25]. The previous conditions and characteristics represent mainly the warranty management of a complex industrial asset in its phase of operation by the user. However, the proposed framework is also relevant in the preoperative phase, when the product is still in a phase of design and development, so it is possible to forecast the supposed behavior of industrial assets once they are launched in the market.

References

1. Crespo Márquez A, Moreu P, Gómez J, Parra C, González-Prida V. 2009. The maintenance management framework: A practical view to maintenance management. Taylor & Francis Group, London, ISBN 978-0-415-48513-5. Pp. 669-674.
2. Crespo Márquez A. 2007. The maintenance management framework: models and methods for complex systems maintenance. Springer-Verlag London Limited. ISBN-13: 9781846288203
3. Murthy DNP, Rausand M, Osteras T. 2008. Product Reliability - Performance and Specifications. Springer Verlag, London. Pp 284+xvii.
4. Murthy DNP, Rausand M, Virtanen S. 2009. Investment in new product reliability, Reliability Engineering & System Safety. 94:1593-1600.
5. Murthy DNP, Hagmark PE, Virtanen S. 2009. Product variety and reliability. Reliability Engineering & System Safety. 94:1601-1608.
6. Murthy DNP, Osteras T, Rausand M. 2009. Component reliability specification. Reliability Engineering & System Safety. 94:1608-1617
7. Gómez J, Crespo Márquez A, Moreu P, Parra C, González-Prida V. 2009. Outsourcing maintenance in services providers. Taylor & Francis Group, London, ISBN 978-0-415-48513-5. Pp. 829-837.
8. International Std. 2004. ISO/DIS 14224, ISO TC 67/SC /WG 4. Petroleum, petrochemical and natural gas industries - Collection and exchange of reliability and maintenance data for equipment. Standards Norway.
9. González-Prida V, Crespo Márquez A, Moreu P, Gómez J, Parra C. 2009. Availability and reliability assessment of industrial complex systems: A practical view applied on a bioethanol plant simulation. Taylor & Francis Group, London, ISBN 978-0-415-48513-5. Pp. 687-695.
10. Murthy DNP, Blischke WR. 2005. Warranty management and product manufacturing. Springer-Verlag London Limited. Pp 302+ xviii. ISBN 1852339330.
11. Parra C, Crespo Márquez A, Moreu P, Gómez J, González-Prida V. 2009. Non-homogeneous Poisson Process (NHPP), stochastic model applied to evaluate the economic impact of the failure in the Life Cycle Cost Analysis (LCCA). Taylor & Francis Group. London. ISBN 978-0-415-48513-5. Pp. 929-939.
12. European Std. 2008. EN-ISO 9000 standards. Quality Management System. European Committee for Standardization, Brussels.
13. González-Prida V, Parra C, Gómez JF, Crespo Márquez A. 2010. Reference framework proposal for the management of a warranty program. Euromaintenance 2010. 20th International Industrial Maintenance Congress in the Verona Conference Centre, Italy. Pp 100-105.
14. Feldmann CG. 1998. The Practical Guide to Business Process Reengineering Using IDEF0. Dorset House Publishing.

15. Giaglis GM. 2001. A Taxonomy of Business Process Modelling and Information Systems Modelling Techniques. The International Journal of Flexible Manufacturing Systems. 13:209-228.
16. IEEE Std. 1998. IEEE Standard for Functional Modelling Language -Syntax and Semantics for IDEF0. 1320.1. New York: IEEE.
17. Kim SH, Jang KJ. 2002. Designing performance analysis and IDEF0 for enterprise modelling in BPR. Int. J. Production Economics 76:121-133.
18. González-Prida V, Barberá L, Gómez JF, Crespo Márquez A. 2011. Contractual and quality aspects on warranty: best practices for the warranty management and its maturity assessment. International Journal of Quality and Reliability Management, IJQRM.
19. Duffua SO, Raouf A, Campbell JD. 2000. Planning and control of maintenance systems. Indianapolis: John Wiley and Sons.
20. López M, Crespo Márquez A. 2010. Modelling a Maintenance Management Framework Based on PAS 55 Standard. Quality and Reliability Engineering International. John Wiley & Sons Ltd.
21. Feldmann CG. 1998. The Practical Guide to Business Process Reengineering Using IDEF0. Dorset House Publishing.
22. Giaglis GM. 2001. A Taxonomy of Business Process Modelling and Information Systems Modelling Techniques. The International Journal of Flexible Manufacturing Systems. 13:209-228.
23. Shoung-Hie Kim, Ki-Jin Jang. 2002. "Designing performance analysis and IDEF0 for enterprise modelling in BPR". Int. J. Production Economics 76 121-133.
24. http://www.raytheon.com/
25. Department of Defense, USA, "Requirements for a Logistics Support Analysis Record" MIL-STD-1388-2B

Chapter 6
Auditing the Initial Case Study

In this chapter, we will proceed to analyze the case study proposed in Chap. 3 as an audit, suggesting improvement actions in accordance with the reference framework described in Chap. 5 [1, 2]. It is intended to check the support provided from distinct methodologies and tools proposed as stages in the reference framework [3, 4]. With the purpose of outlining in greater detail a complete development of a framework, further researches in this field will be able to focus on the identification of those good and bad practices followed by different companies in several sectors.

6.1 Considerations to the Current Situation

This brief exercise is helpful to check whether our organization is considering the nowadays available techniques to manage properly these kinds of issues or there are still some fields left on the decision-making scope, which could be targeted for improvement. For that purpose, Tables 6.1, 6.2, 6.3 and 6.4 present in few words some weak points of the case study and the referred suggestions according to the application of the proposed framework.

6.2 Improvement Actions

Once identified those areas to be enhanced, it is possible to propose some practical actions to improve the warranty program management and, in general, its decision-making for our defined study scenario.

V. González-Prida Díaz and A. Crespo Márquez, *After-sales Service of Engineering Industrial Assets*, DOI: 10.1007/978-3-319-03710-3_6,

Table 6.1 Effectiveness stage checking

Phase	Initial scenario	Proposed framework
1. *Balance score card*	Decisions are usually taken in accordance with problems arising. There is an after-sales strategy at the beginning of the project, and the following changes are related to short-term reachable objectives. There is a reactive management behavior; thus, managers act when a problem appears but do not take an advantage to avoid its appearance	Strategical formulation for the warranty program must be translated to middle-term reachable objectives and the development of action plans. A periodical review of performance and strategy is basic just to quantify the progress and take into account any necessary changes in the organization's support infrastructure
2. *Criticality analysis*	Customers' complaints are usually prioritized following a temporal scheme (as a FIFO method) in the case scenario. That is due to the fact that the concept of inoperance is not really clear between client and manufacturer. In addition to this, managers who take decisions on this area have to fight normally against business objectives	A systematical decision about which assets should have priority related to the warranty management can avoid discrepancies with the existing program strategy. Taking into account the probability of an event and the impact that this event would cause, it can allow aligning after-sales actions with the business targets
3. *Failure root analysis*	It is about repetitive and systematical failures that are specially treated. However, the concept of systematical failure is not clear, and therefore, special interventions on such cases are usually not fast	The failure root cause analysis should run a protocol of systematical actions which can allow a soon intervention, readapting the considerations taken initially on the balanced scorecard

Table 6.2 Efficiency stage checking

Phase	Initial scenario	Proposed framework
4. *Maintenance techniques adapted to warranty*	The maintenance criteria about capacity planning, the spare parts provisioning, the warranty task schedule, skill level of the technicians, etc., is applied during the whole warranty program	An initial maintenance plan, applied to the warranty time horizon, can suppose a good first approach. Nevertheless, it is crucial for its adaptation to the warranty necessities
5. *Warranty policy risk-cost-benefit*	There is no risk–cost–benefit analysis; however, a costs provision is supposed at the beginning of the project, and during the development of the warranty program, there is a follow-up and a readaptation of the initial budget assigned to warranty assistance	An integrated model for total warranty costs depends on the information available. Moreover, the complexity of the issue is usually high and considerations must be taken under certain assumptions in order, either to simplify the analytical resolution, or to reduce sometimes the computational needs

6.2.1 Case Study Scenario Effectiveness

For each department implicated in the warranty management process, a balanced scorecard model is suggested to avoid discordances between local objectives and the complete business goal. Objectives, measures, targets, and initiatives mainly for the after-sales department, but also for logistics, manufacturing, quality, purchasing, and engineering departments, must be developed under the supervision of the management team and periodically reviewed in order to quantify the middle-term progress and to formulate new strategic goals and action plans. In accordance with this program strategy, it is crucial to have clearly defined between client and manufacturer concepts as systematical failure or inoperance degree, in order to enable special interventions when needed. The Criticality and the Failure Root Cause Analysis must be therefore performed using those data about incidences, damaged materials, repairs, etc., gathered by after-sales, quality, and engineering departments and saved in the database, having also in mind those considerations taken initially on the balanced scorecard. These analyses and the database compiling must be followed up and reviewed by the quality department.

6.2.2 Case Study Scenario Efficiency

As soon as the warranty assistance began to be applied, the current warranty needs related to the actual organization's support infrastructure appear much clearer. With these data, a reliability analysis, together with a risk-cost-benefit analysis (adapted both to warranties) can be performed, making easier on one hand those decisions

Table 6.3 Assessment stage checking

Phase	Initial scenario	Proposed framework
6. *RAMS analysis*	The after-sales support is offered in the case scenario while the production line is still open, the product engineering and manufacturing improves with the data feedback from the warranty program. However, in this case study, it is supposed that every vehicle shows the same level of reliability	Although this analysis must be carried out at the very early stages of any product's life, in the case where the product is already launched, this analysis can help to take measures soon and to correct or improve the product, foreseeing also probable claims from the users due to a low reliability on the first products. Therefore, this action increases reliability progressively, with the consequent client's satisfaction
7. *Life cycle cost*	This analysis is carried out in the scenario at the beginning of the project, together with the rest of the areas of the product life cycle, being warranty costs simply considered as a percentage of the complete project budget	Life cycle cost regarding warranty issues is highly influenced on the values for reliability and failure rate, cost of spares, repair times, and component costs. As commented along the paper, a low budget for product engineering leads to high warranty costs in the future. Therefore, a complete LCC analysis can help to determine an appropriate budget for the possible warranty needs in the future

Table 6.4 Continuous improvement stage checking

Phase	Initial scenario	Proposed framework
8. *e-Technologies*	High levels of knowledge, experience, and training are required in the study scenario to the warranty technicians, in order to work with special tools and test equipment useful to check the status of each vehicle. However, it is not implemented in new technologies as e-monitoring, e-diagnosis, e-prognosis, etc., included in the concept of "e-warranty"	The implementation of new technologies for a warranty program is crucial as a support. It includes those resources, management, and services, necessary to enable proactive decision, in order to reach a successful goal which supposes higher level of product quality and program effectiveness
9. *Customer Relationship Management*	The implementation of new technologies for a warranty program is crucial as a support. It includes those resources, management, and services, necessary to enable proactive decision, in order to reach a successful goal which supposes higher level of product quality and program effectiveness	Identify products to be attended or to open faster information channels; a high improvement will be also to enable the identification of clients' groups in order to develop warranty strategies and to optimize assistance times. The main part of the strategy is the database, which provides also interesting information about quantitative and qualitative analysis, statistical results, etc.
10. *Six Sigma*	To lubricate all these machinery. There is a team in the case study (management team), not necessarily multidisciplinar, which focuses the communications with the client and among the different departments, distributing orders, and taking decisions	The proposed methodology requires however a multidisciplinary full-dedicated team, together with improvement tools, mainly statistical, in order to engage the complex mechanisms inside a company

about capacity planning, spare parts provisioning, warranty task schedule, skill level of the technicians etc., and, on the other hand, the readaptation of the initial budget assigned to warranty assistance. These facts are of course alive during the development of the warranty program. That means the different departments must follow up those data which affect their scopes and responsibilities, according to the event evolutions and the boundary conditions for each moment. The management team should gather all these information and, together with the after-sales department, periodically update the required resources to cover properly the actual warranty necessities.

6.2.3 Case Study Assessment

The warranty database, fed with those information coming from after-sales, Engineering, and Manufacturing areas and kept by the quality department, is a good tool to perform a Reliability, Availability, Maintainability, and Safety analysis, as well as a life cycle cost study. The results from the RAMS analysis will be expressed in terms of probability, taken into consideration the mechanisms and relationships between failure cause and effect. These results feed back the Engineering and manufacturing department, so they can apply changes in the design and assembly, improving consequently the product reliability, and will be useful for a later phase related to the Six Sigma application. The reliability values and the failure rates influence highly the cost of spares, the repair times, etc. This information must be taken into account on the one hand by the management team in order to determine an appropriate budget for the possible warranty needs in the future and on the other hand by the after-sales department in order to attend soon those claims related to important or costly failures. By these actions, the after-sales department should also identify those preventive actions which can help to avoid or decrease future customer complaints, started by similar causes.

6.2.4 Case Study Improvement

The huge and diverse amount of data requires of course an automatic processing and a systematical treatment. The use of a database and other computing tools to save and store all these information is therefore essential. The help of the IT department to make easier the data processing and the information flow can be very useful. In addition to this, the use of new e-technologies has been also proposed along this paper. These new e-tools can be for instance (and when possible) the vehicle condition monitoring system. This system can check whether systematical failures have been eliminated and solved. The above-mentioned warranty database, together with a customer relationship management system, helps the management team to (among others) identify customer's groups or, in the case of a fleet of vehicles, to determine different vehicles' behaviors when they are located in different destinations, in order to plan a proper warranty schedule and to analyze with the after-sales department possible reasons and causes to localized failures. Finally, the management team has been here defined as a full-time dedicated team to coordinate and deal with all those warranty issues. Nevertheless, according to the Six Sigma methodology, this team will be much more effective if included among its members are engineers, technicians, and managers from each implicated department. By this way, the management team can not only have a complete view of the warranty process, but also know the casuistry of each involved area and how decisions can affect them. As an example, Table 6.5 tries to summarize how much the different department can be involved in the described phases of the proposed framework. In any case, this involvement degree will depend of course on the company policy for its warranty program management. Other references are [5–7].

6.3 Quality and Maturity Aspects in Warranty Management

6.3.1 Standards and Best Practices for the Quality of Warranty Management

Now, it is the moment to observe what companies are doing nowadays on this field, their best practices in warranty management concerning business process and covering quality and contractual aspects. There are some standards which capture best practices from the industry, whose application can be very useful for their warranty management in order to find competitive advantages in this field (Table 6.6). Some of these standards are presented as follows:

- The Baldrige National Quality Program and the associated Award helps to improve organizational performance practices, capabilities, and results, sharing information on best practices among US organizations of all types and being used as a working tool for managing performance and for guiding planning and opportunities for learning [8].
- Total quality management (TQM) is a management philosophy which tries to improve every function in a company (marketing, finance, design, engineering, production, customer service, etc.) in order to focus them on one company target, meeting customer needs [9].
- European Foundation for Quality Management (EFQM) helps organizations to continually improve and achieve higher levels of performance in management by processes and quality, providing networking and mutual learning experiences, as well as supporting the implementation of best in class tools [10].

Table 6.5 Involvement degree of each department in the different phases

		Logistics	Manufacturing	Administration	After sales	Quality	Purchasing	Engineering
Effectiveness	Balance scorecard	**	**	***	***	**	**	**
	Criticality analysis	*	**	*	***	***	*	***
	Failure root cause analysis	*	**	*	***	***	*	***
Efficiency	Maintenance design tools adapted to warranty	***	**	**	***	**	***	**
	Warranty Policy Risk-Cost-Benefit Analysis	**	**	***	**	**	***	**
Assessment	Reliability, availability, maintainability, and safety	*	***	*	***	***	*	***
	Life cycle cost	*	*	***	***	*	***	**
Improvement	E-technologies	*	*	**	***	**	*	*
	Customer relationship management	*	**	***	***	**	*	**
	Six Sigma	**	**	***	**	**	**	**

where
* Department less involved in the corresponding phase of the framework
** Department partially involved in the corresponding phase of the framework
*** Department highly involved in the corresponding phase of the framework

Table 6.6 Standards and international practices (adapted from [14])

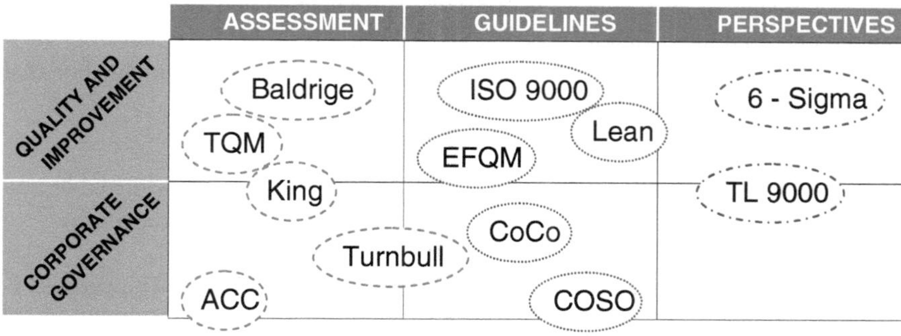

- ISO 9000 is a family of standards for system management by processes and quality, which is maintained by the International Organization for Standardization. It proposes a set of procedures that cover all key processes in the business in order to ensure that such processes are effective, reviewing periodically the quality system itself for effectiveness and facilitating continual improvement [11].
- Committee of Sponsoring Organizations of the Treadway Commission (COSO) is a recognized organization of the private sector, which provides guidance on critical aspects of organizational governance, business ethics, internal control, enterprise risk management, fraud, and financial reporting [12].
- TL 9000 is a quality management practice created to focus on supply chain directives, mainly for the international telecommunications industry [13].

Unfortunately, no single model works in every condition. Standards and reference frameworks should facilitate and improve business processes and productivity in general [15]. Therefore, the proposed framework has to capture features of these standards, fitting our problem better. According to [16], a process is a set of tasks logically related to run for attaining a certain result of business. In addition to this, Ref. [17] considers a process as a logic organization of persons, materials, energy, equipment, and procedures in designed activities to produce a final specified result.

Particularizing to our case, warranty management should be carried out as a redesign [18] using an integrated methodology in order to achieve an improvement on processes' performance, such as cost, quality, service, and speed [19]. There is, consequently, a double view of each process: a strategic view (for its contribution to business) and an operational view. The warranty management framework proposed at the beginning of this work is focused on both previous views [20]. The framework is based on activities which help to assure the quality on the warranty service [21], facilitating the decision-making and finding improvement points in a quick and easy way. Summarizing, it tries to be oriented to customers, processes, and delivery of services. In addition to all the above-mentioned standards and references, the Capability Maturity Model Integration (CMMI) is a framework to improve processes oriented to those customer services as warranty assistance can be. It provides a list of essential elements of management in order to reach a determined level of maturity in management, making efficient processes [22, 23]. The CMMI is therefore now proposed to evaluate those areas involved in the warranty management. For that purpose, maturity is quantified in five progressive levels:

- *Level 1—Initial:*
  It represents a process with unpredictable result. The process is unstable and unorganized. It is defined by who performs it without explicit procedures.
- *Level 2—Repeatable and reactive:*
  It represents a process characterized by repeatable performance. The process is planned, implemented, monitored, and checked according to predefined objectives.
- *Level 3—Defined and proactive:*
  It represents a process characterized by a proper implementation program in the company. The process is based on well-defined methodologies, techniques, and supporting technologies. Proper procedures are established to drive this process.
- *Level 4—Managed:*
  The process is controlled, adopting quantitative techniques and, if this is the case, statistical analysis. The business objectives are checked by the comprehension of the results of the quantitative analysis.
- *Level 5—Optimized:*
  Focused on continuous improvement aligned with business objectives, a corporate policy is established to manage the quality of the process, based on quantitative data and feedbacks about the processes. If this is the case, also new methodologies, techniques, and technologies are tested.

6.3.2 Best Practices Applied to the Warranty Management Framework

In order to support our framework on the selected international frameworks, we have selected mainly widely extended and general practices:

- EFQM (2006): The European model of excellence from the European Foundation for Quality Management is a model used as a reference to maximize the continuous improvement of the management toward internal and external efficiency, coordinating activities and resources of the company.
- ISO 9001 (2008): The family of ISO 9000 is a list of procedures about "quality management" used to develop the proposed reference framework, as a guide for quality and processes orientation.
- TPM: Total productive maintenance by Japanese Institute of Plant Maintenance (JIPM) is a maintenance program centered in machinery to reduce quality failures, breakdowns, and accidents and to improve the productivity and motivation without quality reduction [24, 25]. TPM is used as a guide in

maintenance, giving a special attention to the personnel knowledge which can be adapted for the warranty management.

Based on this, the after-sales department, as any other, has to operate according to a business plan and to contribute to the scope of company objectives complying requirements [26, 27]. When the strategy is defined, the following step is to implement it, considering the restrictions of the organization. For that goal, it is necessary to set all those functions required to perform the strategies and classifying them in processes and activities. Once the strategy has been defined, the next step is to set it into practice, taking into account the organization's restrictions. It is therefore necessary to define all functions to be able to carry out strategies and classify them in processes and activities. In order to categorize activities by their strategic value (for instance, their usage in cost and resources, or in reference to their contribution to quality, taking into account the type of knowledge in which it is based, as well as the control possibilities). The overall warranty management process can be divided into blocks, as the ones proposed for maintenance by Refs. [28, 29]. Consequently, Table 6.7 defines a process where it can be distinguished as the following blocks:

- Strategy organization (SO), in order to coordinate and integrate warranty management activities for the achievement of the department/business objectives.
- Control (Ctr.), in order to deal with alarms and what to do about it, in a proactive sense (e-warranty).
- Incidents management (IM), in order to face complaints and how to correct them.
- Warranty analysis (WA), in order to improve continuously the whole warranty management.
- Incidents prevention (IP), in order to decrease the rate of complaints, conserving customers.
- Capacity management (CM), in order to assign properly the available resources.

Table 6.7 describes basic and common activities, indicating their relation to the process which leads them. Nevertheless, the technical actuation for each activity can be different from one company to another. The table here shown is supported in the indicated doctoral thesis in Ref. [30].

After this, there is a necessity to measure and guide the evolution of warranty management toward higher levels of performance. In order to achieve this goal, a warranty control system should be established and the knowledge has to be managed continuously, because any incident must be documented to follow a continuous improvement. In these terms, a methodology for the warranty management has been developed as a framework proposal. Its implementation can help for example to develop plans and take decisions with clearer goals and planned objectives as well as to find a better choice of strategies. The proposed framework facilitates the management of changes due to the fact that it can help in the decision-making, linking the effects with their root causes. For that target,

Table 6.7 Basic activities in warranty assistance

Activity	Description	SO	Ctr.	IM	WA	IP	CM
Scheduling	Plan on all the warranty activities	***	**	**	**	**	**
Documentation	Elaboration and compilation of processes, procedures, reports, etc.	**	**	**	**	**	***
Budgetary control	Accountability of budget assigned to warranty	***	*	**	*	**	***
Resource management	Control of resources, tools, staff, vehicles, etc.	**	**	**	**	**	***
Training	Internal and external learning/training	**	**	***	**	**	**
Logistics	Spare parts and warehouse management	**	**	**	**	***	**
Safety	Prevention of labor risks with regulations on safety and health	***	**	**	**	**	**
Legal communications	Contacts with administration, government, or other companies	**	**	***	**	**	**
Monitoring	Monitoring alarms about product status	*	***	*	**	*	**
Configuration control	Installing of new parts	*	***	**	**	**	**
Incidence management	Management of all kinds of incidences	*	*	**	**	**	*
Element modification	Verification of product changes	*	**	*	*	**	**
Measures	Evaluation of product quality	*	*	*	**	**	**
Facilities	Environment conservation, e.g., cleaning, disinfection, etc.	*	*	*	*	**	*
Support in field	Support in field on demand to other complaint	*	*	*	*	**	*
Predictive analysis	Predictive analysis to avoid or minimize future impacts	*	*	*	**	*	**
Problem management	Analysis of repetitive failures or with unknown causes	*	*	*	**	*	*
Impact analysis	Analysis of possible impacts and risks on products	**	**	**	**	***	**
Preventive analysis	Improvement plans or tasks to optimize product performance	*	*	*	**	*	**
Change management	Coordination activities which can affect warranty assistance	*	**	*	**	**	*
Capacity management	Capacity analysis in infrastructure	*	*	*	*	*	**

where
* Activity less involved in the corresponding process block
** Activity averagely involved in the corresponding process block
*** Activity highly involved in the corresponding process block

warranty measurements have to be structured hierarchically for each organizational level with performance indicators. It is possible to increase control and knowledge in warranty with the use of the following:

- Modeling techniques as simulation and future forecasting.
- Information system facilities, "e-warranty."

Table 6.8 Framework stages regarding processes

Steps of the proposed framework	SO	Ctr.	IM	WA	IP	CM	Purpose
Balanced scorecard (BSC)	***	**	**	**	**	**	Measurement of warranty performance through indicators
Criticality analysis	**	**	***	**	**	**	Resources and task structure according to priority criteria
Failure root cause analysis	**	**	*	***	*	*	Determination of root cause of failures
RA and MDT	***	*	*	**	*	*	Improvement of the life cycle of the parts in terms of performance
Risk–cost optimization	**	*	*	**	*	***	Decrease in risks and their consequences
RAMS analysis	**	**	*	*	***	*	Prevision of product behavior
Life cycle cost analysis	***	*	*	**	*	*	Improvement of the life cycle of the parts in terms of costs
E-warranty	**	***	*	**	**	**	Remote online communication to facilitate the decision-making and knowledge creation
CRM	***	*	**	*	**	**	Management of complaints, communications, and customers' requirements
Six Sigma	***	*	**	*	***	**	Integration of human factor with statistical tools, to engage the whole mechanisms inside the company

where
* Step less involved in the corresponding process block
** Step averagely involved in the corresponding process block
*** Step highly involved in the corresponding process block

There are many management tools, some from quality management [31–33] and others taken from reviewed models for maintenance [25, 28, 34–36]. However, the application is more suitable in some processes than others. In Table 6.8, it is described where the uses of different management tools proposed for the warranty management framework are more profitable.

As it has been previously stated, this technical performance valuation (more or less involved) in the corresponding process differs from one company to the other [30]. The different degree of participation will depend on the company's policy for its warranty management program, the different faces of the business (and particularly after-sales service) that intend to promote or boost, as well as the own vision of technical staff, engineers, and administrative staff on their level of involvement in their area together with the company's. After this, the Table 6.9 describes some other tools and methodologies not included in the framework proposal for warranty management, but also useful for its improvement in general terms.

As stated in this chapter, it is suggested to consult next section where aspects about quantification of maturity are dealt with.

Table 6.9 Other tools and methodologies in warranty management

Tools and methodologies	Purpose
Human performance management	Evaluation of human resources according to the objectives
Competences-based management	HR management according to the capacities of the personnel
Knowledge management	Knowledge generation to improve the warranty assistance
Outsourcing	Strategic decision to center the efforts in the key activities
Centralized and automated management	To improve the control and reduce unnecessary tasks
Quality function deployment (QFP)	Link between performance and customer perception
Logistic and inventory tools	Resources management
Work flows	Sequence of task representation
Arrow diagram (AD), activity network diagram (AND), PERT, and CPM	Activities planning
Linear, nonlinear, goal, and multiobjective programming	In order to facilitate the decision-making in circumstances of poor information
Tree diagram	To plan tasks and decisions in a hierarchical way
SWOT—Strengths, weaknesses, opportunities, threats	Analysis of the organization, internal and external
Interrelationship digraph (ID) and Isikawa diagrams	In order to link causes with their consequences
Expert and support systems	Artificial Intelligence in order to make decisions
Game theory	Choice among different alternatives in warranty
Simulation modeling	Modeling and checking alternative conditions and consequences
Histograms and pareto diagrams	Resource or task structure according to the frequency
Multidecision criteria methods	Resource or task structure based on several variables
Process decision program chart (PDPC)	In order to make decisions related to risks and consequences
Brainstorming and quality circles (QC)	Continuous improvement and innovation
Affinity diagram	Link between innovation and different information
Statistical prediction techniques	Prediction and improvement of warranty activities, equipment life cycle, and quality
Queuing theory	Simulation of resources or tasks
Temporal series	Prediction according to historical data
Casual models	Prediction related to different information
Matrix diagram	Information linking in order to define priorities
Markovian process	Time estimation for product part availability
Deterministic models	In order to improve warranty assistance
Check list	In order to check and save the performed tasks

6.4 Case Study on the Maturity Assessment

In this section, we will present a methodology to evaluate the level of maturity that an organization has when it comes to a warranty management program. For that, we will depart from the described scenario in Chap. 2, where the initial problem of after-sales management was proposed, and we are going to apply the techniques and best practices developed previously in this chapter.

6.4.1 Valuation of the Maturity of the Initial Scenario

This brief exercise is very useful to value quantitatively the way in which our organization is considering available techniques to properly manage after-sales service or, on the contrary, whether there are still some improvements related to the decision-making process. Till now, we have analyzed all of those aspects related to best practices and their application to our reference framework; it is required now to quantify how a company manages warranty assistance. In other words, the specific warranty management needs quantification for its maturity in terms of the evolution developed by the organization and its capabilities; in order to make decisions, managers need to know the present situation and the potential future of the organization and make the pertinent decisions in the weaker areas. Nowadays, maturity analyses have become frequent and these need to be carried out on a certain frequency to know whether the application of measurements has provided the desired effect. To achieve this, the organization must know which are the best practices and the competitive advantages, and also their weaknesses, in the sector where the company performs. The evaluation of maturity can be based on existing management practices as CMMI levels [23] and EFQM enablers [10]. For our case, the method to determine the maturity level of a company regarding the warranty management and taken as reference our framework proposal will be assigning certain rates to the involved departments in reference to their influence in the already commented management process blocks. That is, we will be basing it on the degree of involvement that is shown in Table 6.10.

As it has been mentioned, this evaluation on technical involvement (major or minor) in the correspondent process differs from one company to another. The different degree of involvement will depend on business policies for their warranty management program, as well as the vision of the company on technicians, engineers, and administrators in general, related to their level of involvement in their area and in relation to the entire company. From that table, giving a value for each degree (for example, * = 1, ** = 2, and *** = 3), we get the matrix represented in Table 6.10. In the same way, we assigned certain ratios to the influence of management blocks with respect to the departments of our case study (Chap. 3). That is to say, based on the degree of involvement that is shown in Table 6.5 and

Table 6.10 Values assigned to the involvement degree acc. to framework processes

Phases of framework	Strategy organization	Control	Incident management	Warranty analysis	Incident prevention	Capacity management
BSC	3	2	2	2	2	2
CA	2	2	3	2	2	2
FRCA	2	2	1	3	1	1
RA & MDT	3	1	1	2	1	1
RCB	2	1	1	2	1	3
RAMS	2	2	1	1	3	1
LCC	3	1	1	2	1	1
EW	2	3	1	2	2	2
CRM	3	1	2	1	2	2
Six Sigma	3	1	2	1	3	2

giving a value for each degree (for example, * = 1, ** = 2, and *** = 3), we get matrix represented in Table 6.11.

As it has been discussed, the degree of involvement will depend on the politics of the business regarding its warranty management and the vision that every employee has in their field in relation to the whole of the company. We see that Table 6.10 is a matrix 10 × 6, and Table 6.11 is a matrix 10 × 7. By exchanging the first one and multiplying both matrices, we get as a result a new matrix with the 6 × 7. Now, we apply the following operations:

1. Dividing each element by the sum of its corresponding row, it is possible to get the quantitative value of how the involvement services are distributed in each of the departments.
2. Dividing the previous result by the sum of the column corresponding to after-sales department, we can compare the degree of involvement of each department under a common reference
3. Multiplying by 100, we get the percentage values of the previous result, and they are shown in Table 6.12.

Evidently, we observe that the after-sales service has a 100 % involvement in the management and decision-making that affect a warranty program. Nevertheless, we can also see how other departments have quite a high degree of involvement in such decisions, such as the quality or engineering department.

6.4.2 CMMI and EFQM Practices

In order to achieve a determined level of maturity, CMMI defines several generic and specific practices that have to be fulfilled in each level [22] in accordance with the defined issues of the framework: mission, objectives, responsibilities, etc. Moreover, combining it with the EFQM model, it is possible to express the

Table 6.11 Values assigned to the degree of involvement according to departments

Stages of framework	Logistics	Manufacturing	Administration	After sales	Quality	Purchasing	Engineering
Balance scorecard	2	2	3	3	2	2	2
Criticality analysis	1	2	1	3	3	1	3
Failure root cause analysis	1	2	1	3	3	1	3
MDT adapted to warranty	3	2	2	3	2	3	2
Risk–cost–benefit analysis	2	2	3	2	2	3	2
RAMS	1	3	1	3	3	1	3
Life cycle cost	1	1	3	3	1	3	2
E-TECHNOLOGIES	1	1	2	3	2	1	1
Customer relat. management	1	2	3	3	2	1	2
Six Sigma	2	2	3	2	2	2	2

Table 6.12 Matrix of maturity according management processes

	Logistics	Manufacturing	Administration	After sales	Quality	Purchasing	Engineering
Strategy organization	9.1649	11.0448	13.6298	16.4498	12.4548	11.0448	12.6898
Control	8.3812	11.4289	12.1908	17.5243	14.0957	9.5241	13.3337
Incident management	8.6874	11.4515	13.0311	16.5850	13.4259	9.8720	13.4259
Warranty analysis	8.9119	10.8924	12.5427	16.8337	13.2029	10.8924	13.2029
Incident prevention	8.6147	11.9281	12.9221	16.5668	13.5848	9.6088	13.2535
Capacity management	9.0663	11.1586	13.9482	16.0404	12.9021	10.8099	12.5534
% of involvement	52.8265	67.9043	78.2648	100.0000	79.6662	61.7519	78.4592

fulfillment percentage inside each level. Further research can deepen in such aspects for different companies with diverse configuration in their modules or departments. When the analyzed business reaches a certain degree of consecution in every department and in every process, the framework is considered totally implemented. The degree of progress for this consecution is characterized by the following:

- 100 % when the framework is totally implemented.
- 66 % when the framework is widely implemented.
- 34 % when the framework is partially implemented.
- 0 % when the framework is not implemented.

Following with our case, with these obtained weights (proportional to the involved departments in relationship to each process), it is possible to deduce the actual contribution that such company is really applying to achieve the excellence. This excellence according to the maturity level is the contribution of each maturity level to the achievement of the mentioned excellence as shown in Table 6.13, expressed also in reference to the EFQM enablers.

The EFQM enablers are crucial due to the following:

1. *Policy and Strategy*: Company develops and keeps the mission and vision of the organization through a clear strategy focused on the client, and supported by relevant policies, plans, objectives, targets, and processes.
2. *Leadership*: Company develops and facilitates the achievement of the mission and vision through appropriate actions and behaviors, leading to the effective management of the organization and its relationships.
3. *People*: Company develops and manages the knowledge and full potential of its people, not only individual but also team based.
4. *Partnerships and Resources*: Company plans and manages its external partnerships and internal resources, in order to manage its processes effectively according to its policy and strategy.
5. *Processes*: Company designs, manages, and improves its processes in order to generate an increasing value for its customers.

These enablers are also susceptible to be assessed. Therefore, it is possible to combine the previously mentioned weights for departments or modules in a generic company reference, with the percentages or rates of such enablers. Consequently, the goal of all these matters is to check the potential of the whole company regarding the warranty management, analyzing each department from the point of view of itself:

1. Performance in the processes.
2. Quality of its activities.
3. Attention to customers.
4. Utilization and development of technologies.
5. Capacity to improve and innovate.

Table 6.13 Excellence according to maturity levels

Level	Policy and strategy	Leadership	Personnel	Partnership and resources	Processes
1	Reactive execution and purpose	Lack of coordination, without reference nor defined responsibilities	Reactive and variable execution according to personal initiative disorganization	Requests on demand and without control	Unstable and unpredictable situation
2	Oriented to customer attending to the management of requirements	Commitment of the interest groups and definition of the responsibilities for the projects	Management of personnel according to the results, monitoring the efficiency and security	Management of partnerships and resources. Specific management of information	Repetitive management and planning according to results
3	Orientated to customer attending to the integrated performance to accomplish the requirements. Standardization of the organization	Identification of personnel involvement and environment preparation for continuous improvement. Assignment of the responsibility based on processes	Development of knowledge and personnel skills, facilitating decision-making activities	Unique and integrated definition of the information, attention to maintain the operation of the resources depending on risks	Unified and coherent management with processes and objectives. Prediction through qualitative techniques
4	Guide the organization for the analysis with the purpose of improving the objectives, with the agreement of customers and internal staff	Defining and implementing the mechanisms for the qualitative analysis	Quantitative prediction and evaluation of needs and improvements of the human resources	Quantitative analysis of the operation for the resources, their procurement, and logistics	Estimation of future efficiency and possible variables from the actual situation of the process
5	Innovation in a sustainable way of the processes and the technologies for the customers satisfaction and the social perception	Implement the concepts of continuous improvement and proactivity	Take the potential of the personnel to optimize the efficiency of the organization	Eliminate the causes that produce variations of the operations of the resources	Continuous improvement of the efficiency, through proper adjustment of the processes

Summarizing, all this work helps basically to establish properly a reference framework for the management of the warranty assistance, increasing the value of the whole organization. The proposed methodology will allow us to measure the evolution regarding standards and regulations and the comparison with other companies of the sector. To determine the maturity is necessary to apply external references as objectives [37, 38]. In this sense, the effective management needs to be based on quantitative business models that integrate warranty with other decisions like production, etc. [39]. Consequently, measurements of warranty performance and company performance will be useful to characterize the performance of the warranty management framework in the achievement of the objectives.

References

1. González-Prida V, Gómez JF, Crespo Márquez A. 2009. Case study: warranty costs estimation according to a defined lifetime distribution of deliverables. World Congress on Engineering Asset Management. WCEAM Athens. ISBN 978-1-84996002-1.
2. González-Prida V, Parra C, Gómez JF, Crespo Márquez A. 2010. Audit to a specific study scenario according to a proposed reference framework for the improvement of the warranty management. ESREL/ESRA Rhodes (Greece).
3. González-Prida V, Gómez JF, López M, Crespo Márquez A, Moreu P. 2009. Warranty cost models State-of-Art: A practical review to the framework of warranty cost management. ESREL Prague. ISBN 979-0-415-55509-9. Pp. 2051-2059.
4. González-Prida V, Parra C, Gómez JF, Crespo Márquez A. 2010. Reference framework proposal for the management of a warranty program. EURENSEAM, Congress Euromaintenance Verona, Italy.
5. Crespo Márquez A, Moreu P, Gómez J, Parra C, González-Prida V. 2009. The maintenance management framework: A practical view to maintenance management. Taylor & Francis Group, London, ISBN 978-0-415-48513-5. Pp. 669-674.
6. González-Prida V, Crespo Márquez A, Moreu P, Gómez J, Parra C. 2009. Availability and reliability assessment of industrial complex systems: A practical view applied on a bioethanol plant simulation. Taylor & Francis Group, London, ISBN 978-0-415-48513-5. Pp. 687-695.
7. González-Prida V, Barberá L, Crespo Márquez A. 2010. Practical application of a RAMS analysis for the improvement of the warranty management. IMS 10th IFAC Workshop on Intelligent Manufacturing Systems. Lisbon, Portugal.
8. Shetty YK. 1993. The Quest for Quality Excellence: Lessons from the Malcolm Baldrige Quality Award. SAM Advanced Management Journal. Vol. 58.
9. Harrington HJ. 1995. Total improvement management: the next generation in performance improvement. New York. McGraw-Hill. Pp. 488.
10. EIPM (European Institute of Purchaising Management), EFQM (European Foundation for Quality Management). 2006. EFQM Framework for Management of External Resources.
11. Intenational Organization for Standardization. 2008. ISO 9000. Quality management system in production environments.
12. COSO (Committee of Sponsoring Organizations of the Treadway Commission.) http://www.coso.org/
13. TL9000, QuEST Forum. http://www.tl9000.org/
14. Gartner, Inc., www.gartner.com

15. Peters TJ,.Waterman RH. 1982. In search of excellence: Lessons from America's best-run companies. Harper & Row. New York. ISBN 0060150424.
16. Davenport T, Short J. 1990. The New Industrial Engineering: Information Technology and Business Process Redesign. Sloan Management Review 31(4):11-27.
17. Pall GA. 1987. Quality Process Management. Prentice Hall.
18. Davenport T. 1993. Process Innovation: Reengineering Work Through Information Technology. Harvard, Business School Press.
19. Hammer, Champy. 1993. Reengineering the Corporation. Harper Business.
20. King WR. 1994. Process Reengineering: The Strategic Dimensions. Information Systems Management.
21. Klein MM. 1994. The most fatal reengineering mistakes. Information strategy: The Executive's J. lO (4)21-28.
22. Chrissis MB, Konrad M, Shrum S. 2003. CMMI®: Guidelines for Process Integration and Product Improvement. Addison Wesley, ISBN: 0-321-15496-7.
23. Mutafelija B, Stromberg H. 2003. Systematic Process Improvement Using ISO 9001:2008 and CMMI®. Artech House, Boston London.
24. Nakajima S. 1992. Introduction to TPM. Productivity Press.
25. Wireman T. 1991. Total Productive Maintenance. Industrial Press.
26. Porter M. 1985. Competitive Advantage. Free Press.
27. Pall GA. 1987. Quality Process Management. Prentice Hall.
28. Crespo Márquez A, 2007. The maintenance management framework: models and methods for complex systems maintenance. Springer-Verlag London Limited. ISBN-13: 9781846288203.
29. Duffua SO, Raouf A, Campbell JD. 2000. Planning and control of maintenance systems. Indianapolis: John Wiley and Sons.
30. Gómez J. 2010. Marco de referencia para la gestión y evaluación del mantenimiento de redes de distribución de servicios. Tesis doctoral dirigida por Adolfo Crespo Márquez. Universidad de Sevilla.
31. Ishikawa K. 1985. What is Total Quality Control? Prentice-Hall Inc., Englewood Cliffs, NJ.
32. Nayatani Y, Eiga T, Futami R, Miyagawa H. 1994. Seven New QC Tools: Practical Applications for Managers. Productivity Press. ISBN-10: 4883190048.
33. Mach P, Guaqueta J. 2001. Utilization of the seven Ishikawa tools (old tools) in the sixsigma strategy. Electronics Technology: Concurrent Engineering in Electronic Packaging. 24th International Spring Seminar. Pp. 51-55.
34. Pintelon LM, Gelders LF. 1992. Maintenance Management Decision Making. European Journal of Operational Research. 58(3):301-317.
35. Wireman T. 1990. World Class Maintenance Management. Industrial Press.
36. Garg A, Deshmukh SG. 2006. Maintenance Management: Literature Review and Directions. Journal of Quality in Maintenance Engineering. 12(3):205.
37. Campbell JD. 1995. Outsourcing in maintenance management: A valid alternative to self-provision. Journal of Quality in Maintenance Engineering. 1(3):18-24.
38. Tsang A. 2002. Strategic Dimensions of Maintenance Management. JQME, 8(1):7.
39. Murthy DNP, Atrens A, Eccleston JA. 2002. Strategic maintenance management. Journal of Quality in Maintenance Engineering. 8(4):287-305.

Part IV
Development of the Framework Stages

Chapter 7
Management Effectiveness

7.1 The Balanced Score Card in After-sales Management

7.1.1 Alignment of Strategy with Global Objectives

The balanced score card (BSC) is a methodological tool with a multidimensional focus that makes possible to integrate corporative strategies with the company's own operation in order to determine the organizational objectives, assessing business performance through management indicators [1]. This way, the BSC helps implementing strategic lines set by management [2] and permits, as well, to align the targets of the departments or operational units with the strategic global objectives, controlling their deviations. The BSC can be understood as a communication, information, and training system (of the company strategy), which does not substitute the traditional process of strategic planning [3], but on the other hand, it complements it and helps its communication and implementation [4]. This methodology transforms both vision and strategy in a compound of objectives and grouped actuation indicators based on four perspectives (dimensions) considered as key elements for management and control [5] (Fig. 7.1):

1. *Perspective of learning and growth.* The objective is to assure the capacity of adaptation and long-term renovation of the company (in response to the various changes in the field), as well as to keep knowledge of the areas considered as basic competencies.
2. *Perspective of internal processes.* It considers quality, productivity, and costs of distinct objective processes developed by the company, among them, the processes of maintenance management and warranty reclamation management (number of defective units, productive life cycle, idle capacity of equipment, etc.
3. *Perspective of customers.* It evaluates the way in which the company generates value for the customers. It seeks to measure the impact and satisfaction that the company generates on its customers.
4. *Financial perspective.* It measures survival, growth, and development of the company in financial and value generation terms.

V. González-Prida Díaz and A. Crespo Márquez, *After-sales Service of Engineering Industrial Assets*, DOI: 10.1007/978-3-319-03710-3_7,

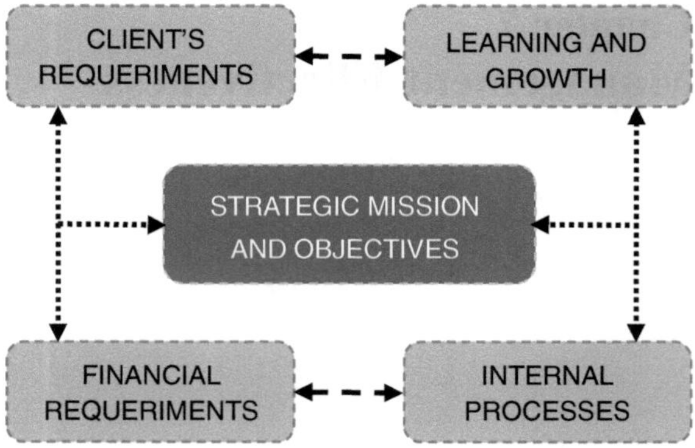

Fig. 7.1 BSC perspectives

The BSC maintains the technical and financial measurement [5], but at the same time carries out a series of more general and integrated measurements which link internal processes, employees, and system actuation to the company's long-term success. In this way [6], the BSC complements the financial indicators as well as it clarifies, translates, and transforms vision and strategy, allowing us to identify, plan, and determine strategic initiatives. These indicators must be defined to measure in the clearest way possible the objective they are associated with. This permits their strategic follow-up and the assessment of their consecution by an appointed member of staff responsible for such task. Each initiative, indicator, and objective will have someone in charge of controlling its degree of compliance. Afterward, these initiatives or strategic actions are defined so that they will permit to reach the objectives and proposed targets. The establishment of actions requires considering the implantation effort and the benefits that are obtained from it. Lastly, a periodical follow-up system must be established according to the degree of consecution of the strategic objectives [6] to be able to make decisions and opportune corrections in the defined strategy. The concerning indicators in the BSC methodology must be relevant, practical, measurable, and implementable [7]. There are two types of indicators in the BSC frame [8, 9]: indicators of results (lag measurements) and indicators of performance (lead measurements). It is necessary that the BSC balances both measurements [10]. *Lag measurements* reflect the results of previous decisions, giving information about what had occurred, but not being able to change the result. On the contrary, *lead measurements* generally assess process performance and allow us to detect what is happening and what actions should be taken to improve results. Lead measurements are more predictive and make quicker adjustments possible. Once those indicators are defined, it is necessary to integrate them with the rest of the company's preexisting information systems [1]. Moreover, the source of data needed to feed the indicators and its appropriate periodicity must be identified.

7.1.2 BSC Principles in Global Warranty Management

The BSC is a dialogue and communication process in all areas of a company including the after-sales service. As this communication process works, a greater participation, alignment, and synergy are achieved. The management of technical and financial indicators allows the company to use the same language for after-sales services. The financial perspectives, customers, processes, and learning suggest, for example, calculating the availability in terms of the repair average time and failure average time, improving the relation between parameters such as production, sales, costs, and availability. One of the targets of the BSC applied to warranty services is to transform strategic objectives in the sales area into specific action plans based on key and contrastable management indicators, developed based on the four perspectives of the methodology (Table 7.1).

The process consists of setting measures, targets, and the action plans to achieve them. This way, it is possible to align ways of management to/with the objectives of the company. This alignment is also possible because of the development of key indicators through a series of functional measures, closer to the obtained results in different processes and therefore easier to measure and control. The BSC in all makes a total implementation of maintenance strategies at all levels possible, encouraging participation of all the involved people in the consecution of strategic objectives and achieving the strategic alignment of the entire organization based on the strategy transformation of specific action plans.

7.1.3 Application to Improve Effectiveness in Warranty Management

As it was previously stated, this methodology aligns local objectives of a department or a business unit, with the global objective of a company. In the specific case of warranty management, these BSC methodology principles are intended to be used within a constant improvement framework (refer to Chap. 5, Fig. 5.9) in order to avoid possible contradictions between a warranty program and the global strategy of the company. That is why, the application of this BSC is proposed. Basically, this methodology considers targets and objectives based on four points of view, as it has been previously mentioned. When these viewpoints are applied to a warranty management program, the BSC methodology includes the following steps:

1. Strategic formulation of the warranty program (as the possible externalization [12] of warranty services or the development of warranty equipment with polyvalent technicians.
2. Translation of after-sales strategy to reachable short-term objectives (such as the identification of measures destined to push personnel to a global vision.

Table 7.1 Application of the BSC methodology to warranty management (adapted from [11])

	Strategic objectives	Measures (KPIs)	Targets	Action plans	Perspective
Mission and strategy	To improve the effectiveness of warranty costs	Warranty costs per sold unit	Current (X) % Objective (X − 1) %	To assure the correct acquisition of data and analysis of criticality of equipment	Financial
	To improve repair time and service quality	Repetitive failures MTTR	Number of repetitive failures < x Reduce the MTTR in a (Y) %	Failure analysis program and after-sales support improvement	Customers
	Improvement in the service process and its documentation	In compliance with quality standards	Certification of services before a determined date	To develop procedures of pending technical inspections	Internal process
	To assure the correct levels of training to accomplish the mission	Level of training according to type of service	Definition of precise training levels	To carry out training and evaluation	Learning

3. Development of action plans (in order to reach the objectives identified in the previous step and taking into account the necessary changes in the business infrastructure of support).

7.2 Hierarchical Analysis in the Decision-Making Process During Warranty

Once the warranty program objectives and strategies are defined, it is necessary to determine the criticality of the product. In this context, criticality analysis is understood as the impact customer's claims can have on the global strategy and expected results. Those decisions and actions taken from the warranty program will involve the possibility of a certain deviation from business objectives in terms of profit losses, redirection of resources, possible delays, etc., or the use of assembly pieces as spares, among others. Therefore, it is required to apply techniques that can systematically determine which goods have a greater priority in the warranty management process, according to the existing strategy. One of those techniques could be based on, for example, a risk–cost evaluation. The criticality analysis based on this assessment would combine the probability of an event with the impact this event could cause. As a result, a criticality matrix is obtained offering a vision of priorities in regard to probable events related to warranty issues and moreover permits us to align after-sales actions with the targets and objectives of the company. This way, it is possible to forecast where to apply available resources and minimize risks, reducing warranty costs and customer dissatisfaction. Lastly, when the defined strategy has been set and the probable events have been prioritized, it is time to focus on those customer reclamations related to repetitive or chronic failures that represent high-priority events. If it is possible to find and eliminate the causes of these failures through a prompt intervention, this will provide an important initial advantage and a quick strategy of a warranty management program. In any case, there are already different developed methods to carry out these analyses on weaker points. This section intends to synthesize in a simple way how the analysis hierarchical process (AHP) at production and sale stage can help to make more appropriate decisions on strategic actions, for example, in regard to spare parts to properly assist customer claims [13]. The analytic hierarchy process (AHP) method establishes a series of scales from comparisons. Inputs in this methodology can be measured as the price, weight, time, provisioning, etc., or the subjective opinion on how the satisfaction and preference sentiments can be. AHP shows an approximate and realistic assessment of the best decision, including certain and small judgment incoherencies when subjective opinions are adopted, evidently due to the fact that human judgment is not always consistent.

7.2.1 General Features of the Analytic Hierarchy Process

The "AHP" is a methodology developed by Thomas Saaty in 1970, based on facilitating the understanding of a complex problem through a breakdown in parts ordered hierarchically ranked (approaches and alternatives), quantifying and comparing variables through addition of views with geometrical average to synthesize a solution [14, 15]. The process has been used to assist numerous corporate and government decision makers. Some examples of decision-making problems are as follows: choosing a telecommunication system, choosing a product marketing strategy, etc.

In short, problems are decomposed into a hierarchy of criteria and alternatives (see Fig. 7.2). There are several published articles about AHP method on decision-making [16, 17]. In the decision matrix (Fig. 7.3), it is synthesized decision maker information with resulted elements in pair compared criteria with a normalized and reciprocal scale of relative importance (see Table 7.2).

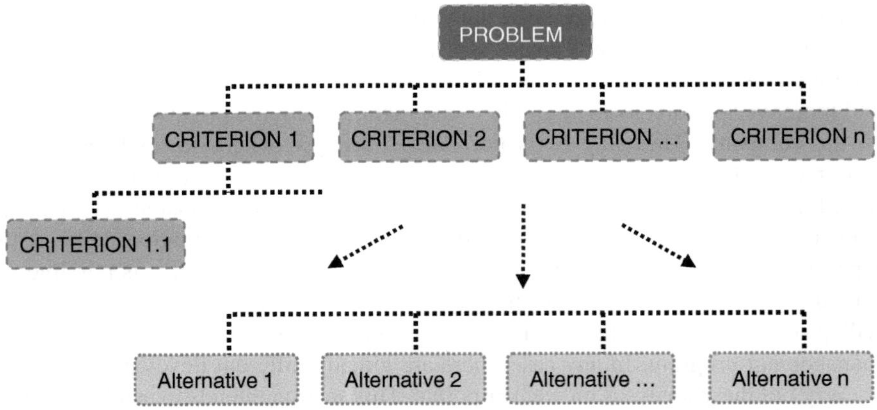

Fig. 7.2 Problem decomposition into criteria and alternatives

Fig. 7.3 Decision matrix AHP

$$W = \begin{vmatrix} w_{11} & w_{12} & \dots & w_{1n} \\ w_{21} & w_{22} & \dots & w_{2n} \\ \dots & \dots & \dots & \dots \\ w_{n1} & w_{n2} & \dots & w_{nn} \end{vmatrix}$$

The elements of decision matrix fulfill

- Reciprocity : $w_{ij} = 1/w_{ji}$ for all $i, j = 1, \ldots, n$.
- Consistency : $w_{ij} = w_{ik}/w_{jk}$ for all $i, j, k = 1, \ldots, n$.
- $\sum j = 1, \ldots, n \quad w_j = 1$.

The method has the following axioms:

1. Reciprocity. The importance of the relation W_{ij} is the inverse of W_{ji}, so $W_{ii} = 1$.
2. Homogeneity. Compared elements in relation to the same property have the same order of magnitude.
3. Dependence. To control the dependence among the elements forms the same level and consecutive levels.
4. Performance of expectations. The structure of criteria and alternatives has to represent the expectations in a ranking.

As it has previously stated, in the application of this method, subjective values can be used. This subjectivity is presented, implying a degree of uncertainty or lack of reliability. That makes it necessary to measure the sensitivity in changes in parameters. To measure here the reliability (CR), we use the ratio of consistency

Table 7.2 Saaty scale to compare criteria in pairs

Saaty scale		Reciprocal Saaty scale	
1	Equal importance of both elements	–	–
3	Weak importance of one element over another	1/3	Slightly less importance of one over another
5	Essential or strong importance of one element over another	1/5	Less importance of one over another
7	Demonstrated and very strong importance of one element over another	1/7	Far less importance of one over another
9	Absolute importance of one element over another	1/9	Absolute less importance of one over another
2, 4, 6, 8	Intermediate values	1/2, 1/4, 1/6, 1/8	Intermediate values

rate (CI) of a comparison array into pairs to the value of the same index of a comparison array into pairs randomly generated:

$$CR = \frac{Cl}{Cl_{random}} \leq 0.1$$

Reliability is sufficient if CR is less than or equal to 0.10; otherwise, it must be reviewed to improve its consistency. The application of this method in the decision-making process involves the acquisition of the following objectives:

- Improve organization, structuring, and documentation (feedback) of the process.
- Use a rational and logical analysis, which minimizes emotional burden of trials.
- Classify and compare alternatives.
- Employ quantitative and qualitative criteria, accurate or measurable, and inaccurate or estimated.
- Consensus and satisfaction in group decision-making.
- Predict at times likely results.

As it has been hinted at, analytical hierarchy process has been applied to multiple business situations, as well as political or personal issues, especially when it is necessary to synthesize the knowledge of different specialists to support decisions:

1. Personal situations, where everything is looking on organizing and reflecting internal preferences, for example acquisition of resources and identification of tracks.
2. Political situations in public administrations. Generally is orientated to consensus achievement or forecasting future, for example identification of public transport routes and public services allocation.
3. Business situations in private companies. Orientated mainly by competitiveness and improvement, it is used in all situations to achieve objectives: organization, structuring projects, resource allocation, prediction, etc.

In our case, AHP method is orientated to continuous improvement, having taken into account the following points [18–20]:

- If the number of alternatives grows, comparisons grow exponentially, and the use of method can be made cumbersome [21].
- It does not consider variation in criteria ranges.
- It is more a comparing tool for management than a statistical method [22].
- Valuations of comparisons can be interpreted differently by different subjects.
- Individually comparison may lead to conflicts, because if $A > B$ and $B > C$, it may occur $A < C$.
- Inclusion of a new irrelevant approach may affect management of two relevant criteria [23]. This would contradict axiom of multi attribute value theory (MAVT) on irrelevant alternatives [24].
- Asymmetrical inconsistency in eigenvectors by Saaty scales [25, 26].

To correct this lack of consistency, the modified AHP method was developed [25], Donegan et al. [26], but also it has its criticisms [27], indicating that improvements in real cases are not as crucial as Saaty AHP original method.

7.3 Root Cause Analysis in Warranty Management

7.3.1 Root Cause Analysis for the Identification of Physical Causes

The general objective of incrementing company benefits, from a maintenance perspective and for the extension of the after-sales services, involves also increasing equipment reliability including the prolonged life span [28] of the company's physical assets as well as those products launched to the market. The main target in industrial asset management will be to predict possible alterations in the process as well as unplanned detection in the production process, minimizing at the same time possible losses. This way, the priority would be to define a sequence of efficient actions, which assure minimal production losses and maximize company's profits [29]. Nowadays, the objectives of any realistic model of maintenance management and after-sales services should be set and dependable on the business plan of the company. Thus, strategies must always be aligned with the company's business plan [30] since this will depend on the consecution of the objectives of after-sales service and the company's own business plan. Subsequently, the development of new technologies and their practice in management requires a skilled technical assistance team in both technical and management matters [31]. Therefore, this justifies the use of more complex tools that permit the generation of more certain solutions in the after-sales service that minimize uncertainty. In the specific case of maintenance, the implementation of methodologies significantly reduces costs through focusing on the root causes of failures.

To confront and overcome these challenges, it is necessary to utilize these maintenance tools: total productive maintenance (TPM) and reliability-centered maintenance (RCM) [32]. More specifically, RCM analyzes functions and failures of a system and identifies the consequences in order to implement preventative measures using standardized and more logical procedure [33], even though this analysis does not lead us to a profound research to identify the mechanisms of failure and their real causes [34]. In another respect, the proactive maintenance uses tools such as root cause failure analysis (RCFA), failure modes and effects analysis (FMEA), critical analysis (CA) and acceptance testing and age exploration (AE). Some authors even make a distinction and identify a sub area in this maintenance, calling it radical maintenance (RM). This involves the detection and prediction of the root causes of failures, to later take the appropriate measures to eliminate the root causes or the conditions that created them [35].

In general, there is a wide variety of tools and methods to determine the root causes of specific events or failures [36]. These may vary in complexity, the quality of information required, and applicability of results. Generally, the most used are the 5 reason analysis, change analysis, current reality tree (CRT), FMEA, failure tree analysis (FTA), Pareto analysis, Bayesian inference, and the Ishikawa diagram. These methodologies have substantial differences, and some of them can be classified as qualitative (5 reason analysis, Ishikawa diagram, HAZOP, among others) and quantitative (Bayesian Inference, Pareto analysis, FTA, among others) [29, 37, 38]. While qualitative methodologies are carried out generally in the form of brainstorming, the quantitative ones can require complex mathematical methods. The importance of the use of root cause analysis (RCA) tools for maintenance and after-sales services lies in the necessity to understand the main causes of failure on which the administration, management, or operation can have some impact in a way that chronic and recurrent failures can be avoided through a determined action plan. In this sense, it is not sufficient to only find the origins of the failures, but it is necessary to generate corrective and essentially preventative actions.

Here is where the use of these types of tools plays a fundamental role. It should be also mentioned the palliative maintenance between possible strategies, given that it is a possible maintenance to temporarily solve problems of customers under a pending long-lasting warranty of repair. It can also be added that opportune maintenance (preventative action as a consequence of a corrective action) provides a better protection to the organization that manages the warranty. Likewise, the Bayesian network can be used as a support for decision-making based on a probabilistic reasoning, considering that it allows us to calculate the probability of future events and has the capacity to adapt to changes [39]. Moreover, the monitoring and diagnosis objective is to be able to integrate previous knowledge of the processes with the present observed physical evidence. This way, a more plausible explanation of the behavior of the process can be given. The Bayes theorem incorporates this type of predictive support to the diagnosis [40]. The flowchart presented in Fig. 7.4 (based on the projects of [34] and [36]) shows the location of

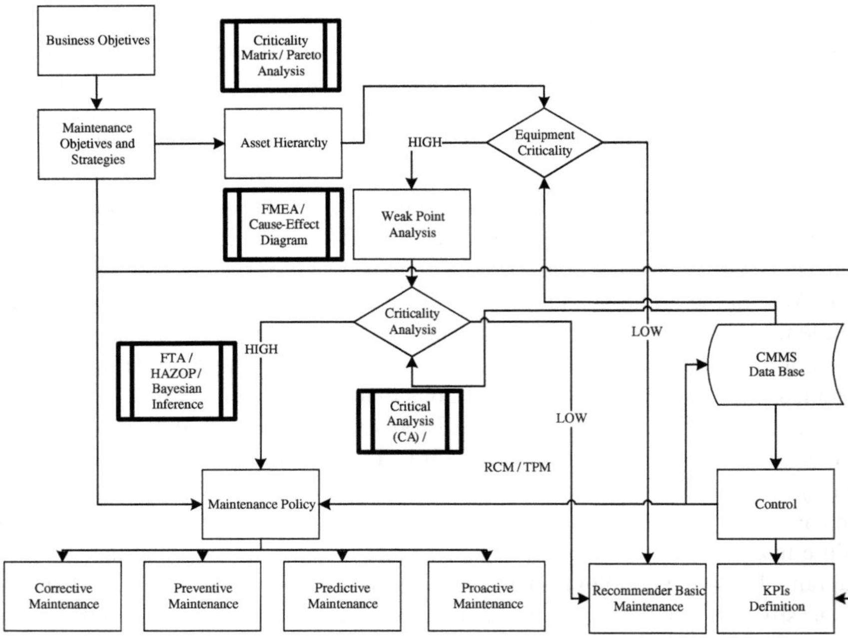

Fig. 7.4 Location of the RCA methodologies focused on maintenance management

the distinct methodologies of RCA in a model of maintenance management in stages. It is the base of the defined in the previous chapter for a warranty management program.

7.3.2 Root Cause Analysis Models

Even though the hierarchical analysis process has been introduced in this chapter, the Pareto analysis is also mentioned located in the hierarchical stage of the critical equipment so that, along with the criticality matrix, it can help to determine which equipments are critical in a systematic level. The FMEA can be used at the stage of weak point analysis of critical equipment, where an evaluation of causes, failure modes, and effects could be considered relevant. The CA helps to determine whether the weak points of critical equipment are significant in the system's performance. The ETA or the Bayesian inference can be used to carry out more complex analysis in the definition of root causes of failures in equipment and the critical weak points of the system as well as for the correct development of an action plan defined on the adapted after-sales/maintenance strategy. All the previously mentioned, it is important to point out that the applicable functioning of the methodologies is achieved when they are properly used for a determined

Table 7.3 Classification of the RCA groups based on their focus

Root Cause Analysis groups	Description
Deductive	Approach that involves a general-to-specific reasoning (for example, failure tree analysis)
Inductive	Approach that involves reasoning from individual cases to general conclusions, giving a general focus (cause-and-effect diagram and HAZOP analysis)
Morphological	Method based on a structure of the system studied. It focuses on the potentially dangerous elements, focusing in factors that have more influence on the system security (for example, evolution of accidents, barriers, health and safety at work)
Non-orientated system techniques	Concepts and techniques non-orientated to systems such as the previous ones (for example, analysis of change and study of probability of human error)

requirement of a specific stage within a reference management framework, in terms of the characteristics, information requirements, and resources. The root cause of a failure can be defined as the most basic cause which can be reasonably identified and controlled [41]. The literary review supporting this focus explains that there are three levels of root cause failure of a system:

- Physical Root Cause: Failure on equipment caused by physical reasons.
- Human Root Cause: Failure on equipment caused by human intervention.
- Latent Root Cause: Failure on equipment caused by decisions at an organizational level that leads to an event of failure.

The failure analysis (FA) or RCA consists of examining in detail the items that are in a failure state in order to define the root cause of the failure and improve the system reliability [42]. This process identifies the factors that caused the failure using a structured focus with designed techniques that make the proper identification and problem solution possible. Its execution eliminates or minimizes those root causes that can generate recurrent failures, not focusing on the consequences of such failure [43]. Within the RCA methods, four groups can be distinguished [44] (Table 7.3).

Following, there is a brief explanation of the most used RCA methodologies in reliability engineering, standing out their advantages and limitations. These are

- *Failure Modes and Effects Analysis*:

The failure mode analysis, its effects (FMEA), and its criticality (FMECA) is a qualitative method that identifies potential modes of failures and analyzes the consequences of all possible failures on systems that could affect a component, proposing measures to avoid or minimize the consequences of these failures on the system [45]. The method systematically analyzes all failure modes at a component, equipment, and subsystem level and assesses the effects and criticality (FMECA) in the system and the probability of occurrence [34]. Basically, it identifies those zones that need improvement in order to assure that the functions of the system work in a reliable and secure way (at a global level). The method follows an inductive approximation, starting from the failure of a component and its produced effects through the system, seeking out all possible consequences. In its limitations, the low performance in complex problems stands out by not being able to show the related causes based on evidence beyond the specific mode of failure that is analyzed [37]. Besides, its effectivity is limited by the experience of the group of work that is developing it.

- *Failure Tree Analysis*:

FTA is a deductive and quantitative method that begins by searching an unwanted event "superior event" with the aim of analyzing the causes for such event and quantifying the probability of occurrence. The cause analysis is carried out through a logical diagram, where the combination of diverse elemental events leads to the superior event. It is a graphical representation of events of a hierarchical order, and it makes it possible to identify and classify possible events (graphically represented in a tree-type scheme) that can cause a failure in the system and carry out estimations of failure occurrence probability of the system. It is widely spread for its easiness of use and due to the fact that presents a high and intuitive level of abstraction of the system. The diagram is drawn using conventional logical symbols, so that cause relations can be identified with the relations "Y" and "O" or several combinations. With this detailed information, the efforts to improve the system's security and reliability are more centered and adapted to such system. Moreover, the FTA can assist to predict failure occurrence considering that it provides data that describe the circumstances in which they can occur, defining the importance of each critical element of the system. It is an applicable technique on static and dynamic complex systems, and it provides the objective basis to analyze system design and its common failures and makes it possible to check the requirements of security compliance, as well as to justify changes and complements. One of its limitations is the fact that it does not work correctly as a RCA, given that it cannot manage functional or sequential dependencies between components. Furthermore, it requires the use of specific information related to known failure reasons of components. However, it is frequently used as a support for RCA, requiring the use of that specific information related to known failure reasons of components [37].

- *Cause-and-Effect Diagram (CED)*:

CED is a tool that splits potential causes into more detailed categories, in a way that can be organized and related according to the factors which help to identify the root causes [43]. It is also known as the Ishikawa diagram or fish bone diagram. Its limitations lie on the fact that it cannot show all cause relations between the primary effect and the root cause, and also, it does not provide evidence to support the cause factors [37].

- *Pareto Analysis*:

This analysis is a statistical focus for problem-solving, which uses a database to identify the predefined number of cause factors occurred in the system. It is based on the Pareto principle, which assumes that the 80 % of the problems are caused by the 20 % of the causes. Sometimes, it is used as an RCA method; however, the Pareto analysis functions more appropriately in defining the beginning of an analysis. They are also commonly used to set maintenance priorities ordering failure codes of the equipment according to its relative cost or its contribution to downtime. It is limited by the accuracy of the information analyzed, and losses cause connection in the cause-and-effect principle [37].

- *HAZOP*:

This methodology, which acronym stands for HAZard and OPerability, has become the cornerstone of risk studies of processing plants. The HAZOP study makes it possible to determine in a qualitative way which are the consequences on a system when the operational conditions or design varies [38]. This study is traditionally carried out as a structured brainstorming exercise, provided and led by a HAZOP study leader, and uses the participant collective experiences. To achieve the study's goal, basically some questions are set and analyzed such as "what deviation can occur?" "what parameters of the process are relevant to measure?" "why do those deviations occur?" (causes) and "how are they expressed" (consequences).

In a HAZOP study, these questions are asked once the total system is divided into its constituent parts, such as sections and nodes. These questions are related to the objective of the system, while the process represents the means to reach those objectives. The restrictive factor of the HAZOP study is the consideration of time and resource requirements, so that it can be carried out throughout the plant life cycle (which is mainly the focus of this methodology). One step further in the study would be to do a quantitative search of the deviation range limits of the nominal values of design, in which the equipment could maintain its aptitude and the system could continue operating properly.

- *Bayesian Inference*:

The Bayesian nets (BN) are one of the most widely applied techniques in probabilistic inferences, given that provide a flexible structure for evaluating and modeling uncertainty. This way, the BNs (graphically) are formed by two nodes that represent random variables connected through arcs that quantify a causal relation between the nodes and in which each one represents a random variable that can take two or more discrete values. Process monitoring and failure diagnosis require to assume as accurate numerous information sources which could be not certain or not totally reliable or incomplete, to be able to infer when the status of an equipment process has changed and to be able to identify the root causes of this change. Such inferences are necessarily imperfect and are valuable if the conclusions are supported by some quantitative measurements in spite of the uncertainty [40].

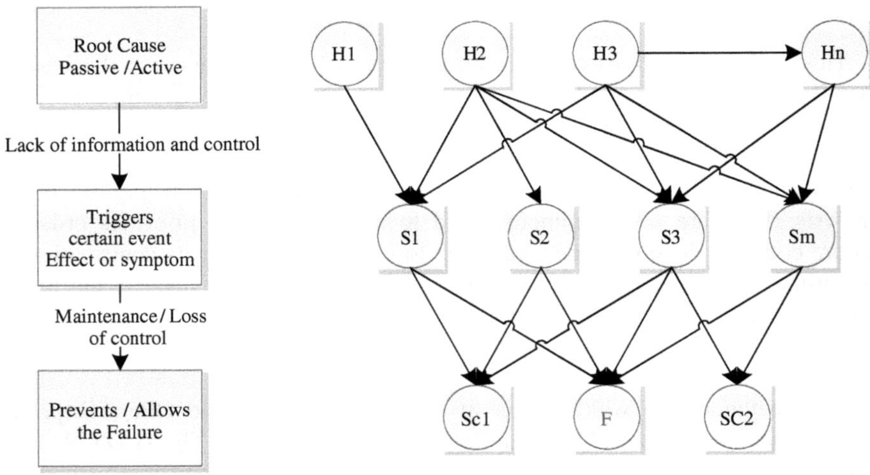

Fig. 7.5 Concepts of a BN in a root cause analysis (adapted from [29])

Two types of inferences can be supported and considered through the BNs: the diagnosis focus and the predictive focus [46]. When there are abnormal changes in the conditions during the operation of a process, and those changes cannot be identified nor corrected, some unwanted events, known as failures, can be generated. A causal representation of the facts through BNs generates a chain of events and transitions, which are relevant to the RCA under uncertainty and for the purpose of decision-making support about proper corrective actions [29]. Figure 7.5 represents a model of BNs for an RCA, in which the root causes (H_i) group contains all conceivable hypotheses of failure sources, which can possibly produce failure events or modes (S_j) that represent all those symptoms preceding a failure (F).

These symptoms refer to changes in operation conditions, which affect the equipment performance and final result of operation. BNs have proven to be useful for a wide variety of predictive purposes and monitoring. Similar applications have been documented in the medicine field and image processing, among other fields [40]. In the manufacturing industry, it has been also been used as a monitoring and diagnosis method in real time to identify failures of components in multistage processes [47]. There is a high necessity for decision-making in reference to maintenance management and after-sales service extension. BNs are an applicable methodology combined with other supportive decision-making methods, such as the decision support (DS). DS is a computer-assisted information system that supports all the company's activities regarding decision-making. A properly designed DS is the interactive software that compiles useful information from a combination of data, documents, personnel knowledge, and identified business models and helps solving problems and making decisions. An alteration analysis (that which includes the RCA and DS) is an industrial control process that has the global aim to extract from maintenance and after-sales databases the necessary information for an early evaluation of abnormalities, as well as to locate and solve problems in the operation, maintenance, and after-sales assistance processes [29]. Other tools that also could be added are the APR (risk preliminary analysis), PDR (Nets of Petri Estocastica), CM (Markov chains), MSC (Monte Carlo simulation), etc.

7.3.3 Advantages and Inconveniences

The previously described tools and methodologies present advantages and inconveniences intrinsic to each of them. Depending on the type and analysis in-depth carried out, it is necessary to evaluate each methodology in order to select and use that which better adapts to the raised necessities. Even though all methodologies have the capacity to define an analyzed problem, the CEDs are not able to show every causal relation between primary effects and root causes. Neither they are able to deliver a clearer route toward the root causes since they only categorized or order isolated causes in groups that produce a primary effect.

However, they do possess a low requirement level of information and resources and are relatively easy to use [37]. The HAZOP study is a structured one in the form of brainstorming and developed by people involved in the problem to solve; therefore, it is highly dependent on the experience of those in charge of carrying it out and should be conducted in several sessions which require time and other resources.

Its advantage lies in the developed plans to prevent the recurrence [38]. The FMEA is effective to find failure causes of a component and however loses capacity to solve complex problems when it fails to establish causal relations further than the failure mode that is being analyzed. The FTA is a quantitative method that works extremely well in engineering problems. It finds causes related to the original design of the system, determining possible scenarios and selecting appropriate solutions, as long as human factors are not included [37]. Equally, the Bayesian inference (in spite of requiring greater resources and having less simplicity in its use) presents large capacities to establish causal relations for a wide number of variables and it is apt as a support for decision-making and recurrence prevention [29]. Its structure provides the combination of previous knowledge, either obtained in a causal form or by observed data. The BNs can be used to know the casual relations and facilitate comprehension and the optimum form of problem analysis and future event prediction [46, 48]. Then, Table 7.4 shows a summary comparative chart based on a group of criteria for methodologies commonly used in RCA analysis.

The difficulty in obtaining solid results through some of these methodologies lies in the usage of only one focus, given that the qualitative analysis (brainstorming type) can leave aside a relevant amount of information contained in quantitative data, while another mere numeric focus is feasible of having a turn since it does not take into consideration aspects such as experience or relevant qualitative information. The use of an isolated method can derive in an incomplete

Table 7.4 Comparison of some RCA methodologies [34, 37, 38, 40, 43, 45, 46]

RCA methodologies/ characteristics		Capacity to define the problem	Simplicity of use	Information requirements	Dependence on the experience of directors	Time and resource consumption	Definition of all the casual relations	Deliver of a route towards the root causes	Explains how solutions prevent recurrence	Capacity to include human failures	Prediction of future events
Cause-and-effect diagram	High	✔	✔		✔					✔	
	Low			✔		✔	✔	✔	✔		✔
HAZOP	High	✔		✔	✔	✔	✔		✔		
	Low		✔					✔		✔	✔
Bayesian inference	High	✔		✔		✔	✔	✔	✔		✔
	Low		✔		✔					✔	
FMEA	High	✔	✔		✔	✔		✔			
	Low			✔			✔		✔	✔	✔
Failure tree analysis	High	✔	✔	✔			✔	✔	✔		
	Low				✔	✔				✔	✔

analysis. Thus, in some specific cases, the integration of RCA tools can be convenient specially when dealing with complex systems, obtaining this way better results [49]. In fact, one of the most common combinations to support an RCA analysis is that of FMECA and FTA [34]. All in all, there is evidence regarding the integration of the distinct types of RCA to generate more solid results, that is to say, each tool unitarily possesses its own limitations. However, the integration among them allows them to eliminate their own and individual limitations of each and every one of them. The present research proposes a representation model of a system malfunctioning, and the identification of the physical causes of such failures by the integration of BNs, failure trees, FMEA, and HAZOP studies based on the variables determined status that, given their dependencies, can unchain a state or failure event. The integration of these methodologies represents a more solid result in the identification of the main causes of a deviation of performance from the generation of the causality chain.

Following this chapter, it is suggested to consult next sections where aspects about the balance score card are dealt with as well as a hierarchical analysis procedure focused on what was here elaborated.

7.4 Case Study on Balanced Score Card

7.4.1 Application from a Financial Perspective

This practical example is based also on the case that will be described in Chap. 13, which deals with a manufacturing company that also assembles farm machinery [50]. In this particular case, the principles described in this chapter (about the balanced score card) will be applied. Those that mainly constitute as business activities and form a part of considerable profits in such company are the production, sales and repair of harvest machinery, and its harvest heads, whether they are produced or previously used. One key element for the company is to improve and optimize the technical support services, that is to say, to improve after-sales processes with the aim of increasing benefits coming from this area and retaining clients so that future sales are assured. After-sales activities consist of supplying spare parts, warranty service as well as maintenance contract with their customers. The company's management intends to compare from a financial point of view, the four aspects of the business (sale of new and secondhand harvesters as well as new and secondhand harvesters' harvest heads) that are the most efficient in regard to warranty costs. By doing this, and from a global viewpoint, the company can reassign objectives and assure an appropriate and satisfactory customer service in all their products, detecting which product provides the most, and the least, benefits. The inactivity of a harvest machine could cause great losses to the customer. Therefore, during harvest season, these machines should be at their highest performance capacities, and in the event of a breakdown, a quick repair should be guaranteed either by the parent company or one of the authorized assistance centers. Because of this, the company must take on a new strategy to reduce the number of technical assistance interventions, especially during harvest season since these corrective repair services are totally unexpected and difficult to control. To achieve this, the company must decide to take measurements or KPI (key performance indicators): sales price of the product; the corresponding warranty cost; the percentage in respect to sales price; along with the objectives (percentage and monetary) of what the warranty cost should be. Information for each type of product is listed in detail in Tables 7.5, 7.6, 7.7 and 7.8, according to the type of machine.

In the case of the data related to new or secondhand attachments (Tables 7.6, 7.8), the price to consider should just take into account the price of the harvest heads as reference and not the price of the rest of the machine (practically speaking, this case study has considered an approximate base price to be 180,000 € for the machine). It is clearly observed that the data in this case study have been distributed by projects. Nevertheless, this is an equivalent form to what could be temporary evolution, in which supposedly manufacturing processes are improving and the arisen problems during the warranty period should tend to decrease. According to Ref. [2], maintenance can be defined as all those actions that control the process of deterioration that can lead to failures in the system (preventive maintenance), as well as the restoration of a system to its operative state through corrective actions after failure (corrective maintenance).

When the manufacturer must provide warranty assistance to a customer, both parties must have a clear idea of the what the service consists of and what the circumstances should be, whether assistance must be provided or not. In the sales of farm machinery (new or secondhand), just as many other sectors, there is a possibility of performing maintenance contracts and/or warranty extensions. This is not the aim of this study, but could be an interesting object to evaluate, applying the balanced score card methodology, how the introduction of preventive maintenance contract can impact the company's after-sales performance.

In other words, carry out this case study from three different scenarios (as it will be mentioned in Chap. 13):

- Scenario A: No preventive maintenance contracts are applied, neither under the warranty period nor not.
- Scenario B: Users hire preventive maintenance services only during the warranty period.
- Scenario C: Users hire preventive maintenance services not only during the warranty period but also once it expires throughout the product's life cycle.

This case study is based on scenario A. That is, the study only considers warranty costs. However, the results of this analysis could differ if we consider the influence of maintenance hire.

Table 7.5 KPI relative to Harvesters Model 554

Project	Machine	Sales price	Warranty (€)	Warranty/sales price (%)	Objective % warranty/sales price	Obj. improvement % warranty/sales price	Warranty objective (€)	Warranty objective improvement (€)
PF 102	554 BCT	156,309	528	0.3	2	1.8	3,600	3,240
PF 103	554 BCT	180,000	406	0.2	2	1.8	3,600	3,240
PF 114	554 BCT	171,000	918	0.5	2	1.8	3,600	3,240
PF 115	554 BCT	147,090	1,818	1.2	2	1.8	3,600	3,240
PF 116	554 BCT	150,000	191	0.1	2	1.8	3,600	3,240
PF 117	554 BCT	150,000	226	0.2	2	1.8	3,600	3,240
PF 118	554 BCT	182,000	314	0.2	2	1.8	3,600	3,240
PF 119	554 BCT	176,724	1,065	0.6	2	1.8	3,600	3,240
PF 120	554 BCT	147,500	532	0.4	2	1.8	3,600	3,240
PF 121	554 BCT	173,748	1,686	1.0	2	1.8	3,600	3,240
PF 122	554 BCT	175,000	4,561	2.6	2	1.8	3,600	3,240
PF 123	554 BCT	165,000	243	0.1	2	1.8	3,600	3,240
PF 125	554 BCT	165,000	946	0.6	2	1.8	3,600	3,240
PF 128	554C	170,791	901	0.5	2	1.8	3,600	3,240
PF 129	554B	190,790	–	0.0	2	1.8	3,600	3,240
PF 130	554B	191,000	2,268	1.2	2	1.8	3,600	3,240
PF 131	554C	193,000	2,239	1.2	2	1.8	3,600	3,240
PF 132	554C	188,476	3,227	1.7	2	1.8	3,600	3,240
PF 133	554C	179,740	–	0.0	2	1.8	3,600	3,240

Table 7.6 KPI relative to harvest heads model LM

Proyec.	Machine	Sales price	Warranty (€)	% Warranty/ sales price	Target % warranty/ sales price	Improvement obj. % warranty/sales price	Warranty target (€)	Improvement target warranty (€)
MC 00	LM 600	55,524	4,678	8.4	2	1.8	1,200	1,080
MC 01	LM 500	67,000	2,581	3.9	2	1.8	1,200	1,080
MC 02	D225+LM600	206,750	5,279	2.6	2	1.8	1,200	1,080
MC 03	D210+LM600	237,154	978	1.7	2	1.8	1,200	1,080
MC 04	DX180+LM500	215,000	190	0.5	2	1.8	1,200	1,080
MC 05	LM 600	74,000	947	1.3	2	1.8	1,200	1,080
MC 06	LM 500	58,000	1,396	2.4	2	1.8	1,200	1,080

Table 7.7 KPI relative to used harvesters

Project	Machine	Sales price	Warranty (€)	% Warranty/ sales price	Target % warranty/ sales price	Improvement obj. % warranty/ sales price	Warranty target (€)	Improvement target warranty (€)
OP 02	CAT 574	115,000	439	0.4	2	1.8	2,600	2,340
OP 09	CAT 574	85,000	297	0.3	2	1.8	2,600	2,340
OP 36	EL 574 B	162,000	2,259	1.4	2	1.8	2,600	2,340
OP 51	TJ 1210Y	60,000	570	0.9	2	1.8	2,600	2,340
OP 63	P Bisón	90,000	3,840	4.3	2	1.8	2,600	2,340
OP 67	P Bisón	70,000	933	1.3	2	1.8	2,600	2,340
OP 70	CAT 574	75,000	3,540	4.7	2	1.8	2,600	2,340
OP 72	CAT 574	70,000	–		2	1.8	2,600	2,340

Table 7.8 KPI relative to used headings

Project	Machine	Sales price	Warranty (€)	% Warranty/ sales price	Target % warranty/ sales price	Improvement obj. % warranty/ sales price	Warranty target (€)	Improvement target warranty (€)
OP 66	CAT 550	110,000	–	0.0	2	1.8	3,000	2,700
OP 10	CAT 570	121,052	258	0.2	2	1.8	3,000	2,700
OP 15	CAT 580	172,133	1,334	0.8	2	1.8	3,000	2,700
OP 18	CAT 550	120,000	–	0.0	2	1.8	3,000	2,700
OP 21	V 921.1	— €	4,666		2	1.8	3,000	2,700
OP 25	CAT 550	94,828	1,293	1.4	2	1.8	3,000	2,700
OP 27	CAT 580	155,000	7,366	4.8	2	1.8	3,000	2,700
OP 35	DX 225	220,444	469	0.2	2	1.8	3,000	2,700

7.4.2 Results in the Practical Application

Following our particular case, in Fig. 7.6, four graphics can be observed as the result of the previous tables.

Graphics in Fig. 7.6 represent the following values:

- Percentage values of the warranty cost in terms to the sales price, dots in the form of rhombus and colored in dark gray;
- The tendency of the warranty, continuous line in light gray;
- The objective value, dash and dot line;
- And the "challenge" or proposed improvement in the objective, gray dotted line.

In the upper right corner of each graphic, the average total cost of the warranty is shown in respect to the sales price. In short, these graphics show the tendencies of warranty costs: upward trend (in the case of used harvesters) and on a downward trend (harvest heads LM) or oscillating in a high and low degree. Those products that have tendencies which end up rising are marked by a dark gray triangle, whereas the inverse is marked by a light gray rectangle. Likewise, the product whose mean percentage in warranty costs exceeds the objective value is

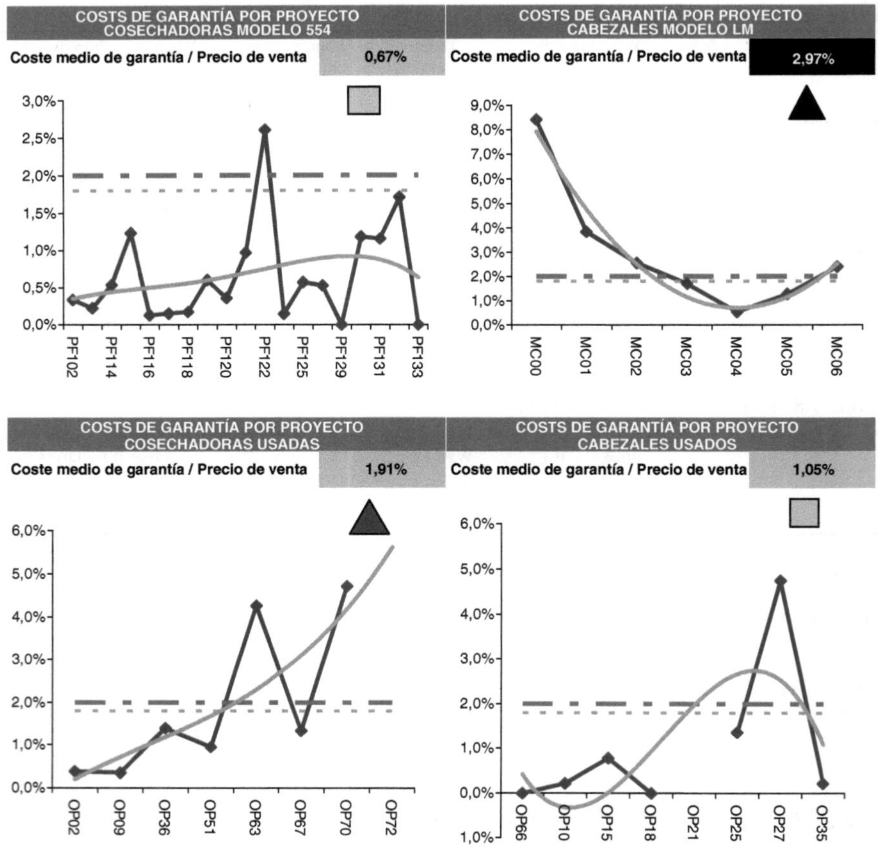

Fig. 7.6 BSC, a financial perspective

marked in dark gray, whereas the opposite (those who improve the objective) are marked in light gray. In conclusion, we can observe how the sale of harvest heads model LM does not present an appropriate behavior in the after-sales services from a financial point of view. In the other extreme, we can find the case of the Harvesters Model 554. On the other hand, it can be observed in the case of sales of used or secondhand products that, in spite of achieving in both cases the warranty costs' objectives, the after-sales service costs of used harvesters will increase in a near future, exceeding the objective value. These values, tendencies, or drastic changes observed in different graphics also show the derivated results in the decision-making process (personnel restructuration, tools update, tools repair, etc.). In any case, the methodology of BSC helps in a simple way to detect problems, which could later be addressed in a management council or in the prevision of possible circumstances that generate them. The following are other interesting references concerning the balanced score card: [3–5], [7] and [51] and about financial perspective: [1], [6] and [8].

7.5 Case Study on an Analytic Hierarchy Process

7.5.1 Practical Application of AHP

In Chap. 3, a scenario was described on which the present case study will be based. It is the moment to apply the methodology also represented as AHP previously developed in this same section. This application will be performed in a simple way and following these next three steps:

1. *State the objective*: to obtain a spare part for the warranty assistance with minimal effects on the assembly line.
2. *Define the criteria*:
    - Criterion 1: Stock extra (the higher the extra stock the better).
    - Criterion 2: Terms
        - Criterion 2.1: Supply term (the shorter the supply time the better).
        - Criterion 2.2: Repair term (the shorter the repair time the better).
    - Criterion 3: Reliability (the higher is the reliability, or lower is the failure frequency, the better).
    - Criterion 4: Costs
        - Criterion 4.1: Supply cost (the lower is the supply cost the better).
        - Criterion 4.2: Repair cost (the lower is the repair cost the better).
    - Criterion 5: Vehicles amount
        - Criterion 5.1: vehicles still to deliver (the higher is the amount of vehicles to deliver the better).
        - Criterion 5.2: vehicles under warranty (the lower is the amount of vehicles under warranty the better).
3. *Selection of alternatives*:

- Spare part is obtained from warehouse.
- Spare part is obtained by cannibalization.
- Spare part is obtained by purchasing.

This information can be arranged in a hierarchical tree or in a chart. In any case, the information is then synthesized to determine relative rankings of alternatives. Both qualitative and quantitative criteria can be compared using informed judgments to derive weights and priorities. In order to define the relative importance of the criteria, it will be used judgments, determining by this way the ranking of the mentioned criteria. Using pairwise comparisons, the relative importance of one criterion over another can be expressed (see Table 7.2): 1 equal, 3 moderate, 5 strong, 7 very strong, and 9 extreme. In order to turn this matrix (see Table 7.9) into ranking of criteria, in other words, to obtain a ranking of priorities from a pairwise matrix, Dr Thomas L. Saaty demonstrated mathematically that the

Table 7.9 Matrix of pairwise comparisons

	Extra stock	Supply term	Repair term	Reliability	Supply cost	Repair cost	Vehicles still to deliver	Vehicles under warranty
Extra stock	*1/1*	3	3	2	4	4	5	5
Supply term	1/3	*1/1*	1/2	2	1/3	1/3	1/4	1/4
Repair term	1/3	2	*1/1*	2	1/3	1/3	1/4	1/4
Reliability	1/2	1/2	1/2	*1/1*	3	3	4	4
Supply cost	1/4	3	3	1/3	*1/1*	1/2	1/3	1/3
Repair cost	1/4	3	3	1/3	2	*1/1*	1/2	1/2
Vehicles still to deliver	1/5	4	4	1/4	3	2	*1/1*	1/2
Vehicles under warranty	1/5	4	4	1/4	3	2	2	*1/1*

eigenvector solution was the best approach [52]. To calculate the eigenvector, a short and simple way is to raise the pairwise matrix to powers that are successively squared each time, calculating the sum of each row and normalizing.

This process is finalized when the difference between these sums in two consecutive calculations is smaller than a prescribed value. The computed eigenvector gives us the relative ranking of our criteria (see Table 7.10). In order to review its consistency, we apply the already-stated formulas CI y CR to $\lambda_{\max}$, which is obtained from the previously indicated eigenvalues (see Table 7.10) and the matrix of pairwise comparisons (Table 7.9), once it is normalized [26].

Therefore,

$$\mathrm{CI} = \frac{\lambda_{\max} - n}{n - 1} = 0.103948072$$

Consequently,

$$\mathrm{CR} = \frac{\mathrm{CI}}{\mathrm{CI}_{\mathrm{Rando}}} = \frac{0.103948072}{1.41} = 0.073722037 < 0.10$$

The value $\mathrm{CI}_{\mathrm{Rando}}$ depends on the number of criteria being compared (in this case, eight criteria). Saaty proposed the use of an appropriate consistency index (also called RCI "random consistency index"). He randomly generated reciprocal matrix using scale 1/9, 1/8,..., 8, 9 (similar to the idea of bootstrap [53, 54]), obtaining a random consistency index in order to check whether it is about 10 % or less. The average random consistency index of sample size 500 matrices is shown in Table 7.11.

In terms of the alternatives, a pairwise comparison determines the preference of each alternative over another (see Table 7.12).

Table 7.10 Ranking of defined criteria

Ranking order	Criteria	Eigenvalues
1	Extra stock	0.2797
2	Reliability	0.1777
3	Vehicles under warranty	0.1338
4	Vehicles still to deliver	0.1162
5	Repair cost	0.0854
6	Repair term	0.0732
7	Supply cost	0.0705
8	Supply term	0.0636

Table 7.11 Random consistency index (RCI)

N	1	2	3	4	5	6	7	8	9	10
RCI	0	0	0.58	0.9	1.12	1.24	1.32	1.41	1.45	1.49

As mentioned earlier, AHP can combine both qualitative and quantitative information. Particularly in our study case, it has been here applied a qualitative point of view, trying to describe a generic context which has considered:

- A high extra stock, reliability, as well as a high amount of vehicles still to be delivered;
- A short term in the supply and/or the repair of a piece; and
- A low cost in the supply and/or the repair of a piece, as well as a low amount of vehicles under warranty.

With the previously mentioned context, computing the eigenvector will determine the relative ranking of alternatives under each criterion (see Table 7.13).

Multiplying in terms of eigenvalues, the relative ranking of alternatives and the relative ranking of criteria, we can obtain the highest ranked alternative under the described context (Tables 7.14, 7.15, 7.16).

Therefore, the best decision under the previously described context is to obtain the spare parts from the warehouse, the second option for the manager is to obtain the piece by cannibalization, and finally, the worst decision would be to obtain the spare parts by purchasing. In summary, the AHP provides a logical framework to determine the benefits of each alternative. Other deductions are possible to obtain from the numerical results.

In this particular example, the executive manager can check the influence, for instance, of purchasing additional units, which would change the weight of the criterion extra stock but also the supply term and the supply cost, or performing the repair of the failed piece although it would have a longer repair term. In other words, this tool allow us to use it as a sensitivity analysis, showing how projected choices can change with qualitatively or quantitatively variations in the input of key assumptions on which the decision-making is based and showing in some way a criticality analysis of the issue.

Table 7.12 Eigenvalues of the defined alternatives

	Warehouse	Cannibalization	Purchasing	Eigenvalues		Warehouse	Cannibalization	Purchasing	Eigenvalues
Extra stock					*Supply cost*				
Warehouse	1.0000	3.0000	3.0000	0.5936	Warehouse	1.0000	0.5000	0.3333	0.1571
Cannibalization	0.3333	1.0000	2.0000	0.2493	Cannibalization	2.0000	1.0000	0.3333	0.2493
Purchasing	0.3333	0.5000	1.0000	0.1571	Purchasing	3.0000	3.0000	1.0000	0.5936
Supply term					*Repair cost*				
Warehouse	1.0000	2.0000	0.3333	0.2493	Warehouse	1.0000	3.0000	3.0000	0.5936
Cannibalization	0.5000	1.0000	0.3333	0.1571	Cannibalization	0.3333	1.0000	2.0000	0.2493
Purchasing	3.0000	3.0000	1.0000	0.5936	Purchasing	0.3333	0.5000	1.0000	0.1571
Repair term					*Vehicles still to deliver*				
Warehouse	1.0000	0.3333	2.0000	0.2493	Warehouse	1.0000	3.0000	3.0000	0.5936
Cannibalization	3.0000	1.0000	3.0000	0.5936	Cannibalization	0.3333	1.0000	2.0000	0.2493
Purchasing	0.5000	0.3333	1.0000	0.1571	Purchasing	0.3333	0.5000	1.0000	0.1571
Reliability					*Vehicles under warranty*				
Warehouse	1.0000	0.3333	2.0000	0.2493	Warehouse	1.0000	3.0000	3.0000	0.5936
Cannibalization	3.0000	1.0000	3.0000	0.5936	Cannibalization	0.3333	1.0000	0.5000	0.1571
Purchasing	0.5000	0.3333	1.0000	0.1571	Purchasing	0.3333	2.0000	1.0000	0.2493

Table 7.13 Ranking of alternatives under each criterion

Objective		Obtaining a spare part							
Criteria		Extra stock	Supply term	Repair term	Reliability	Supply cost	Repair cost	Vehicles still to deliver	Vehicles under warranty
Alternatives	Warehouse	1	2	2	2	3	1	1	1
	Cannibalization	2	3	1	1	2	2	2	3
	Purchasing	3	1	3	3	1	3	3	2
Best situation of each criterion		High	Short	Short	High	Low	Low	High	Low

Table 7.14 Eigenvalues according to alternatives (matrix 1)

	Extra stock	Supply term	Repair term	Reliability	Supply cost	Repair cost	Vehicles still to deliver	Vehicles under warranty
Warehouse	0.5936	0.2493	0.2493	0.2493	0.1571	0.5936	0.5936	0.5936
Cannibalization	0.2493	0.1571	0.5936	0.5936	0.2493	0.2493	0.2493	0.1571
Purchasing	0.1571	0.5936	0.1571	0.1571	0.5936	0.1571	0.1571	0.2493

Table 7.15 Eigenvalues according to criteria (matrix 2)

Criteria	Eigenvalues
Extra stock	0.0316
Supply term	0.0812
Repair term	0.0935
Reliability	0.2322
Supply cost	0.0959
Repair cost	0.1163
Vehicles still to deliver	0.1624
Vehicles under warranty	0.1870

Table 7.16 Result of the alternative ranking

[Matrix 1] × [Matrix 2]	Warehouse	0.4117
	Cannibalization	0.3367
	Purchasing	0.2516

Although costs have been here included, in many complex decisions, costs should be set aside until the benefits of the alternatives are evaluated. Otherwise, it could happen that the general costs of the warranty program were too high, taking not care about its benefits. In other words, discussing costs together with benefits can sometimes bring forth many political and emotional results.

References

1. Kaplan RS, Norton DP. 1996. The Balanced Scorecard: translating strategy into action. Harvard Business School Press.
2. Kaplan R, Norton D. 2005. The office of strategy management. Harvard Business Review. 83(10):72.
3. Kaplan R, Norton D. 2006. Alignment: Using the Balanced Scorecard to Create Corporate Synergies. Boston. Harvard Business School Press.
4. Abran A, Blugione L. 2003. A multidimensional performance model for consolidating Balanced Scorecards. Advances in Engineering Software. 34:339-349.
5. Fleisher C. Mahaffy D. 1997. A balanced scorecard approach to public relations management assessment. Public Relations Review. 23(2):117-142.

6. Michalska J. 2005. The usage of The Balanced Scorecard for the estimation of the enterprise's effectiveness. Journal of Materials Processing Technology. 162-163 751-758.
7. Kaplan R, Norton D. 1992. The balanced scorecard: Measures that drive performance. Harvard Business Review. 70(1):71-79.
8. Macdonald M. 1998. Using the balanced scorecard to align strategy and performance in long-term care. Healthcare Management Forum. 11(3):33-38.
9. Kumar U, Ellingsen HP. 2000. Development and implementation of maintenance performance indicators for the Norwegian oil and gas industry. Proceedings of the 14th International Maintenance Congress (Euro-Maintenance 7-10 March, Gothenburg). Pp. 221-8. JQME 13,3 256
10. Parida A, Chattopadhyay G. 2007. Development of a multi-criteria hierarchical framework for maintenance performance measurement (MPM). Journal of Quality in Maintenance Engineering. 13(3):241-258.
11. Crespo Márquez A. 2007. The Maintenance Management Framework. Models and methods for complex systems maintenance. Springer Series in Reliability Engineering. Springer-Verlag. ISBN: 978-1-84628-820-3
12. Gómez J, Crespo Márquez A, Moreu P, Parra C, González-Prida V. 2009. Outsourcing maintenance in services providers. Taylor & Francis Group, London, ISBN 978-0-415-48513-5. Pp. 829-837.
13. González-Prida V, Crespo Márquez A, Moreu P, Gómez J, Parra C. 2009. Availability and reliability assessment of industrial complex systems: A practical view applied on a bioethanol plant simulation. Taylor & Francis Group, London, ISBN 978-0-415-48513-5. Pp. 687-695.
14. Saaty TL. 1995. Decision Making for Leaders. RWS Publications, New York.
15. Saaty TL, Vargas LG. 1982. Logic of Priorities. Boston: Kluwer-Nijhoff Publishing.
16. Bevilacqua M, Braglia M. 2000. The Analytical Hierarchy Process Applied To Maintenance Strategy Selection. Reliability Engineering and System Safety.
17. Belton V. 1986. A comparison of the analytic hierarchy process and a simple multi-attribute value function. European Journal of Operational Research.
18. Harker PT, Vargas LG. 1990. Reply to "Remarks on the analytic hierarchy process. J.S. Dyer". Management Science.
19. Saaty TL. 1994. Fundamentals of Decision Making. RWS Publications, Pittsburgh.
20. Saaty TL. 1994. Highlights and Critical Points in the Theory and Application of the Analytic Hierarchy Process. European Journal of Operational Research.
21. Moffett A, Garson J, Sarkar S. 2005. A software package for incorporating multiple criteria in conservation planning. Environmental Modelling and Software. 20:1315-1322.
22. Dyer J. 1990. Remarks on the Analytic Hierarchy Process. Management Science. 36:249-258.
23. Dyer J. 2005. MAUT-multiattribute utility theory. Multiple criteria decision analysis: state of the art surveys. Ed. by J. Figueira, S. Greco and M. Ehrgott. Pp. 265-294. Kluwer, Dordrecht.
24. Arrow K, Raynaud H. 1986. Social choice and multicriterion decision-making. The MIT Press, Cambridge.
25. Donegan HA, Dodd FJ, McMaster TBM. 1992. A new approach to AHP decision-making. The Statiscian 41.
26. Donegan HA, Dodd FJ, McMaster TBM. 1995. Theory and methodology inverse inconsistency in analytic hierarchies. European Journal of Operational Research.
27. Zanakis SH, Solomon A, Wishart N, Dublish S. 1998. Multi-attribute decision making: A simulation comparison of select methods. European Journal of Operational Research.
28. Eti M, Ogaji S, Probert S. 2006. Reducing the cost of preventive maintenance (PM) through adopting a proactive reliability-focused culture. Applied Energy. 1235-1248.
29. Weidl G, Madsen A, Israelson S. 2005. Applications of object-oriented Bayesian nets for condition monitoring, root cause analysis and decision support on operation of complex continuous processes. Computers and Chemical Engineering. Pp. 1996-2009.
30. Bertolini M, Bevilacqua M. 2006. A combined goal programming - AHP approach to maintenance selection problem. Reliability Engineering and System Safety. Pp. 839-848.

31. Smidt-Destombes K, Heijden M, Harten A. 2004. On the availability of a k-out-of-N system given limited spares and repair capacity under a condition based maintenance strategy. Reliability Engineering and System Safety. Pp. 287-300.
32. Eti M, Ogaji S, Probert S. 2006. Development and implementation of preventive-maintenance practices in Nigerian industries. Applied Energy. Pp. 1163-1179.
33. Moubray J. 1997. Reliability-centred maintenance. New York. Industrial Press Inc.
34. Li D, Gao J. 2010. Study and application of Reliability-centered Maintenance. Journal of Loss Prevention in the Process Industries. Pp. 622-629.
35. Gao J. 2005. Informatization and intellectualization of the engineering asset. National Conference for Device Management.
36. Barberá L, González-Prida V, Crespo Márquez A. 2012. Review and evaluation criteria for software tools supporting the implementation of the RCM methodology. (Sent to ESREL Helsinki).
37. Gano D. 2007. Apollo Root Cause Analysis - A New Way of Thinking (3rd ed.).
38. Rossing N, Lind M, Jensen N, Jørgensen S. 2010. A functional HAZOP methodology. Computers and Chemical Engineering. Pp. 244-253.
39. Box G, Kramer T. 1992. Statistical process monitoring and feedback adjustment: A discussion, technical report, center for quality and productivity improvement. Technometrics. Pp. 251-285.
40. Dei S, Stori J. 2005. A Bayesian network approach to root cause diagnosis of process variations. International Journal of Machine Tools & Manufacture. Pp. 75-91.
41. Paradies M, Busch D. 1988. Root Cause Analysis at Savannah River Plant. Conference on Human Factors and Power Plants. Pp. 479-483.
42. Sikos L, Klemes J. 2010. Reliability, availability and maintenance optimisation of heat exchanger nets. Applied Thermal Engineering. Pp. 63-69.
43. Doggett A. 2004. A statistical comparison of three root cause analysis tools. Journal of Industrial Technology. Pp. 1-9.
44. American Institute of Chemical Engineers. 1992. Guidelines for Investigating Chemical Process Incidents. New York. AIChE.
45. Cai X, Wu C. 2004. Application manual of modern machine design method (1st ed.). Beijing: Chemical Industry Press.
46. Ben-Gal I. 2007. Bayesian Nets. F. Ruggeri, F. Faltin, & R. Kennett, Encyclopedia of Statistics in Quality & Reliability. Wiley & Sons.
47. Wolbrecht E, D'Ambrosio B, Paasch R, Kirby, D. 2000. Monitoring and diagnosis of a multi-stage manufacturing process using Bayesian Nets. Artificial Intelligence for Engineering Design. Analysis and Manufacturing. Pp. 53-67.
48. Zitrou A, Bedford T, Walls L. 2010. Bayes geometric scaling model for common cause failure rates. Reliability Engineering and System Safety. Pp. 70-76.
49. Hitchcock L. 2006. Integrating Root Cause Analysis Methodologies. Engineering asset management. Pp. 614-617.
50. Legnani E, Cavalieri S, Crespo Márquez A, González-Prida V. 2010. System Dynamics Modeling for Product-Service Systems - A case study in the agri-machine industry, In: Proc. of APMS. International Conference on Advances in Production Management Systems, Cernobbio, Como, Italy.
51. Djamaludin I, Murthy DNP, Kim CS. 2001. Warranty and preventive maintenance. International Journal of Reliability, Quality and Safety Engineering. 8(2):89-107.
52. Saaty TL. 1990. How to make a decision: The analytic hierarchy process. European Journal of Operational Research.
53. Chernick MR. 2008. Bootstrap methods: a guide for practitioners and researchers. Wiley-Interscience. ISBN 978-0-471-75621-7.
54. Qi Luo (Ed.) 2010. Advancing Computing, Communication, Control and Management, Springer. ISBN 978-3-642-05172-2.

Chapter 8
Management Efficiency

8.1 The Logistic Support in the Technical Assistance

This part of the chapter will deal with some aspects about the logistic support and its influence in warranty management. As an introduction, this section will examine issues such as:

1. The classification of elements' candidates for a logistic support analysis (i.e., those components of the industrial asset, which, due to its importance according to certain factors, deserve to be especially analyzed for the warranty management).
2. The choice of repair echelons (i.e., those levels where it is more efficient to perform the assistance, in order to take appropriate measures for the development of a warranty program).
3. The definition of tasks (i.e., methods that define maintenance and warranty tasks for a component of a specific product).

In order to launch a new product to the market with a suitable warranty program, it is necessary that during the early stages of its product life cycle, the logistic support has to be identified and assessed for that product. Some references and standards already deal with this issue [1, 2] mainly for the military sector. According to these references, the data required for a logistics support analysis can be categorized under three different groups:

- Data related to product configuration and its requirements;
- Data related to possible failures; and
- Data related to maintenance and components acquisition.

Logistic support analysis is an iterative process that may be useful for the company management, in order to size the customer-service department, the staff qualifications, the materials needed, etc., with the goal of providing the appropriate technical assistance during the product warranty period to the customer [3].

V. González-Prida Díaz and A. Crespo Márquez, *After-sales Service of Engineering Industrial Assets*, DOI: 10.1007/978-3-319-03710-3_8,

Table 8.1 Logistic support influence

Disciplines or areas	Logistic support influence
1. Reliability and maintenance	LORA, FMECA, RCM/failure rates/MTTR, MTBM
2. Maintenance planning	Maintenance tasks
3. Manpower/technicians	Required skills/staff knowledge
4. Deliveries support	Spare parts
5. Tools and equipments	Special tools/test equipments
6. Technical documentation	Catalog of spare parts/maintenance manuals
7. Simulation and training	Staff skills
8. Computer-based support	Hardware/software reliability
9. Facilities	Required equipments and location
10. Transport/dimensions	Mass, size, weight/volume, transport, type
11. Design	Texts with recommendations
12. Standardization	Elements control

The following chart (Table 8.1) shows the interaction with other disciplines and how logistic support can influence them.

This table includes items that are not typical on logistic support. For example, the reliability (corresponding to a product characteristic), maintenance, design, and standardization (which are also not elements of logistical support, etc.). However, they are considered here due to their involvement and their importance in logistic support. In other words, the product identification and evaluation from the point of view of logistic support are to analyze the parts, types, and quantities, the technical requirements and skills levels of technicians, the tools and technical documentation for the operation and repair. All of them should be aimed for implementing a technical assistance program in order to assure a specific product-market position. It will therefore be necessary a database that records those sources of logistic support as well as the identification and acquisition of elements.

8.1.1 Classification of Elements Candidates for a Logistic Support Analysis

Those logistic support candidates to be considered in a logistic support analysis process [1, 4] can be defined as follows:

- *Logistic Support Full Candidate*: Components requiring that their logistic support is fully analyzed (fiability, maintainability, disponibility, required skills, tools, manpower, etc.).
- *Logistic Support Administrative Candidate*: These are components that do not require a complete analysis of their logistic support, but they are necessary to be taken into consideration to perform the complete analysis of full candidates. In other words, an administrative candidate is the one which has to be manipulated in order to access other candidates of the product.

In Fig. 8.1, a procedure to select the corresponding types of candidates mentioned above is proposed in the form of a flowchart. Knowledge of logistic support in the product key elements is essential to plan, schedule, and organize a program of maintenance and warranty support. In this framework, component reliability is considered as the first factor. Those components presenting a high failure rate will be selected to be included in the full candidate list for logistic support. In order to limit or restrict the amount of components in the list, threshold limits can be set for the failure rate (in operating hours, kilometers, etc., depending on the type of product). The second factor considered is maintenance; those components requiring intensive preventive maintenance, whose maintenance tasks require specific staff training or those components presenting a high technical maintenance complexity and documentation will be also full candidates. According to the product maintenance, the candidates to consider will be those components that require scheduled preventive maintenance in order to function properly during the life cycle of the entire system, as well as those cases when it will be needed to implement corrective tasks. The third factor depending on the components needs of support equipment (for diagnosis) or special tools (for repairs) to carry out the development of maintenance, either preventive or corrective, as well as warranty tasks. In addition to this, a fourth factor is the criticality. In this case, candidates will be those components whose failure affects the safety of the product or the user, altering the fulfillment of the product function.

There are four standardized levels to measure the criticality:

1. *Catastrophic*: failure can cause the loss of the entire system.
2. *Critical*: failure can cause serious damage to the system, so the product cannot achieve the success of its functions.
3. *Marginal*: failure can cause damage to the system, so there is a delay or loss of product availability.
4. *Minor*: failure is not serious enough to cause damage to the system, but an unscheduled repair will probably be required.

Finally, the fifth factor considered is the accessibility of the components. We consider that a component will be an administrative candidate when (although it is not fulfilling the requirements for full candidates) it is necessary to be removed to have access to other product full candidates or when the access to this element must be done frequently. To conclude, if a component does not meet any of the above criteria, it will be considered as noncandidate.

8.1.2 Choice of Repair Levels

Once logistic support candidates have been selected a level of repair analysis (LORA) can be carried out [5]. Notice that the choice of the repair level presents a high influence on the logistic support in terms of cost-effectiveness. The aim of the analysis is indeed to choose the proper action, not only during the development of

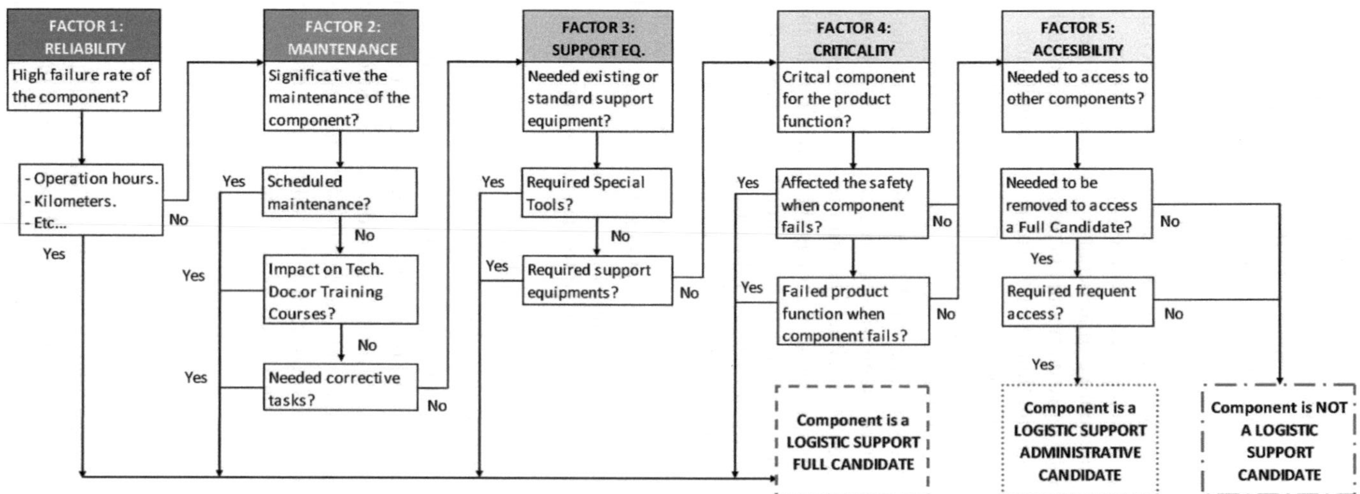

Fig. 8.1 Procedure to select logistic support candidates

Table 8.2 Repair levels (adapted from [5])

Repair level	Significance	Cost
Echelon 1	Developed by the user	C_1
	Preventive maintenance tasks by the operator	
Echelon 2	General changes for parts and adjustments	C_2
	Preventive tasks and correction (e.g., duration 4 man/hour)	
Echelon 3	In-place repair/change of damaged assemblies	C_3
	Preventive tasks and correction (e.g., 4 man/hour < duration 50 man/hour)	
Echelon 4	General repair of damaged sets and subsets	C_4
	Preventive tasks and correction, determined by maintenance manuals when duration > 50 man/hour	

a warranty program, but also during the rest of the product life cycle. The repair levels used in this paper have been extracted from the military sector [5]. Other terms commonly used are also line repairable unit (LRU), which are immediately replaced on ground, and shop repairable unit (SRU) or technical level of intervention (TLI). The proposed framework suggests a way to evaluate and determine how and where a maintenance or warranty task should be executed, in order to afford the lowest cost. Using the already-developed notation for the military sector (Table 8.2), it is possible to define the levels or echelons at which maintenance tasks must be performed according to a cost minimization criteria. Repair cost at each level (C1, C2, C3, and C4) can be calculated knowing the different costs (direct labor, material, handling, etc.) for the different maintenance echelon. Then, it is possible to minimize an objective function in order to determine which echelon is the most recommendable to face a repair. A fifth echelon related to reconstructions and major changes performed by the own industrial maintenance (overhaul) is sometimes considered.

Future researches in this field can be focused on the calculation of different costs (which are not within the scope of this study) or, once the warranty period has expired, the combination of extended warranties with maintenance assistance.

8.1.3 Task Definition

In literature, many methods are described to define maintenance and warranty tasks when components of specific product fail [6]. To achieve this goal, the failure modes, effects, and criticality analysis (FMECA) provides a good basis for determining those maintenance tasks that are necessary and, consequently, their extrapolation to the case of assistance under warranty. In a warranty management context, the reliability-centered Maintenance (RCM) becomes a dependability analysis for the definition of those tasks to be performed during the warranty period, in accordance to the forecast of possible customers' complaints. Within these technologies, the criticality of a failure mode can be assessed using the safety hazard severity code (SHSC) as catastrophic, critical, marginal, and minor failure (already mentioned) or through a more quantitative assessment which is the case of the military field [7].

$$C_m = \lambda_p \cdot \alpha \cdot \beta \cdot t$$
$$C_r = \sum (C_m)_n \quad \text{where}\, n = 1, \ldots, N$$

where [8] is as follows:

- C_m (*modal criticality number*): It is calculated for each failure mode of each logistic support item.
- C_r (*item criticality number*): It is calculated for each logistic support item.
- λ_p (*failure rate*): It is usually obtained from failure rate predictions [9, 10].
- α (*failure mode rate*): It is usually obtained from failure mode database sources such as [11].
- β (*conditional probability*): It is the analyst's best judgment that the failure will occur, based on the item severity classification.
- t (*mission phase duration*): In military or aerospace sectors, it is an average data of the usual system functioning.

The product analysis allows identifying preventive tasks to be performed. As commented above for the FMECA and RCM, the results can be extrapolated to the case of technical assistance in the development of a warranty program. Future research in this field can focus on the combination of RCM and FMECA with customer relationship management (CRM), which is also considered as one of the stages in the proposed framework for warranty management. This integration of RCM and FMECA solutions with an enterprise resource planning (ERP) is generally known as reliability-centered maintenance and optimization (RCMO), which can improve not only the definition of new strategies for customer service, but also the continuous improvement in the efficiency of warranty management, and (consequently) customer satisfaction. After determining the tasks to be applied during the warranty period (and its possible extensions or combinations with maintenance contracts), it is time to define the frequency of these tasks. In general terms, the evaluation of these intervals depends on the experience and product engineering, as well as the forecast of possible claims by the customer depending on product reliability. However, in order to calculate, this frequency is applied usually (although not exclusively) the following formulation:

$$\text{TF} = (1/\text{MTBF}_{\text{tec}} + 1/\text{MTBM}_{\text{in}} + 1/\text{MTBM}_{\text{nd}}) \times \text{AOR}$$

where

TF	Task Frequency
MTBF_{tec}	Mean Time Between Failure Technical
MTBM_{in}	Mean Time Between Maintenance Induced
MTBM_{nd}	Mean Time Between Maintenance No Defect
AOR	Annual Operating Requirements

In order to obtain an in-depth knowledge on this formula, we suggest consulting the Ref. [1], which defines the parameters for calculating the frequency. However, simplifying:

- *Mean time between technical failure* ($MTBF_{tec}$): Indicator of system reliability that is calculated from known failure rates of various system components and documented by technical characteristics. Technical parameters reflect the technical reliability that the system/equipment must demonstrate. In determining these parameter values, all failures and resultant actions to restore the item will be considered.
- *Mean time between maintenance induced* ($MTBM_{in}$): One of the categories of maintenance events contributing to the mean time between maintenance actions (MTBMA) value. Induced malfunctions are those induced in the system/equipment under analysis from external sources (i.e., other equipment, personnel, etc.).
- *Mean time between maintenance no defect* ($MTBM_{nd}$): One of the categories of maintenance events contributing to the MTBMA. These events consist of removals, replacements, and reinstallations of equipment due to erroneous failure indication. The $MTBM_{nd}$ shall be developed using historical data and field feedback information from similar items to establish the number of maintenance events that are the result of erroneous failure indication. An alternative procedure approved by the requiring authority may be used in lieu of the above procedure.

Once obtained the frequency for a specific task, it is possible to determine the required spare parts. For that purpose, it is necessary to know previously the following parameters:

- Task frequency (TF) per year and product unit
- Spares quantity per task (QT)

With these data, it is possible to obtain the spare parts frequency per year and product unit:

$$\Lambda = \sum (\mathrm{TF} \times \mathrm{QT})$$

Summarizing, this entire section tries to transmit that in order to attend warranties with a minimum of waste, expenses or unnecessary effort, it is required to design an appropriate plan for the warranty program. The plan for a specific product will need to identify its functions, how these functions may fail, and then, a set of efficient tasks based on safety considerations of the concerned product and on the assistance economy. As discussed in Chap. 5, an initial maintenance plan applied to the warranty period can be a good first approach for capacity planning under warranty, spare parts provisioning, scheduling of tasks for assistance, the training level of technicians, etc. The program planning and improvement applied to after-sales management can, of course, enhance the efficiency of warranty program policies. This improvement will depend on the time horizon of the analysis. Other references in this field are [12–14].

8.2 Costs-Risks–Benefits Analysis in the After-sales Management

Once described the main lines of the logistic support from the point of view of warranty tasks (logistics support components, their repair levels…), another important aspect to consider in this stage (the efficiency of after-sales service) is to analyze the costs, risks, and benefits applied to the development of a program of warranty assistance. In order to achieve this, firstly, it is necessary to remark that the warranty support service itself does not generate any profit, in the usual sense of profit definition. In general terms, the "benefit", as stated here, refers to a lower cost. That means, a cost that is intended to decrease, so that this "benefit" reverses to the warranty as a parameter related to competition and as a marketing tool, which can increase the sale of a more robust and higher-quality product compared with other market competitors. In terms of cost and failure modeling, a wide range of different approaches, both analytical and empirical, have been developed in other fields. An interesting reference in this area is [15]. Generally, in order to carry out an after-sales policy to analyze risks, costs, and benefits, it will depend on the available information, being also the complexity normally high in this case, so simplifications are usually required to make easier the resolution of such analysis, or just to reduce the computers needs. This section will try to summarize briefly the main concepts related to costs, risks, and earned value (EV), taken these concepts from the general theory of project management and in order to adapt it to the customer service for a better management and control.

8.2.1 The Costs Management in the Technical Assistance

A project cost management includes those processes required to ensure that the project is completed within the approved budget [16, 17]. Regarding warranty tasks, it will be primarily concerned with the cost of the resources required to fulfill the complete activities related to the after-sales service in order to provide to the customers a proper technical assistance. It is possible to observe from different points of view the warranty cost management (Fig. 8.2). On the one hand, the general cost management should consider the effect on the cost of the product of decisions taken for the whole project.

For example, limiting the number of design reviews may reduce the cost of the project, but at the expense of an increase in service and customer's operating costs. A broader view of warranty cost management should be referred also to the life

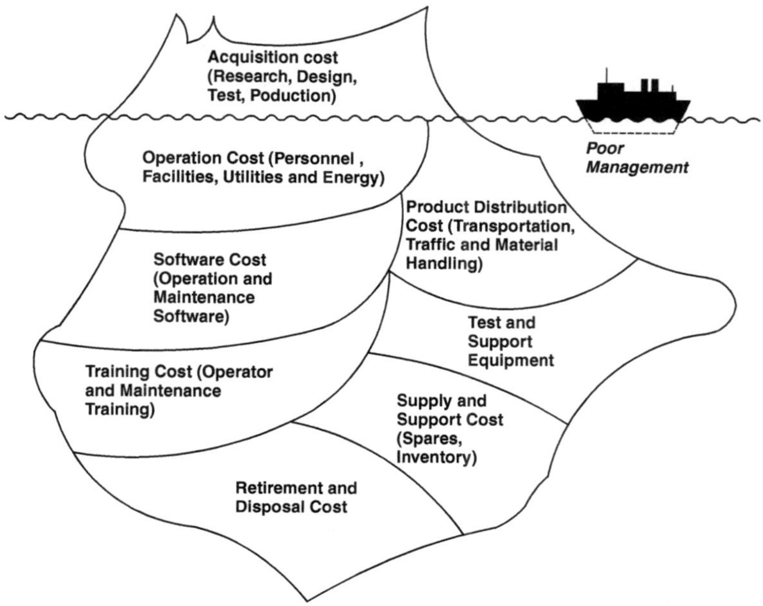

Fig. 8.2 Costs iceberg related to logistic support

cycle cost, considering therefore acquisition, operating, and disposal costs when evaluating with the balance scorecard the different project strategies. Consequently, a creative approach applied to optimize life cycle costs will:

- Save time
- Increase profits
- Improve quality
- Expand market share
- Use resources more effectively
- Solve problems not only for the warranty assistance but for the whole business.

This approach is usually called as value engineering which is used together with the life cycle cost analysis as techniques to reduce cost and time, improve quality and performance, and optimize the decision-making. On the other hand, the particular cost management does not require considering all those decisions taken during the whole product life cycle. The particular costs for warranty refer just to those activities needed to resolve a problem during a specific period of time at the beginning of the phase of use or product operation. Such activities will depend just on the product quality. Frequently, estimating and analyzing the financial performance of a product launched to the market does not consider warranty issues. In this case, controllable and uncontrollable costs should be estimated and budgeted separately to ensure that the expected benefits will be obtained from the actual performance of the product.

In order to influence in warranty costs, it will be easier at the early stages of the project during the prelaunch stage, due to the fact that a soon definition and identification of the expected product requirements are critical to reduce future costs during the warranty period. The process of determining what resources (people, equipment, materials) and what quantities of each (when needed) should be used to perform warranty activities is dealt as part of the product logistic support but supposes, of course, an important part to take into account when the budget to these activities must be established. In addition to this, the process of developing an approximation of the costs for the resources needed to perform the warranty tasks should be preferably done prior to budget request rather than after budgetary approval is provided, developing an assessment of the likely quantitative result and how much will it cost to the organization to provide this product service. The product pricing will be a business decision where the above-mentioned costs should be taken into account, together with other factors. As inputs for the warranty cost approach can be included:

- Resource requirements
- Estimated activities duration
- Historical information
- Used structure of the organization
- Commercial available data on cost estimating, and risks (either as threats or opportunities)…

All these factors have a significant impact on cost which the management team should consider to budget properly the future technical assistance. The following methods can be used to estimate the warranty costs:

Analogous estimating: It uses the actual warranty cost of a previous similar product as the basis for estimating the cost of the current product. It is frequently used when there is a limited amount of detailed information about the product in the early product phases. Generally less costly than other estimating techniques, but it is also generally less accurate. Most reliable when the previous product is similar in fact and not just in appearance (that includes a similar market target), and the staff to prepare such estimation have the needed expertise (because the approach will be considered as a form of expert judgment).

Parametric modeling: It uses product characteristics (parameters) in a mathematical model to predict product failures and consequently the costs of spare parts, repair time, logistic… as well as staff training, technical documentation, tools and equipment etc. Models can be simple or complex, but they are more reliable when the historical information used to develop the model is accurate, and the parameters are quantifiable.

Bottom–up estimating: It involves the cost estimation of individual activities or work packages related for instance to the product maintenance, adapted and summarized to warranty tasks. Its cost and accuracy will depend on the size and complexity of the individual activity or work package taken into consideration. Therefore, the management team must weigh the accuracy against the cost.

In order to simplify the use of the mentioned methods, there are many software applications, simulation, or statistical tools widely used to assist with cost estimating. Mayhew [18], and Doyle [19] are suggested references regarding cost modeling in other fields, although there are plenty of other different methods for cost estimating. All of them should finally express a quantitative assessment of the likely costs of the resources required to complete the warranty activities (labor, materials, and supplies). The estimation is expressed in currency units to facilitate comparisons with other budgets of the whole business or with the warranty costs of other products. For instance, it is very useful for large companies with a specific range of products in the market (e.g., the automotive sector), to control multiple warranties cost baselines to measure different aspects of the warranty performance or different degrees in demands or requirements of specific customers' profiles. In any case, the warranty cost estimation must be refined during the course of the warranty period, adding new details which can be available during this time, making all this exercise more accurate. The estimation report usually includes other supporting details, as follows:

- Description of the estimated scope of work,
- How such an estimation has been developed,
- Justification of the assumptions considered, and
- Indication of the range of possible results.

Once estimated the costs, it will be needed to be expressed in a management plan for warranty costs, which will be an additional element in the overall product plan.

8.2.2 Risk Management in the Technical Assistance

The warranty risk management is the systematic process of identifying, analyzing, and responding to risk likely appearing during a program of warranty assistance. Reference [20] deals with issues regarding risk and spare parts estimation, and Ref. [21] addresses risk and maintenance decisions. Warranty risk is understood as an uncertain event or condition that, if occurs, has a negative effect on the program objective. Nevertheless, it brings the opportunity to improve the product quality. Any risk has a cause and, if it occurs, a consequence.

Therefore, known risks are those that have been identified and analyzed and may be possible to plan for their occurrence and mitigation. On the other hand, unknown risks cannot be managed or their probability of occurrence cannot be assessed.

Nevertheless, executive managers may address them by applying a general contingency based on past experience with similar programs, in our case, related to the after-sales service. Those known risks that are threats to the warranty program may be accepted if they are balanced with the reward that may be gained by taking the risk. Likewise, risks that are opportunities may be pursued to benefit the program's objectives. One measure of an organization's commitment is its dedication to gathering high-quality data on warranty and product risks and the characteristics of these risks. Figure 8.3 illustrates the risk management process, divided into different steps described in Table 8.3. Other references within this scope are, for example, [22–24].

8.2.3 The Earned Value Management Applied to the Warranty

Once described the influence of factors like cost, risk, and time in the warranty management, we melt all these features in a technique which allows us the economical follow-up of the after-sales service [15, 25]. An accounting system developed for general projects' management, applied, and well balanced to warranty management is now presented. This system implies the monitoring of cost performance to detect and understand variances from the warranty plan, ensuring that all appropriate changes are recorded accurately in the cost baseline, in order to prevent incorrect or inappropriate changes (out of acceptable limits) which can affect negatively to the business goal. The inputs required for this control are, for instance:

- Warranty cost baseline,
- Performance reports (regarding budget issues),
- Change requests, and
- Cost management plan, etc.

It may also alert the management team to warranty issues that may cause problems in the future and should be integrated in the general budgetary control system. Other references within this scope are, for example, [26–28]. The acronyms used (internationally accepted) are in the Table 8.4 which shows their meaning. It is important here to underline that the best warranty service is the one that is not needed to be applied. Nevertheless, it does not systematically happen in the real world, so the company must be prepared for any eventuality that may occur. The following account ability system referred to warranty management is for the case when the performances of the client and of the sold product are foreseen (degree of complexity, type of complaints, etc.). As commented before, the concept of "benefit" or "EV," etc., must be understood in the case of warranty management as a lower cost in the incurred value applied in the technical assistance, keeping high levels of service quality.

Earned value management system (EVMS) or earned value analysis (EVA) is an integrated management system that allows the progress monitoring, combining (for our specific case) the technical assistance planning with the warranty cost. The traditional practice of project management (Fig. 8.4) tends to compare incurred costs (actual costs) with planned costs (planned expenditure) [29]. In this figure,

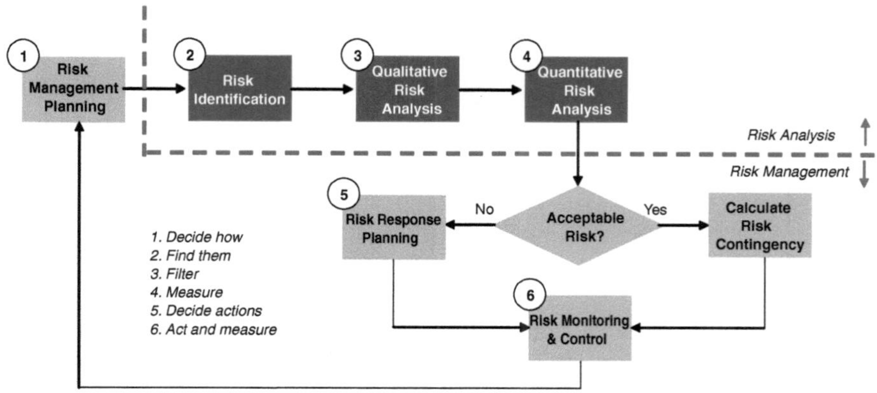

Fig. 8.3 Risk management workflow

the abscissa axis refers to the percentage of the whole-program time. This program time refers to the warranty period since a new product is launched to the market, till the warranty expiration of the last sold unit. What is here suggested is that the level of incurred cost is probably not the right choice to determine the progress of the warranty program. The level of incurred cost in warranty issues is not necessarily a good measure of progress of the technical assistance tasks. Therefore, EVA applied to after-sales provides a third reference point which represents an objective overview of the status of the warranty contract:

- The value of work completed to date and
- The earned value.

This EV is difficult to understand in warranty terms, since the warranty application does not imply any direct income. On the contrary, it is usually considered as an expenditure center. Consequently, the philosophy here is to consider as EV all this benefit obtained by saving resources but applying at the same time a proper service to the customer. This proper service means fast, effective, and without any additional eventuality to the customer which supposes a decrease in the confidence in the company. In the case of the after-sales service, one can ask whether:

- There is an excess in the incurred warranty cost,
- Whether the product is behaving in an unexpected (and negative) way, or
- Whether the product launch is advanced in comparison with the scheduled product positioning in the market.

What is here suggested is that the level of incurred cost is probably not the right choice to determine the progress of the warranty program or the reliability of the product [30–32].

The concept is to compare the budgeted cost and planning with the actual planning at any time of the program, integrating the program's operational

Table 8.3 Inputs/output for the risk management in a warranty program

Warranty risks management	Input	Output
1. Risk management planning	• Warranty program chart • Company's risk management policies and plan templates • Defined roles and responsibilities • Risk tolerances in the warranty program • Work-breakdown structure for warranty assistance	• Risk plan for the warranty management
2. Risk identification	• Risk plan for the warranty management • Scope and objective of the warranty program • Risk categories: Technical or quality risks Warranty management risks Organizational risks External risks Historical information	• List of identified risks • Risks symptoms or warning signs • Insufficient detail in the inputs of other steps
3. Qualitative risk analysis	• Risk plan for the warranty management • List of identified risks • Status of the warranty program • Product type • Data precision • Probability and impact scales • Warranty assumptions	• Risk ranking for the whole warranty program • List of prioritized risks • Trends in the results of the qualitative analysis

(continued)

Table 8.3 (continued)

Warranty risks management	Input	Output
4. Quantitative risk analysis	• Risk plan for the warranty management	• Prioritized list of quantified risks • Probabilistic analysis of the warranty program (for achieving expected costs and times)
	• List of identified risks • List of prioritized risks • Historical information • Expert technician's judgment	• Trends in the results of the quantitative analysis
5. Risk response planning	• Risk plan for the warranty management	• Response plan for warranty risks • Residual risks • Secondary risks • Contractual agreements • Contingency reserves • Alternative strategies
	• Risks ranking for the whole warranty program • List of prioritized risks • Prioritized list of quantified risks • Probabilistic analysis of the warranty program • List of potential responses • Common risk causes in warranty assistance • Trends in the results of the qualitative and quantitative analysis	• Revision to the warranty risk plan
6. Risk monitoring and control	• Risk plan for the warranty management • Response plan for warranty risks • Program data and records • Additional risk identification and analysis • Scope changes	• Work plans (for emerging risks) • Corrective actions • Changes request on the warranty program • Update to the risk response plan • Risk database for the warranty assistance

Table 8.4 Acronyms meaning

Acronym	Meaning	Acronym	Significance of acronym
EVMS	Earned Value Management System	BCWS	Budgeted Cost of Work Scheduled
EVA	Earned Value Analysis	BCWP	Budgeted Cost of Work Performed
EV	Earned Value	ACWP	Actual Cost of Work Performed
ITD	Incurred To Date	Cv	Cost variance
ETD	Estimate To Complete	Sv	Schedule variance
EAC	Estimate At Completion	CPI	Cost Performance Index
BAC	Budget At Completion	SPI	Schedule Performance Index

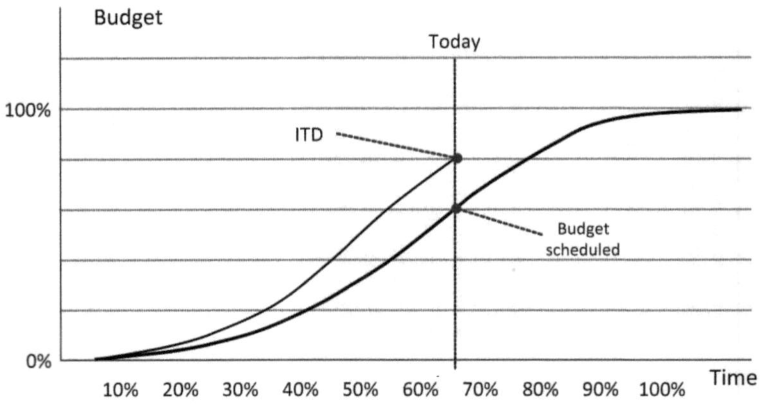

Fig. 8.4 Traditional practice: ITD versus planned costs

progress (earned value). The basis of the EVMS applied to after-sales service is the structured decomposition of the warranty program in work packages. A value must be assigned to each work package linked to the completion of activities such as milestones, deliverables… related to the warranty schedule. That is in formulas as follows:

- EV = Earned Value = [Completed % of the Warranty Schedule] × [Budget for the Warranty Program]
- EV = Σ ([Completed % of the Work Package] × [Budget for the Work Package])

There are other concepts to have in mind during the benefit analysis of a warranty program. Such concepts are for instance the following ones, regarding, of course, the work of the warranty program taken into consideration:

- BCWS: Budgeted Cost of the Work Scheduled
- BCWP: Budgeted Cost of the Work Performed
- ACWP: Actual Cost of the Work Performed

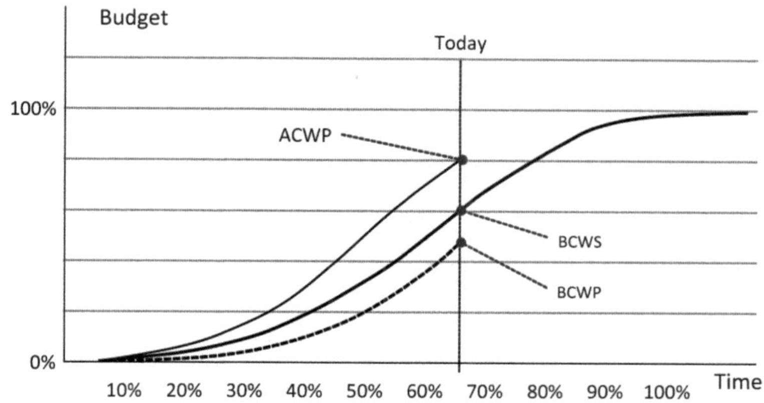

Fig. 8.5 Budgeted costs versus actual costs

Figure 8.5 illustrates the meaning of the above-mentioned concept. As we can see comparing these two concepts, the ACWP corresponds to the incurred to date (ITD), and the BCWS corresponds to the budget scheduled at the beginning of the warranty program. The BCWP which corresponds to the already-defined EV. In order to implement all this, the warranty program must have a defined work-breakdown structure. It does not require being an exhaustive list of tasks; however, this is a tool which helps to organize the total work scope of the warranty program by the subdivision in discrete work elements or tasks of the warranty program. Consequently, by combining the following information, it is possible to calculate the EV, making an early diagnosis of the warranty program evolution:

- Budget for packages of the warranty program incurred cost by packages
- Warranty program planning
- Degree of the operative program advance

Comparing the obtained curves (Fig. 8.6), it is possible to control the reserves and to determine risks along the warranty program. Here, the estimate at completion (EAC) corresponds to the ITD plus the estimate To complete (ETC), and the budget at completion (BAC) corresponds to the originally planned value of the complete warranty program. The commented work packages in terms of warranty are related to that technical assistance which must be given to the corresponding amount of product units that are foreseen to be sold according to a determined schedule. The product delivery schedule can be transformed in terms of warranty costs which are expected to be spent for a specific amount of units in a specific moment. In other words, if we know the units' delivery schedule and the foreseen average of warranty costs per unit sold, it is direct to know the expected warranty cost for the whole program. Reader can find case studies [33] (see also Chap. 2), describing a context very useful to illustrate and facilitate the comprehension of these concepts related to warranty terms. Other interesting references are [34, 35]. Additional concepts related to this benefit analysis are:

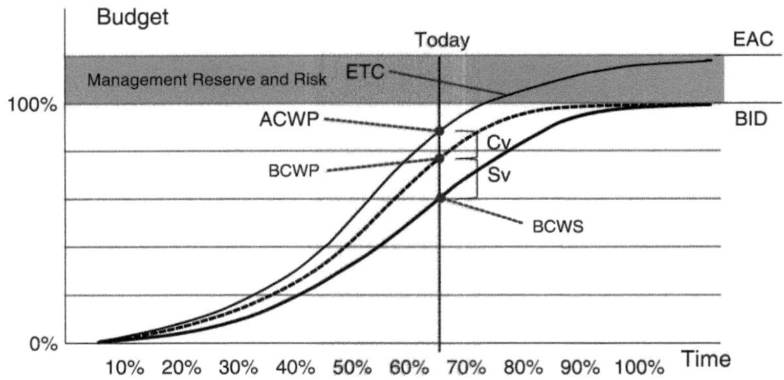

Fig. 8.6 Comparation between EAC and BAC

- Cost variance (Cv) is the difference between the EV and the Actual Cost of the warranty program at any specific time:

$$\text{Cv} = \text{BCWP} - \text{PTCA} = \text{Earned}-\text{Actual}$$

- Schedule variance (Sv) is the difference between the Actual and the planned EV of the warranty program at any specific time:

$$\text{Sv} = \text{BCWP} - \text{BCWS} = \text{Earned}-\text{Budget}$$

- Cost Performance Index (CPI) is the rate about how much costs to generate each € of EV:

$$\text{CPI} = \text{BCWP}/\text{ACWP} = \text{EV}/\text{ITD}$$

- Schedule Performance Index (SPI) is the rate about how fast supposes to generate real value in comparison with the budgeted one:

$$\text{SPI} = \text{BCWP}/\text{BCWS} = \text{EV}/\text{Quoted}$$

The meaning of such indexes is as shown in the following chart (Table 8.5).

However, one of the main advantages of this benefit analysis is the ability to use these indexes to foresee the future direction of the warranty program. This is due to the fact that the calculation of the ETC is adjusted to the performance level achieved to that date:

- $\text{ETC} = (\text{BAC} - \text{ITD})/(\text{SPI} \times \text{CPI})$
- $\text{EAC} = \text{ITD} + \text{ETC} = \text{ITD} + (\text{BAC} - \text{ITD})/(\text{SPI} \times \text{CPI})$

With these results, we can observe that:

- If the past performance is good: CPI & SPI > 1, ETC$\downarrow$ and EAC $<$ BAC
- If the scheduled activities are not been achieving: CPI & SPI < 1, ETC$\uparrow$ and EAC $>$ BAC
- If CPI & SPI $= 1$, EAC $=$ BAC

Table 8.5 Chart about CPI and SPI meaning

	>1 (favorable)	<1 (unfavorable)
CPI	Under the budget	Over the budget
SPI	Ahead of schedule	Behind the schedule

Summarizing, the EV management provides integrated data about cost, operations, and planning regarding (in our case) a warranty program. It is also useful for an early detection of deviations as well as a provision of objective information on the status of the warranty contract. It has the ability to objectively determine the EAC and the warranty program planning, integrating risk management, and the improvement of cost prediction. In continuation to this chapter, it is suggested to consult next section which deals with risk analysis aspects in the management of customer reclamations as it was here focused and developed. Finally, the reader may notice how the present Section offers an interpretation of the content included in the PMBOK guide (A guide to the project management body of knowledge). Those essential matters regarding risk and cost in the PMBOK are here applied to those aspects related to aftersales and warranty topics [36].

8.3 Practical Case on Customer Risk

8.3.1 Application Scenario

Two important aspects about warranty management are represented in this chapter and Chap. 10. Firstly, the risks analysis for the improvement of after-sales management and, secondly, customer relations management. As we can see, the concept of warranty is generally attributed to a physical product sold in the market for its use during an expected life cycle. When the user is not satisfied with the product, discards it, uses it with dissatisfaction, or resells it in the secondhand market [37], all of these together with the uncertainty and the difficulty is assessing the level of customer dissatisfaction in terms of costs. This is not always the situation that can be found in the market. There are also services that are sold as products, and which utility (or distribution) should be guaranteed. This is the case of companies that provide services such as the distribution of water, gas, energy, etc. For this specific case, we will consider a water-distribution campaign, which serves customers through a network of elements with hierarchical structure (see Fig. 8.7). These types of distribution companies have different characteristics from the manufacturers due to the fact that they involve certain particular characteristics [38]:

- Geographical distribution with variable environmental conditions.
- High number and different types of elements.
- High number and different types of clients.
- Elements are interrelated among themselves.

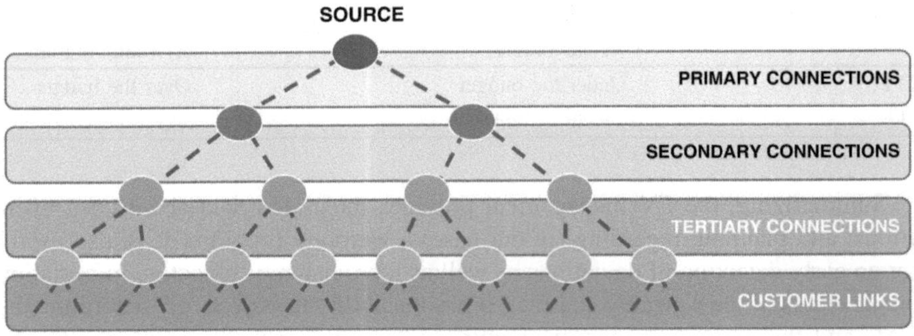

Fig. 8.7 Infrastructure of a distribution company

- Hierarchical structure in the element network with different levels of customer service.
- The infrastructure experiences operational and configuration changes.
- High level of demand in the human resources and material (replacements).

On the other hand, in these types of companies, it is relatively simple to assess customer level of satisfaction and, consequently, the level of warranty assistance provided by technical services. This is due to the fact that the user, once it reaches a certain level of dissatisfaction, can rescind the distribution contract and switch to another competitor company. Nevertheless, this task must include the uncertainty of possible incidents in the service and how it affects quality. According to commercial restrictions, warranty and maintenance services must identify, analyze, predict, quantify, correct, or attenuate the main and possible incidences (risks). Future studies can analyze different methods of statistics and decisions [39] to examine possible risks. Poor quality must be measured in terms of costs, since it is relevant for the company to relate the evolution of quality to customer satisfaction. The preventive and evolution costs are easily quantified and predictable according to recorded tendencies. Included in failure costs, there are some similarly quantifiable, while intangible others are difficult to foresee. That is why, it is necessary to quantify the consequences to determine important risks so that, with the aim of getting an efficient warranty in terms of costs, some factors should be analyzed, such as: magnitude of the incident, appropriate corrective actions, prevision of necessary preventive measures, etc., and the budget that all of these implies. The most important risks from both the business' and customer's points of view are [40]:

- Customer relations: loyalty decrease due to poor and deficient customer service.
- Price and quality: bad-service disposition increases costs and contributes to customer dissatisfaction.
- Internal personnel motivation, image and external prestige of the company.

In short, this analysis is focused on the intangible costs' impact on the customer when dealing with failures and their consequences on the company.

- Loss of production.
- Loss of image and customers.

Owing to the fact that costs are intangible, they will have to be estimated with hypothetical and statistical criteria.

8.3.2 Initial Hypotheses and Quantification of Consequences

In order to show the effect of poor quality, we will take into account three hypotheses:

1. The effect of poor quality on the customer will be based on previously carried out studies regarding this issue [41], in which customer reclamations in more than 400 big service companies in the USA, Japan, and Europe concluded that a dissatisfied customer transmits his own bad experience to an average of 9–10 potential customers, and on the other hand, only a 17 % of dissatisfied customers had the intention to continue with the services provided by the company in question.
2. To assess the relation between customers and businesses, we will use the variable "customer net present value" (CNPV). The CNPV variable will be the updated benefit per client. In other words, it corresponds to the total of obtained income from a client during its relation with the company, minus the direct costs of sales, acquisition, and customer loyalty.
3. As a third hypothesis, there is a certain probability that a dissatisfied customer will abandon the service. This probability will be estimated using warranty and maintenance indicators from statistical and analytical methods ("service level of agreements (SLA)"). These methods analyze the behavior of a group of customers that suffer a failure at a point after a certain period of time [42].

To evaluate the impact of the service on the warranty in terms of costs, we will use the following equation [43]:

$$\text{Risks (R)} = \Sigma(\text{probability of occurrence} \times \text{consequence costs}) \quad i = 1, \ldots, n \tag{8.1}$$

Along all this, intangible costs of loss of production, customer and corporative image, as well as risk assessment are described as follows:

(a) *Production loss*

Any failure of service or interruption in the service causes a profit loss due to the impact in reducing the CNPV per client. That is why, in order to improve the life cycle of the customer (i.e., his/her continuance in the service contract), the warranty should reduce production losses that should be able to be calculated from a point in which a failure is repaired (T_r), this will be the period of time in which the service will be inactive. Therefore, for only one failure, the loss of benefits

(LP) will be the percentage of time the company will have to discount to CNPV for each client (an average value for CNPV is used in the formula). Evaluation time should also be determined, and it will correspond to the average customer continuance in the company (T_c)

$$\mathrm{LP} = \sum_{i=1}^{n} \frac{T_{r_i}}{T_c} \cdot m_i \cdot \mathrm{CNPV} = n \cdot \mathrm{CNPV} \cdot \sum_{i=1}^{n} \frac{T_{r_i}}{T_c} \cdot m_i \qquad (8.2)$$

The value m_i is the number of affected customers per individual failure (i = 1, …, n). Obviously, the higher the hierarchical position of the device in the network, the more significant benefit losses are. This is because the number of affected clients is greater.

(b) *Customer and corporate image losses*

Losses in terms of corporate image are due to bad publicity from a customer's point of view, and it can be classified into three types of possible consequences:

- The company is losing customers.
- The company must reward customer loyalty.
- The company will have to pay more in order to attract more customers.

In other words, these consequences could be translated into a combination of Eqs. (8.3), (8.4) and (8.5). Some parameters are needed for its deduction:

- P_a = probability of customer abandonment;
- p = expected number of people, customers and potential customers that are not directly affected by a failure but just influenced by bad publicity;
- r_p = customers in a part of the population within a geographic area in the network;
- r_{po} = expected proportion of customers as a business' goal in a part of the population within a geographic area in the network;
- s_l = ratio of customer loyalty,
- s_p = ratio of customer publicity,
- s_c = ratio of gaining dissatisfied or influenced customers.

Some of these parameters are difficult to quantify. The ratios s_x refer to those additional costs used with the aim of recuperating a client. Specifically, s_l will be the discount cost or indemnity to the client when the company provides a bad service. Then, s_p will be the cost of publicity by client, and s_c will be the inverted cost to attract new customers. The three costs are here used to describe three different customer's behaviors (see the three equations below):

1. Customers can discard the service; therefore, the company loses net value per client and per year.
2. Customers continue with the service, but an s_l to ensure customer loyalty is required.

3. A potential customer is influenced by bad recommendation. In order to compensate for this, more publicity is needed.

Therefore, these three above-mentioned costs use estimations for the comment:

(b1) *Customer loss cost* (*CLC*):

CLC is the sum of CNPV per every lost customer, multiplied by the customer abandonment probability. As follows:

$$\text{CLC} = P_{\text{a}} \cdot \text{CNPV} \cdot \left(1 + p \cdot r_{\text{p}}\right) \tag{8.3}$$

Equation (8.3) includes abandonment probability of a customer (affected by a failure), and the abandonment probability of others due to a bad experience passed on by the affected customer.

(b2) *Customer loyalty cost* (*CLY*).

In the same way, the CLY costs are calculated as the loyalty costs of influenced customers:

$$\text{CLY} = (1 - P_{\text{a}}) \cdot s_{\text{l}} \cdot \left(1 + p \cdot r_{\text{p}}\right) \tag{8.4}$$

Equation (8.4) integrates the probability to keep the affected customer (in contrast to abandonment probability) and the probability to keep others due to the bad experience passed on by the customer. Just like in Eq. (8.3), not all influenced people are customers, but just a part of them (r_{p}).

(b3) *Capturing new customers cost* (*CCC*):

In this case the equation is:

$$\text{CCC} = P_{\text{a}} \cdot (s_{\text{p}} + s_{\text{c}}) \cdot \left(1 + p \cdot \left[r_{\text{p}} + \left[r_{\text{po}} - r_{\text{p}}\right]\right]\right) = P_{\text{a}} \cdot (s_{\text{p}} + s_{\text{c}}) \cdot \left(1 + p \cdot r_{\text{po}}\right) \tag{8.5}$$

This equation incorporates the capturing cost of lost clients due to the failure and the capturing cost of those clients lost due to the bad experience passed on by the client (r_{p}) through probability of abandonment. Moreover, ($r_{\text{po}} - r_{\text{p}}$) corresponds to the number of potential clients pending whom shall be gained as a result of reaching the objectives.

(c) *Risk assessment*

The study of the probability of abandonment related to reestablishment service time (or repair time) is a record "survival" analysis focused on those customers that "survive" a failure. There are different techniques to solve this type of analysis [44, 45]. In our case, provided that we know the number of clients per network equipment, we can prioritize all those warranty tasks according to those risks for the customer, all this, with the purpose of reducing such risks at a greater speed. Based on the distribution of Weibull and in order to obtain an equation that reflects customer abandonment probability, density probability function will be:

$$f(t_f) = \left(\frac{\beta \times t_i^{\beta-1}}{\alpha^\beta}\right) \times e^{\frac{-t_i\beta}{\alpha}} \tag{8.6}$$

where t_i represents the different times in which a customer abandons the service, α is the scale parameter, and β is the parameter of form (or slope). For obtaining parameters α and β, the best option is to work with the logarithm of credibility function Λ (Eq. 8.7), starting from a list of different t_i.

$$\begin{aligned} \Lambda &= \log L\{t_1, t_2, \ldots, t_N|\alpha \cdot \beta\} = \sum_{i=1}^{N} \log f(t_i|\alpha \cdot \beta) \\ &= \sum_{i=1}^{N} \left[(\ln \beta - \beta \cdot \ln \alpha) + (\beta - 1) \cdot \ln(t_i) - \left(\frac{t_i}{\alpha}\right)^\beta\right] \end{aligned} \tag{8.7}$$

In order to find the maximum, partial derivatives to the credibility logarithm are applied (Eqs. 8.8 and 8.9).

$$\frac{\partial \log L}{\partial \alpha} = \sum_{i=1}^{N} \left[\frac{1}{\beta} + \ln\left(\frac{t_i}{\alpha}\right) \cdot \left(1 - \left(\frac{t_i}{\alpha}\right)^\beta\right)\right] = 0 \tag{8.8}$$

$$\frac{\partial \log L}{\partial \beta} = -n \cdot \frac{\beta}{\alpha} + \frac{\beta}{\alpha^{\beta+1}} \cdot \sum_{i=1}^{N} t_i^\beta = 0 \tag{8.9}$$

Most of the time, these calculations are very complex. Thus, it is necessary to recur to statistical methods such as *Newton–Raphson* to converge rapidly. Basically, the aim is to determine the equations that are able to foresee the future behavior of costs based on hypothesis and general parameters emerging from the market. It is required, therefore, to link them with a probability base. The results shown in this section, the equations that are related to 30 failures on equipment that supports around 250 clients each are resolved. Table 8.6 presents each failure, the number of customers that survive (S) and (F) the time in which customers is lost.

8.3.3 Results of Application

The developed formulas in the last paragraph have been applied to a case of a water-distributing company, similar to the case study that will be studied later in Chap. 11. Table 8.6 shows the data according to the following restrictions:

- $p = 10$, the expected number of people, customers and potential customers, influenced by bad publicity;
- $r_p = 0.33$, relation of customers in a part of the population within a geographical area of the network;

Table 8.6 List of lost and retained clients according to repair time

Customers	Status F or S	Final status time	Subset ID	Customers	Status F or S	Final Status Time	Subset ID
220	S	1	1	1	F	86	23
165	S	7	2	5	F	140	23
187	S	8	3	132	S	140	23
131	S	10	4	1	F	92	24
119	S	12	5	1	F	144	24
1	F	22	6	5	F	162	24
164	S	22	6	137	S	162	24
61	S	24	7	1	F	126	25
2	S	25	8	1	F	139	25
47	S	31	9	7	F	186	25
160	S	35	10	230	S	186	25
80	S	42	11	1	F	201	26
120	S	45	12	55	S	201	26
1	F	52	13	1	F	104	27
175	S	52	13	1	F	132	27
1	F	54	14	3	F	228	27
100	S	54	14	123	S	228	27
160	S	54	15	174	S	231	28
2	F	73	16	1	F	97	29
138	S	73	16	1	F	135	29
1	F	50	17	1	F	159	29
2	F	80	17	1	F	242	29
160	S	80	17	1	F	283	29
1	F	68	18	1	F	314	29
1	F	87	18	1	F	357	29
142	S	87	18	15	F	411	29
1	F	83	19	190	S	411	29
13	S	83	19	1	F	122	30
3	F	95	20	1	F	208	30
136	S	95	20	1	F	242	30
3	F	110	21	1	F	290	30
139	S	110	21	1	F	320	30
1	F	84	22	12	F	367	30
4	F	132	22	192	S	367	30
181	S	132	22				

- $r_{po} = 0.66$, expected proportion of customers, as a target of the company within a geographical area of the network;
- $s_l = 25€$, as the quota to retain customer fidelity,
- $s_p = 25€$, as the quota for customer publicity,
- $s_c = 300€$, as the quota for gaining an dissatisfied or influenced customer,
- $T_c = 21{,}000$ h, the average life span of the customer in the company,
- CNPV = 3,000€.

The meaning of each column is as follows: "customers", the amount of these by subset; "status F or S", F refers to customers lost or that have left, and S refers to customers who remain or survive after a failure; "Final status time" is the period after which customers decide to abandon or remain with the service; "ID of the subset" corresponds to the identification of the faulty device.

Based on this chart and using specific software, the Weibull parameters can be obtained $\beta = 2.24$ and $\alpha = 945.8907389$. If we analyze a concrete failure, for instance, the case of equipment 22, it can be observed that:

- One customer abandons the services after 84 units of stand-by time (repair service time),
- four customers abandon the services after 12 units of stand-by time,
- 181 customers continue with the service after 132 units of stand-by time.

Using the obtained values in Eq. (8.6), the reliability in function of the repair time or abandon probability results in = (1 − 0.985) = 0.0151 where according to Table 8.6, the average repair time is $t_r = 146.25$ h. Along with this and applying the indicated restrictions at the beginning of this section, Eqs. (8.2)–(8.5) result as follows:

$$LP_{per\_client} = CNPV \cdot \frac{T_r}{T_c} = 3.000 \cdot \frac{146.25}{21.000} = 20.90\,€/\text{client} \tag{8.2a}$$

$$\begin{aligned} CLC &= P_a \cdot CNPV \cdot \left(1 + p \cdot r_p\right) = 0.0151 \cdot 3.000\,€ \cdot (1 + 10 \cdot 0.33) \\ &= 149.49\,€/\text{client} \end{aligned} \tag{8.3a}$$

$$\begin{aligned} CLY &= (1 - P_a) \cdot s_1 \cdot \left(1 + p \cdot r_p\right) = (1 - 0.0151) \cdot 25\,€ \cdot (1 + 10 \cdot 0.33) \\ &= 106.62\,€/\text{client} \end{aligned} \tag{8.4a}$$

$$\begin{aligned} CCC &= P_a \cdot (s_p + s_c) \cdot \left(1 + p \cdot r_{po}\right) = 0.0151 \cdot (25 + 300) \cdot (1 + 10 \cdot 0.66) \\ &= 37.30\,€/\text{client} \end{aligned} \tag{8.5a}$$

As a result, intangible costs per customer and failure are estimated at 314.31€ (20.90 + 149.49 + 106.62 + 37.30). Therefore, the warranty activities can be prioritized multiplying that amount by the number of clients related to the equipment in question (bomb, hydraulic substation, etc.). These results are also useful if the after-sales service cost is compared to the prevention and evaluation budget. In short, it has been here developed a simple method to estimate intangible costs of a distribution company, depending on the service level, that is, the repair time of a failure and per client. This practical case is also based on the doctoral thesis [46].

References

1. Department of Defense, USA. Requirements for a Logistics Support Analysis Record. MIL-STD-1388-2B.
2. Murthy DNP, Solem O, Roren T. 2004. Product Warranty logistics: Issues and challenges. Euro. Jr. Oper. Res. 156:110-126.
3. Jones JV. Integrated Logistic Support Handbook. McGraw-Hill and Logistics Management Associates, Irvine, California. ISBN 0-07-033079-4.
4. Department of Defense, USA. Performance Specification Logistics Management Information. MIL-PRF-49506.
5. Department of Defense, USA. Level Of Repair Analysis (LORA). MIL-STD-1390D.
6. Department of Defense, USA. Procedures for Performing a Failure Mode, Effects and Criticality Analysis. MIL-STD-1629 6A.
7. Department of Defense, USA. System Safety Program Requirements. Standard Practice for System Safety. MIL-STD-882.
8. https://www.logsa.army.mil/alc/, Supportability Metrics. Redstone Arsenal, Alabama, USA.
9. Department of Defense, USA. Reliability Prediction of Electronic Equipment. MIL-HDBK-217. Reliability Analysis Center and Rome Laboratory at Griffiss AFB, NY.
10. Department of Defense, USA, "Reliability Prediction Methodology" RIAC 217 Plus. Reliability Analysis Center.
11. Department of Defense, USA. Failure Mode/Mechanism Distributions. RAC FMD–97. Reliability Analysis Center.
12. Blischke WR, Murthy DNP. 2002. Case Studies in Reliability and Maintenance. Wiley, New York. Pp. 661 + xxiii.
13. Rausand M. System Reliability Theory. Department of Production and Quality Engineering. Norwegian University of Science and Technology. Norway.
14. www.relex.com/products/fmeafmeca, Relex Software Corporation, Greensburg, Pennsylvania, USA.
15. Murthy DNP, Blischke WR. 2005. Warranty management and product manufacturing. Springer-Verlag London Limited.Pp 302 + xviii. ISBN 1852339330.
16. Ehrlenspiel K, Kiewert A, Lindemann U. 2007. Cost-Efficient Design. Hundal, Mahendra S. (Ed.). Springer-Verlag jointly published with ASME. XX, 546 p. 368. ISBN: 978-3-540-34647-0. New York, USA.
17. Seuring S, Goldbach M. 2002. Cost Management in Supply Chains. XII, 435 p.109. Springer-Verlag. ISBN: 978-3-7908-1500-9.
18. Mayhew D. 2007. Cost-Justifying Usabilit" Ed. By Randolph Bias, University of Texas at Austin. Elsevier. ISBN-13: 978-0-12-095811-5, ISBN-10: 0-12-095811-2.
19. Doyle D. 2002. Cost Control: A Strategic Guide. Elsevier ISBN-13: 978-1-85971-517-8, ISBN-10: 1-85971-517-6.
20. Ghodrati B, Akersten PA, Kumar U. 2007 Spare parts estimation and risk assessment conducted at Choghart Iron Ore Mine: A case study. Journal of Quality in Maintenance Engineering. 13(4):353-363. ISSN: 1355-2511.
21. Backlund F, Hannu J. 2002. Can we make maintenance decisions on risk analysis results?. Journal of Quality in Maintenance Engineering. 8(1):77-91. ISSN: 1355-2511.
22. Collier PM, Berry AJ, Burke GT. 2006, Risk and Management Accounting. Best practice guidelines for enterprise-wide internal control procedures. Elsevier ISBN-13: 978-0-7506-8040-0. ISBN-10: 0-7506-8040-7.
23. Spedding LS, Rose A. 2007. Business Risk Management Handbook. A sustainable approach. Elsevier. ISBN-13: 978-0-7506-8174-2. ISBN-10: 0-7506-8174-8.
24. Zeng L. 2000. On the Basis Risk of Industry Loss Warranties. The Journal of Risk Finance. Emerald 2000. 1(4): 27-32. ISSN: 1526-5943.
25. Blischke WR, Murthy DNP. 1994. Warranty Cost Analysis. Marcel Dekker. New York.

26. Page JS. 1996. Conceptual Cost Estimating Manual. Elsevier. ISBN-13: 978-0-88415-267-5. ISBN-10: 0-88415-267-7.
27. Murthy DNP, Blischke WR, Kakouros S, Kuettner D. 2007. Changing the way we think about warranty management. Warranty Week.
28. Hines T. 2004. Supply Chain Strategies: Customer Driven and Customer Focused. Manchester Metropolitan University, UK. Elsevie. ISBN-13: 978-0-7506-5551-4, ISBN-10: 0-7506-5551-8.
29. Raby M. 2000. Project management via earned value. Emerald. 49(1): 6-10. ISSN: 0043-8022,
30. Murthy DNP, Rausand M, Virtanen S. 2009. Investment in new product reliability. Reliability Engineering & System Safety. 94:1593-1600.
31. Murthy DNP, Hagmark PE, Virtanen S. 2009. Product variety and reliability.", Reliability Engineering & System Safety. 94:1601-1608.
32. Murthy DNP, Osteras T, Rausand M. 2009. Component reliability specification. Reliability Engineering & System Safety. 94:1608-1617.
33. González-Prida V, Gómez JF, Crespo Márquez A. 2009. Case study: warranty costs estimation according to a defined lifetime distribution of deliverables. World Congress on Engineering Asset Management. WCEAM Athens. ISBN 978-1-84996002-1.
34. Jigeesh N, Bhat MS. 2006. Modeling and simulating the dynamics in project monitoring and earned value analysis. Journal of Advances in Management Research. Emerald. 3(1):26-43. ISSN: 0972-7981.
35. Aneirin O. 2003. Accounting for Business Studies. Manchester Metropolitan University. Elsevier 2003. ISBN-13: 978-0-7506-5834-8. ISBN-10: 0-7506-5834-7.
36. Project Management Institute. 2008. A Guide to the Project Management Body of Knowledge (PMBOK® Guide) - Fourth Edition. ISBN: 978-1-933890-72-2.
37. Akerlof GA. The Market for Lemons: Quality Uncertainty and the Market Mechanism. The Quarterly Journal of Economics. 84(3):488-500.
38. Gómez JF, López M, González-Prida V, Crespo Márquez A. 2010. Network maintenance contribution in quality terms on Distribution Network Services Providers. Reliability, Risk and Safety: Theory and Applications. Pp.1055-1062. Taylor & Francis Group, London, ISBN 978-0-415-55509-8.
39. Ishikawa K. 1985. What Is Total Quality Control?: The Japanese Way?. Prentice Hall, 240 p. ISBN-10: 0139524339.
40. Kaplan R, Norton D. 1996. The Balanced Scorecard: Translating Strategy into Action. Boston, MA: Harvard Business School Press. ISBN 0875846513.
41. Goodman J. 1986. Technical Assistance Research Program (TARP), U.S. Office of Consumer Affairs Study on Complaint Handling in America.
42. Klein J, Moeschberguer M. 1997. Survival Analysis. Techniques for Censored and Truncated Data. Springer–Verlag. New York Inc.
43. Moubray J. 1997. Reliability-centered Maintenance. Industrial Press.
44. Blischke WR, Murthy DNP. 2000. Reliability Modelling, Prediction and Optimization. Wiley.
45. Harrell FE. 2001. Regression Modeling estrategies. Springer.
46. Gómez JF. 2010. Marco de referencia para la gestión y evaluación del mantenimiento de redes de distribución de servicios. Tesis doctoral dirigida por Adolfo Crespo Márquez. Universidad de Sevilla.

Chapter 9
On the Assessment and Control

The reliability and availability of a system depends, first of all, on its design and its installation quality. Also, the reliability and availability conservation will depend on the maintenance to be performed. A good maintenance plan has to be able to analyze all possible failures and has to be designed to avoid them. This means it is necessary to elaborate a detailed failure analysis of all those components that make up the system. Due to the complexity of such a task, it is important to take into account all those indications given by the own manufacturer for the main equipment, performing a reliability analysis for the rest of components. In general terms, if maintenance would be only corrective (attending problems just when they arise), it could be a profitable short-term policy, but not longer. Achieving a good preventive maintenance however, the system life cycle can be increased and, consequently, its profitability in the middle and long term. It is also important to keep in mind that the application of preventive maintenance does not present an immediate consequence, but the effect of taken actions appears after a certain time. In addition to this, these plans should apply the principle of continuous improvement, being updated as a better and deeper knowledge of the system is available. In this chapter, a mathematical formulation to perform a reliability analysis in the design stage of the product (*reliability*, *availability*, *maintainability and safety*, RAMS analysis) is developed and will be focused on customer service planning, as well as the analysis of the new product, once data are gathered related to its behavior when it is launched into the market (LCC analysis). That is, the intention here is to develop two methodologies that allow us to analyze "a priori" the behavior of a future product, and the others would allow us an "a posteriori" analysis of the product's behavior once it is launched into the market and more real data are available (LCC analysis).

9.1 General Characteristics of a RAMS Analysis

Next we are going to get the RAMS variables for two subsystems arranged in series or in parallel and for the case of a system formed by "*n*" identical

V. González-Prida Díaz and A. Crespo Márquez, *After-sales Service of Engineering Industrial Assets*, DOI: 10.1007/978-3-319-03710-3_9,

subsystems in parallel where the system is deemed to fail if "*m*" subsystems or more fails (case "*m*-out-of-*n*"). For these cases, it will be deduced those system features mentioned in the introduction: "ratio of system failure," "mean time between failure," "availability and unavailability of the system," and "mean downtime." However, it should be noted also that sometimes conventional formulas are not used in cases of hybrid configurations, applying for these cases other methods such as a probabilistic formulation.

9.1.1 Subsystems in Series Connection

This section describes a system conformed by two nonidentical subsystems arranged in series (Fig. 9.1).

- *Failure Rate*
  Considering only one subsystem, the failure rate will be λ_1. The probability of failure in dt is $\lambda_1 \mathrm{d}t$. When there are two subsystems in series, the probability of failure in dt is then $(\lambda_1 \mathrm{d}t + \lambda_2 \mathrm{d}t)$. The system failure rate is therefore $(\lambda_1 + \lambda_2)$:

$$\lambda_{\text{series}} = \lambda_1 + \lambda_2$$

The reliability function (assuming an exponential distribution) will be:

$$R(t) = \exp\left[-(\lambda_1 + \lambda_2)t\right]$$

- *Mean Time Between Failures*
  Considering an exponential function for reliability, it follows that:

$$\text{MTBF}_{\text{series}} = 1/(\lambda_1 + \lambda_2) = \frac{\text{MTBF}_1 \cdot \text{MTBF}_2}{\text{MTBF}_1 + \text{MTBF}_2}$$

- *Availability and Unavailability*
  In order to make the system available, each subsystem must be available. Therefore,

$$A_{\text{series}} = A_1 \cdot A_2$$

On the contrary, the unavailability will be:

$$\text{UA}_{\text{series}} = 1 - A_{\text{series}} = 1 - (1 - \text{UA}_1) \cdot (1 - \text{UA}_2) = \text{UA}_1 + \text{UA}_2 - \text{UA}_1 \cdot \text{UA}_2$$

- *Mean Downtime*
  In case of two repairable subsystems, one with MDT_1 and the other with MDT_2, what would the mean downtime be for the two subsystems in series? At any moment, the system will be in one of the following four states:

Fig. 9.1 Subsystems in series connection

- Both subsystems work properly.
- Subsystem #1 does not work, but #2 does.
- Subsystem #2 does not work, but #1 does.
- None of the two subsystems work.

The last three cases are responsible that the entire system would not be operable. Considering an inactive system, the probability of being inactive due to a failure in subsystem #1 is as follows:

$$\lambda_1/(\lambda_1 + \lambda_2)$$

Since subsystem #1 needs a MDT_1 to be repaired, the repair time associated with this subsystem is then:

$$\lambda_1/(\lambda_1 + \lambda_2) \times \mathrm{MDT}_1$$

A similar expression is applied to the case of subsystem #2. Synthesizing both for the complete system, it is obtained as follows:

$$\mathrm{MDT}_{\text{series}} = \frac{\mathrm{MTBF}_1 \cdot \mathrm{MDT}_2 + \mathrm{MTBF}_2 \cdot \mathrm{MDT}_1}{\mathrm{MTBF}_1 + \mathrm{MTBF}_2}$$

9.1.2 Subsystems in Parallel Connection

This section describes a system of two nonidentical subsystems arranged in parallel (Fig. 9.2).

Here, the two subsystems are repairable, with downtime MDT_1 and MDT_2.

- *Failure Rate*
  If the system was only established by the subsystem #1, the ratio of complete system failure would be λ_1 and the probability of failure in $\mathrm{d}t$ would be $\lambda_1 \mathrm{d}t$. Adding the subsystem #2 in parallel, the likelihood of system failure in $\mathrm{d}t$ will be $\lambda_1 \mathrm{d}t$ reduced by the probability that the subsystem #2 is in failure mode. The probability of finding the subsystem #2 in failure mode is given by:

$$\frac{\mathrm{MDT}_2}{\mathrm{MTBF}_2 + \mathrm{MDT}_2}$$

Assuming that $\mathrm{MDT}_2 \ll \mathrm{MTBF}_2$, and using $\mathrm{MTBF}_2 = 1/\lambda_2$, the reduced failure rate for subsystem #1 is given by: $\lambda_1 \cdot \lambda_2 \cdot \mathrm{MDT}_2$. In the same way, De igual modo, the reduced failure rate for subsystem #2 is given by:

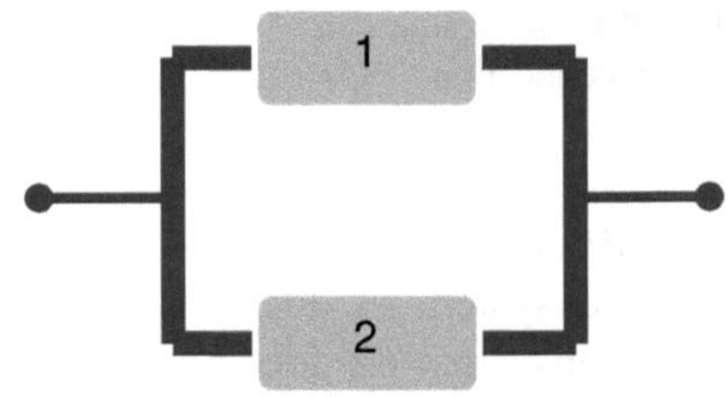

Fig. 9.2 Subsystems in parallel connection

$$\lambda_1 \cdot \lambda_2 \cdot \mathrm{MDT}_1$$

Therefore,

$$\lambda_{\text{parallel}} = \lambda_1 \cdot \lambda_2 \cdot (\mathrm{MDT}_1 + \mathrm{MDT}_2)$$

- *Mean Time Between Failures*
  Considering the approach that the inverse of the failure ratio is MTBF (certain for an exponential distribution), we obtain:

$$\mathrm{MTBF}_{\text{parallel}} = 1/\lambda_{\text{parallel}} = \frac{\mathrm{MTBF}_1 \cdot \mathrm{MTBF}_2}{\mathrm{MDT}_1 + \mathrm{MDT}_2}$$

Note that if the two subsystems are not repairable, then the MTBF for the case in parallel is the sum of the individual MTBFs.

- *Availability and Unavailability*
  For the system becomes available, any of the subsystems should be available. Therefore,

$$A_{\text{parallel}} = A_1 + A_2 - A_3 \cdot A_4$$

On the contrary, the unavailability will be:

$$\begin{aligned}\mathrm{UA}_{\text{parallel}} &= 1 - A_{\text{parallel}} = 1 - (A_1 + A_2 - A_1 \cdot A_2) = (1 - A_1) \cdot (1 - A_2) \\ &= \mathrm{UA}_1 \cdot \mathrm{UA}_2\end{aligned} \tag{9.1}$$

- *Mean Downtime*
  By the definition of:

$$\text{Unavailability} = \frac{\mathrm{MDT}}{\mathrm{MTBF} + \mathrm{MDT}} \approx \frac{\mathrm{MDT}}{\mathrm{MTBF}}$$

The MDT may be obtained for the case in parallel, using expression (9.1) above.

$$\mathrm{UA}_{\mathrm{parallel}} = \frac{\mathrm{MDT}_{\mathrm{parallel}}}{\mathrm{MTBF}_{\mathrm{parallel}}} = \frac{\mathrm{MDT}_{\mathrm{parallel}}}{\frac{\mathrm{MTBF}_1 \cdot \mathrm{MTBF}_2}{\mathrm{MDT}_1 + \mathrm{MDT}_2}};$$

$$\mathrm{UA}_1 \cdot \mathrm{UA}_2 = \frac{\mathrm{MDT}_1}{\mathrm{MTBF}_1} \cdot \frac{\mathrm{MDT}_2}{\mathrm{MTBF}_2}$$

Consequently,

$$\mathrm{MDT}_{\mathrm{parallel}} = \frac{\mathrm{MDT}_1 \cdot \mathrm{MDT}_2}{\mathrm{MDT}_1 + \mathrm{MDT}_2}$$

9.1.3 Case of n Identical Subsystems in Parallel

Considering a system with n identical subsystems, where the entire system fails if m or more subsystems fail (Fig. 9.3). In that case, the formulas for failure rate, MTBF, availability, and downtime for the entire system will be as given below indicated.

- *Failure Rate*
  If the system consists only of subsystem #1, then the failure rate is λ, and the probability of failure in dt is λdt. In order to have a system failure, we need other $(m - 1)$ subsystems in failure mode. The option that any of the subsystems would be in failure mode is given by MDT/(MTBF + MDT), or (MDT/MTBF) assuming that MDT $\ll$ MTBF. In order to find $(m - 1)$ subsystems in failure mode, the probability will be:

$$(\mathrm{MDT}/\mathrm{MTBF})^{m-1}$$

There will be ${}_{n-1}C_{m-1}$ ways for the group $(m - 1)$ out of the $(n - 1)$ subsystems. In addition to this, any subsystem can be selected to be analyzed as subsystem #1. Considering all together we obtain:

$$\lambda_{m\text{-out-of-}n} = \lambda \cdot \left(\frac{\mathrm{MDT}}{\mathrm{MTBF}}\right)^{m-1} {}_{n-1}C_{m-1} \cdot n = \frac{n!}{(n-m)!(m-1)!} \lambda^m \cdot \mathrm{MDT}^{m-1} \tag{9.2}$$

This is the failure rate for exactly m failed subsystems. The failure rate for a number of subsystems higher than m will be reduced by a factor of $(\lambda \cdot \mathrm{MDT})$. For a consistent checking, let us consider $n = m = 2$. This corresponds to two identical subsystems in parallel. When $m = 2$ subsystems fails, the entire system fails. The expression (9.2) for this case is as follows: $\lambda_{\mathrm{system}} = \lambda^2 \cdot (2 \cdot \mathrm{MDT})$, which agrees with the formula of the previous section for two parallel-connected subsystems.

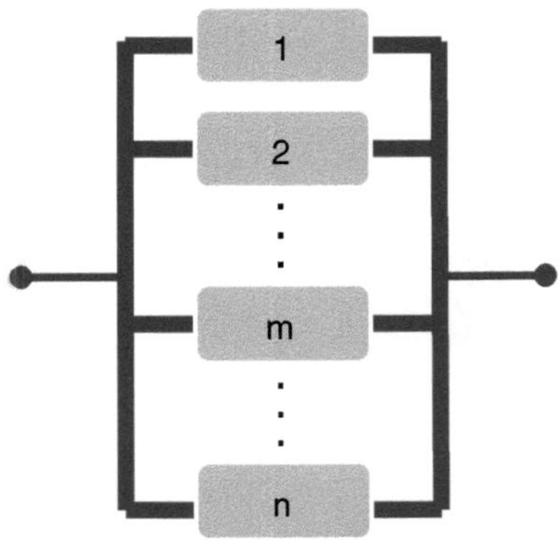

Fig. 9.3 Parallel connection of n identical subsystems

- *Mean Time Between Failures*
  Considering the approach that the inverse of the failure rate is the MBTF (indeed, as mentioned above, certain for an exponential distribution), we obtain:

$$\mathrm{MTBF}_{m-\mathrm{out-of}-n} = 1/\lambda_{m-\mathrm{out-of}-n} = \frac{\mathrm{MTBF}^m}{\frac{n!}{(n-m)!\cdot(m-1)!}\cdot \mathrm{MDT}^{m-1}}$$

- *Availability and Unavailability*
  In order for the system to be available, at least $(n - m + 1)$ subsystems must be available. Therefore,

$$A_{m-\mathrm{out-of}-n} = \sum_{i-n-m+1}^{n} \frac{n!}{(n-i)!\cdot i!} A^i (1-A)^{n-i}$$

Using the next equation:

$$1 = [A + (1-A)]^n = \sum_{i=0}^{n} \frac{n!}{(n-i)!\cdot i!} A^i (1-A)^{n-i}$$

availability can be rewritten as follows:

$$A_{m-\mathrm{out-of}-n} = 1 - \sum_{i=0}^{n-m} \frac{n!}{(n-i)!\cdot i!} A^i (1-A)^{n-i} \approx 1 - \frac{n!}{m!\cdot(n-m)!}(1-A)^m$$

And the unavailability will be given by (again, for MDT $\ll$ MTBF):

$$\mathrm{UA}_{m-\mathrm{out-of}-n} = \frac{n!}{m! \cdot (n-m)!} \mathrm{UA}^m$$

- *Mean Downtime*
  By the definition of:

$$\mathrm{UA}_{m-\mathrm{out-of}-n} = \frac{\mathrm{MDT}_{m-\mathrm{out-of}-n}}{\mathrm{MTBF}_{m-\mathrm{out-of}-n}}$$

MDT is obtained for the case "m-out-of-n," by the application of previous sections for $\mathrm{UA}_{m\text{-out-of-}n}$ and $\mathrm{MTBF}_{m\text{-out-of-}n}$. Therefore,

$$\mathrm{MDT}_{m-\mathrm{out-of}-n} = \frac{\mathrm{MDT}}{m}$$

9.1.4 Summary of Formulas and Results

Once obtained the expressions that link the RAMS variables with the units that make up the industrial assets in terms from the components themselves (Table 9.1), the next step will be to provide a quantitative assessment of these parameters. From the quantitative assessment one can observe that subunit or component which can be critical in the functioning of the entire system.

The obtained results can be useful to establish a reliability ranking for the components, under certain assumptions and boundary conditions. It allows to the possibility of drawing up a preliminary list of recommended spares, useful for the warranty assistance. In reference to this, it should be underlined that the initial data will be approaches and should not be taken as precise values. That is because the intention of this analysis is first at all to provide a starting point for dealing with the possible behavior of the elements that make up the industrial asset. Thus, once the design process will be deeper and the company has experience and a more precise and detailed knowledge of the systems, the values considered in this analysis may be refined and adjusted to reality, giving conclusions more successfully. Interesting references in this field are [1–3]. Once analyzed the product reliability in its design stage, next step will be to perform a similar analysis, but once the product is on the user's hands and historical data on its behavior are available. This analysis will assess whether the initial assertions were correct, and will allow more realistic decisions for future batches of products offered for sale.

Table 9.1 Summary chart of formulas

	Two subsystems in series	Two subsystems in parallel
System failure rate	$\lambda_{\text{series}} = \lambda_1 + \lambda_2$	$\lambda_{\text{parallel}} = \lambda_1 \cdot \lambda_2 \cdot (\text{MDT}_1 + \text{MDT}_2)$
System MTBF	$\text{MTBF}_{\text{series}} = \frac{\text{MTBF}_1 \cdot \text{MTBF}_2}{\text{MTBF}_1 + \text{MTBF}_2}$	$\text{MTBF}_{\text{parallel}} = \frac{\text{MTBF}_1 \cdot \text{MTBF}_2}{\text{MDT}_1 + \text{MDT}_2}$
System availability (A)	$A_{\text{series}} = A_1 \cdot A_2$	$A_{\text{parallel}} = A_1 + A_2 - A_1 \cdot A_2$
System unavailability (UA)	$\text{UA}_{\text{series}} = \text{UA}_1 + \text{UA}_2 - \text{UA}_1 \cdot \text{UA}_2$	$\text{UA}_{\text{parallel}} = \text{UA}_1 \cdot \text{UA}_2$
System mean downtime (MDT)	$\text{MDT}_{\text{series}} = \frac{\text{MTBF}_1 \cdot \text{MDT}_2 + \text{MTBF}_2 \cdot \text{MDT}_1}{\text{MTBF}_1 + \text{MTBF}_2}$	$\text{MDT}_{\text{parallel}} = \frac{\text{MDT}_1 \cdot \text{MDT}_2}{\text{MDT}_1 + \text{MDT}_2}$

9.2 Influence of Warranty on the Life Cycle Cost of a Product

Throughout the last years, specialists in areas of engineering and management of operations have improved the costs quantification processes, including the use of techniques to quantify reliability and failure impact on total costs of an industrial asset throughout its life cycle [4]. These improvements have permitted a decrease of uncertainty in the decision-making process in areas of vital importance such as design, development, after-sales, maintenance, replacement, and production asset acquisition. During this process, many of the decisions and actions should be adopted throughout the industrial asset's period of use. The challenge now is to apply these improvements in a product that is launched in the market. Necessities of support and maintenance depend generally on decisions taken at a design and manufacturing phase [5]. The most part of these actions have a high impact on the asset's life cycle, being of special interest, those decisions related to the improvement process of factors that affect reliability (quality design, technology used, technical complexity, failures frequency, preventive/corrective maintenance, levels of maintainability and accessibility …), considering that all these aspects have a great influence in the total costs of the asset's life and influence to a large extent the possible expectations to extend the product in question's useful life at reasonable costs (refer to, for example [6–10]).

9.2.1 Background to Techniques of Life Cycle Costs Analysis

In this section, we will focus on the life cycle costs analysis (LCCA), which is integrated in the third stage (evaluation and control) of the proposed reference framework. It is relevant to highlight that there are diverse methodologies brought up in the LCCA field, and that relate, for instance, to the environmental impact or to the total costs of production assets, etc. [11].

Even though these methodologies have their particular characteristics, they tend to propose a reliability analysis based on a constant failures rate. Previous records of the LCCA can be consulted in [12]. However, some of the most important contributions throughout these last 20 years are presented in chronological order as follows:

- 1991 (*Woodhouse*). Presents the LCCA as a systematic process of technical–economical evaluation, taking into account economical and reliability aspects [13]. Presenta el LCCA como un proceso sistemático de evaluación técnico-económica, teniendo en cuenta los aspectos económicos y la fiabilidad [13].
- 1993 (*Fabrycky y Blanchard*). LCCA model that includes a structured process to estimate Reliability costs [14].
- 1996 (*Kirt y Dellisolla*). Present the LCCA as an economical estimation technique bound to evaluate in a quantitative form all costs associated to the expected economical useful life span [12].
- 1997 (*Woodward*). Research line that includes all basic aspects of reliability analysis and its impacts on life cycle costs [10].
- 2000 (*Willians y Scott*). Model based on the distribution of Weibull to estimate frequency of failure and its reliability costs impacts [15].
- 2001 (Riddell y Jennings). Commercial software development on LCCA and denominated *APT Life span* [16].
- 2002 (*Durairaj y Ong*). The quick application of costs analysis techniques enables us to evaluate beforehand potential design problems and quantify potential impact in terms of costs throughout the industrial asset's life span [11].
- 2003 (*Parra y Omaña*). The consultants firm *Woodhouse Partnership* and the Venezuelan Technological Institute of Petrol [Instituto Tecnológico Venezolano del Petróleo (INTEVEP)], test an evaluation model of the total costs in the life span of 56 systems of gas compression [17].

9.2.2 Basic Aspects of LCCA

The life span cost of a physical asset is defined through the identification of its applicable functions at each phase of the product's life, the estimation of the costs of these functions and the application of acquired costs during its life span extension. Therefore, this valuation must include all those costs related to designing, manufacturing, and production [18–20]. Part of these costs, in the case of a launched product into the market, has to be assumed by the own buyer. In any case, involved costs in the life cycle of the product are here considered, without specifying who (manufacturer or user), has to cope with such expenses. From a financial point of view, costs generated throughout the life span of an asset can be classified in two types (Fig. 9.4):

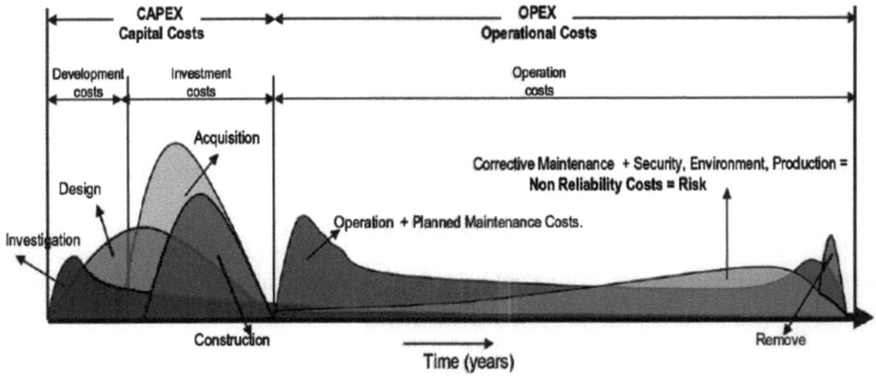

Fig. 9.4 Costs generated throughout the life span of the product

- CAPEX: Capital expenses (design, development, acquisition, installation, personnel training, manuals, documentation, tools, and facilities for repairs, spare parts, product withdrawal, etc.).
- OPEX: Operational expenses (labor, operations, planned maintenance, stock, hiring, corrective maintenance, penalization for failures, or low reliability events).

As it has been previously mentioned, after-sales services are normally offered at the same time that production lines are still open. That is the reason why, product engineering and manufacturing can be improved with feeding data coming from a warranty program, reducing consequently, the general expenses of the life span of the product. With an adequate reliability and availability evaluation, it is possible to prove during the early stages of the product, how the expressed requirements in the initial technical specifications are incompatible or even impossible to reach for determined product configurations [21, 22]. If the product is already launched, this analysis can help to speed up the necessary measures to correct and/or to improve the product, foreseeing also probable claims from the users due to the product unreliability. As referred above, the typical life cycle cost analysis includes costs for planning, research and development, production, commissioning, warranty, operation & maintenance, and disposal [23]. From the consumer's point of view, the asset's life cycle cost, also known as the total cost of ownership, will include the acquisition costs (purchase price), operation & maintenance cost, and disposal cost (including residual value).

In general, the lower the budget for product engineering the higher risk of the warranty programs . In these cases, warranty claims for important or costly failures need prompt and strong analysis to identify, not only further tasks to proceed with the repair, but also preventive actions which can avoid or at least decrease, future claims due to similar reasons. An overall review of all the warranty complaints can be helpful to show, for example, repetitive failures and trends related to vendor/ buyer problems, quality issues, manufacturing conditions, product design, etc.

9.2.3 Development of the Stochastic Model

To evaluate the warranty service, and particularly to estimate the appropriate warranty period, it is necessary to apply convenient measures that must be defined during the strategical phase of the warranty program. These measures in general must make possible the comparison mainly of reliability data and the inclusion of the life span valued cost. Following warranty assistance or on a periodical basis review of product behavior, it is then possible to evaluate and improve the warranty program. The company therefore must establish and apply a standard and repeatable method to gather and analyze data also to interpret results that can be based on business factors rather than industrial ones. The results must be used to support and justify those improvements to be applied. There are different stochastic models that can be considered useful for failures frequency analyses in systems and repairable unit. This chapter is intended also to compare two different stochastic methods, the non-homogeneous Poisson process (NHPP) and the generalized renewal process (GRP). The first one is usually applied to simple repair cases, and the second one is usually implemented when repairs are complex. In particular, the NHPP is a model also called "minimal repair" and assumes that the unit goes back to an "as bad as old" (ABAO) condition after repair. Hence, after the element's repair, there is the assumption that is operative but as old as it was before the failure. These motives, together with its simplicity, are the main reasons for which it has been decided to apply this methodology in repair warranty cases (which obviously, must be minor, with the equipment left in a condition "as bad as old"). The NHPP differs from homogeneous Poisson process (HPP) in the fact that the occurrence failure rate varies with time instead of being constant [24]. In this process, times between failures are not independent nor identically distributed. The NHPP is a stochastic process in which the probability of failure occurrence at any interval $[t_1, t_2]$ has a Poisson distribution and its mean can be calculated as follows:

$$\text{mean} = \int_{t_1}^{t_2} \lambda(t)\mathrm{d}t \tag{9.3}$$

where $\lambda(t)$ is the rate of occurrence of failure (ROCOF) and it is defined as the inverse of waiting time between failures, $1/E[x_i]$ (refer to for example [24, 25]). In other words, the NHPP is a stochastic process in which the ROCOFs n at any interval $[t_1, t_2]$ has a Poisson distribution with the mean:

$$\bar{\lambda} = \int_{t_1}^{t_2} \lambda(t)\mathrm{d}t \tag{9.4}$$

where $\lambda(t)$ is the ROCOFs. Therefore, according to the Poisson process,

$$\Pr[N(t_2) - N(t_1) = n] = \frac{\left[\int_{t_1}^{t_2} \lambda(t)\mathrm{d}t\right]^n \exp\left[-\int_{t_1}^{t_2} \lambda(t)\mathrm{d}t\right]}{n!} \tag{9.5}$$

where $n = 0, 1, 2, \ldots$ corresponds to the expected total number of failures at time intervals $[t_1, t_2]$. The expected total number of failures is obtained from the function of accumulated intensity:

$$\Lambda(t) = \int_0^t \lambda(t)\mathrm{d}t \tag{9.6}$$

One of the most common forms of ROCOF used in reliability analysis of repairable systems is the Potence Model Law [24, 25]:

$$\lambda(t) = \frac{\beta}{\alpha}\left(\frac{t}{\alpha}\right)^{\beta-1} \tag{9.7}$$

This form comes from the assumption that all intermediate times between successive failures follow a Weibull conditional function of density with parameters α and β. The Weibull distribution tends to be used in the area of maintenance due to its flexibility and applicability to diverse failure processes; however, solutions coming from gamma and logarithmic are also possible.

This model involves that failure ith is conditioned by the operative time until the failure $(i - 1)$th.

Figure 9.5 shows the referred condition [26]. This condition also is brought about from the fact that the system kept the "as bad as old" condition after repair $(i - 1)$th. Therefore, the repair process does not extend life to the component or system. With the aim of developing this methodology, the notation will be as follows (see Fig. 9.6):

- Subindex meaning:
  - i: failure event and "as bad as old" repair (from 1 to n)
  - n: last known failure event
  - s: next event after the last known failure (n)
- Times meanings:
  - t_i: duration of time, interval between failures
  - T_i: accumulated duration of intervals

According to the mentioned notation in order to get maximum probability estimators (ML) in the parameters of the potency law, the following conditional probability definition is considered:

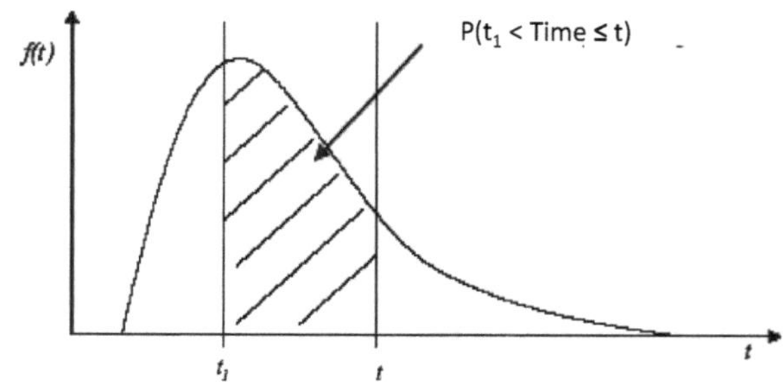

Fig. 9.5 Conditional probability of failure occurrence

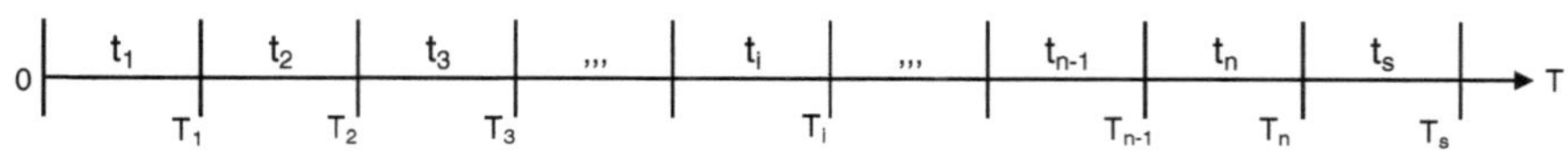

Fig. 9.6 Timeline of failure events

$$P(T \leq Ts|T > tn) = \frac{F(Ts) - F(Tn)}{R(Tn)} = \frac{1 - R(Ts) - 1 + R(Tn)}{R(Tn)} = 1 - \frac{R(Ts)}{R(Tn)} \tag{9.8}$$

where $F(\cdot)$ and $R(\cdot)$ are the failure probability of components and the reliability on the respective times, respectively. Assuming a Weibull distribution from Eq. (9.8) can be obtained:

$$F(T_i) = 1 - \exp\left[\left(\frac{T_{i-1}}{\alpha}\right)^{\beta} - \left(\frac{T_i}{\alpha}\right)^{\beta}\right] \tag{9.9}$$

where in case that the failure event is $i = s$, then $i - 1 = n$. Therefore, the Weibull conditional function of density is as follows:

$$F(T_i) = \frac{\beta}{\alpha}\left(\frac{T_i}{\alpha}\right)^{\beta-1} . \exp\left[\left(\frac{T_{i-1}}{\alpha}\right)^{\beta} - \left(\frac{T_i}{\alpha}\right)^{\beta}\right] \tag{9.10}$$

For the case of NHPP, there can be obtained different function of probability. We will use the expression based on the estimation of a time T after the occurrence of the last failure and before the next one. These expressions can be seen in detail in [27]. In the case of repairable components already finished, the maximum probability function L can be expressed as follows:

$$L = \prod_{i=1}^{n} f(T_i) = f(T_1) \prod_{i=2}^{n} f(T_i) R(T_n|T) \tag{9.11}$$

Therefore,

$$L=\left\{\frac{\beta}{\alpha}\left(\frac{T_1}{\alpha}\right)^{\beta-1}\exp\left[-\left(\frac{T_1}{\alpha}\right)^{\beta}\right]\right\}$$
$$\times\left\{\left(\frac{\beta}{\alpha}\right)^{n-1}\prod_{i=2}^{n}\left(\frac{T_1}{\alpha}\right)^{\beta-1}\exp\left(\sum_{i=2}^{n}\left[\left(\frac{T_{i-1}}{\alpha}\right)^{\beta}-\left(\frac{T_i}{\alpha}\right)^{\beta}\right]\right)\right\} \quad (9.12)$$
$$\times\left\{\exp\left[\left(\frac{T_n}{\alpha}\right)^{\beta}-\left(\frac{T}{\alpha}\right)^{\beta}\right]\right\}$$

Once again, the parameters estimators ML are calculated according to [24, 25]. The results are the following:

$$\hat{\alpha}_n=\frac{T_n}{n^{\frac{1}{\hat{\beta}_n}}} \quad (9.13)$$

$$\hat{\beta}_n=\frac{n}{\sum_{i=1}^{n}\ln\left(\frac{T_n}{T_i}\right)} \quad (9.14)$$

where T_i is the moment when the ith failure occurred. T_n is the total time in which the last failure took place, and n is the total number of failures. The expected total number of failures at times intervals $[T_n, T_s]$, that is to say, the interval t_S, multiplied by the Weibull accumulated intensity function will be [27, 28] as follows:

$$\Lambda(T_n,T_s)=\frac{1}{\alpha^{\beta}}\left[(T_s)^{\beta}-(T_n)^{\beta}\right] \quad (9.15)$$

where T is the time after the last produced failure and T_n is the total accumulated time:

$$T_n=\sum_{i=1}^{n}t_i \quad (9.16)$$

Time till next failure (TNF) from the NHPP [28], taking into account the Weibull parameters and the total accumulated time, will be given by the next expression:

$$\mathrm{TNF}=\left\{\left[\frac{1}{\alpha}+(T_n)^{\beta}\right]^{(1/\beta)}\right\}-T_n \quad (9.17)$$

Assuming a Weibull distribution of NHPP, the reliability function will be according to the next expression [28]:

$$R(T_s)=\exp\left\{-\alpha\times\left[(T_s)^{\beta}-(T_n)^{\beta}\right]\right\} \quad (9.18)$$

Therefore, the failure probability of the system will be as follows:

$$F(T_s) = 1 - R(T_s) \tag{9.19}$$

9.2.4 Proposed NHPP Model for Warranty Assessment

Now and as in the previous section, the steps to estimate the minimum time for the warranty period (t_{MTW}) in accordance with NHPP are described. As an starting point, data will be necessary. It will be necessary to know the intervals of time t_i ($i = 1,\ldots, n$), when the failure events and repair took place. With these data available, the following steps will be required:

1. Calculation of parameters:

Taking into account T_n as the total accumulated time:

$$T_n = \sum_{i=1}^{n} t_i \tag{9.16}$$

Parameters α and β of the Weibull distribution in the case of the nth event will be then as follows:

$$\hat{\alpha}_n = \frac{T_n}{n^{\frac{1}{\beta_n}}} \tag{9.13}$$

$$\hat{\beta}_n = \frac{n}{\sum_{i=1}^{n} \ln\left(\frac{T_n}{T_i}\right)} \tag{9.14}$$

where T_i is the moment in which failure ith occurs, T_n is the total time in which the last failure event occurs, and n is the total number of failures, and deriving by the maximum likelihood approach (maximum likelihood estimation, MLE):

$$LK = L\{\alpha, \beta; \mathrm{data}\} = \left(\frac{1}{\alpha^\beta}\right)^n \cdot \beta^n \cdot \prod_{i=1}^{n} \exp\left(-\frac{t_n^\beta}{\alpha^\beta}\right) \cdot t_i^{\beta-1} \tag{9.20}$$

2. Calculation of the TNF:

Time till next failure, taking into account the parameters of Weibull and the accumulated total time, will be given by the following expression:

$$\mathrm{TNF} = \left\{ \left[\frac{1}{\alpha} + (T_n)^\beta\right]^{(1/\beta)} \right\} - T_n \tag{9.17}$$

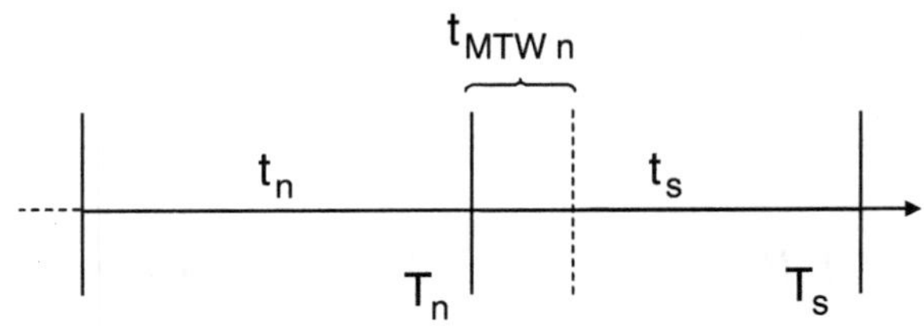

Fig. 9.7 Timeline of failure events (detail)

3. Calculation of the minimum time for warranty (t_{MTW}):

With all the previous calculations, it is possible to obtain now the minimum time for warranty period after the *n*th repair (Fig. 9.7) with the following expression (taking into account the 20 % of TNF as minimum warranty time):

$$t_{\mathrm{MTW}_n} = \mathrm{TNF}_n \times 20\ \% \tag{9.21}$$

4. Calculation of the expected number of failures by t_{MTW}:

The expected total number of failures in the time interval $[T_n, T_n + t_{\mathrm{MTW}n}]$ according to the Weibull accumulated intensity function is [27] as follows:

$$\Lambda(T_n, T_n + t_{\mathrm{MTW}n}) = \frac{1}{\alpha^\beta}\left[(T_n + t_{\mathrm{MTW}n})^\beta - (T_n)^\beta\right] \tag{9.22}$$

where as it has been mentioned before, t_{MTW} is the minimum warranty time after carrying out the last repair, being $(T_n + t_{\mathrm{MTW}n})$ equal to $(T_{\mathrm{MTW}n})$.

5. Calculation of reliability of a system for the recommended warranty period $(T_n + t_{\mathrm{MTW}n})$:

Assuming Weibull distribution, the reliability function will be according to the next expression:

$$R(T_{\mathrm{MTW}n}) = \exp\left\{-\alpha \times \left[(T_{\mathrm{MTW}n})^\beta - (T_n)^\beta\right]\right\} \tag{9.23}$$

6. Calculation of failure probability in the recommended warranty period $(T_n + t_{\mathrm{MTW}n})$:

The probability of failure in the system will be consequently the following:

$$F(T_{\mathrm{MTW}n}) = 1 - R(T_{\mathrm{MTW}n}) \tag{9.24}$$

As we can see in the described process, a balance of risks has been applied for the estimation of the recommended warranty period, and not an objective function. Basically, the percentage considered as minimum for reliability of the warranty period is a parameter to be estimated during the budget estimation process, together with the probability to determine its relation to those costs. It is important to highlight here that NHPP is an effective tool for minor repair cases as the ones in warranty reclamations. Nevertheless, extending this point to maintenance contracts in which the after-sales service of a company must deal with greater repairs (for example overhauls), where the application of a GRP methodology is useful for these types of cases [29], when repairs are complex.

9.2.5 Proposed GRP Model for the Warranty Assessment

Now, through the GRP model, the flexibility is increased in order to report on the five possible system states after a repair, although the complexity is increased too at the same time:

- As good as new
- As bad as old
- Better than old, but worse than new
- Better than new
- Worse than old

The NHPP model accounts for the second state, while the rest repair states have not been involved. Generalized renewal process (GRP) copes with all the five possible states, modeling the failure behavior of a specific system and understanding the effects of the repairs on the system's age [26], introducing the concept of virtual age (V_n) and furthermore an indicator of repair quality (q). V_n represents the calculated age of the element immediately after the nth repair occurs.

(a) Calculation of parameters:

$$V_n = q \cdot \sum_{i=1}^{n} q^{n-i} \cdot x_i = q \cdot (V_{n-1} + x_n) \tag{9.25}$$

The summation (with $V_0 = 0$ for $t_0 = 0$) assumes that the nth repair compensates for the damage accumulated during the time between the $(n - 1)$th and the nth failure, but also damage produced in previous intervals.

The indicator of repair quality (q) represents the state "as good as new" when $q = 0$, while the assumption of $q = 1$ corresponds to a NHPP "as bad as old." The values of q that fall in the interval $0 < q < 1$ represent the after repair states in which the condition of the element is "better than old but worse than new," whereas the cases where $q > 1$ correspond to a condition "worse than old." Similarly, cases with $q < 0$ would suggest a component or system restored to a state "better than new." Therefore, physically speaking, q can be seen as an index for representing the effectiveness and quality of repairs.

(b) Calculation of failure probability and system reliability:

For $V_{n-1} = y$, the system has a time to the nth failure, x_n, which is distributed according to the following cumulative distribution function (cdf):

$$F(\alpha, \beta, q; x_n, v_{n-1}) = \frac{F(x_n + v_{n-1}) - F(v_{n-1})}{1 - F(v_{n-1})} = 1 - \mathrm{e}^{-\left[\frac{(x_n+v_{n-1})^\beta - v_{n-1}^\beta}{\alpha^\beta}\right]} \tag{9.26}$$

where $R(t) = 1 - F(t)$ is the reliability function.

There are different formalizations of the GRP process to model equipment evolution over time. In this case, we will try to simplify the implementation of the GRP method based on the study by Ref. [30] built on the GRP II by Ref. [31], which uses the maximum likelihood approach (MLE), to cover complex systems with multiple repairs.

$$\begin{aligned}\log L\{\alpha, \beta, q; x_1, x_2, \cdots, x_n\} &= n \cdot (-\ln \alpha + \ln \beta) \\ &- \left[\left(\frac{T - \left(\sum_{i=1}^{n} x_i\right) + V_n}{\alpha}\right)^\beta - \left(\frac{V_n}{\alpha}\right)^\beta\right] \\ &- \sum_{k=1}^{n}\left[\frac{(x_k + V_{k-1})^\beta - V_{k-1}^\beta}{\alpha^\beta}\right] \\ &+ (\beta - 1) \cdot \sum_{k=1}^{n} \ln\left(\frac{x_k + V_{k-1}}{\alpha}\right)\end{aligned} \tag{9.27}$$

(c) Calculation of minimal time for the warranty period ($t_{\mathrm{MTW}n}$):

Now, the minimal time for the warranty period after the nth repair based on the risk until the next failure expressed as a probability derived from the reliability is as follows:

Solving $t_{\mathrm{MTW}n}$ for 80, 60, or 40 % of reliability:

$$t_{\mathrm{MTW}n} = \left[{v_{n-1}}^\beta - \alpha^\beta \cdot \ln(R(T_{\mathrm{MTW}n}))\right]^{\frac{1}{\beta}} - v_{n-1}. \tag{9.28}$$

Besides, as a continuation to this chapter, it is suggested to consult next sections where aspects of RAMS are being dealt with as well as those related to the life cycle cost analysis focused on what has been here developed.

9.3 Case Study on a RAMS Analysis

9.3.1 Study Scenario

The case study here presented corresponds to a practical application referred in this chapter, where a formulation has been developed to apply a RAMS analysis. The industrial asset to analyze consists on a device with an electrical and a mechanical part and will be controlled and powered from the outside. According to the examples provided in Appendix A of ISO/DIS 14224 [32], we can simplify the battery limits of the device as shown in Fig. 9.8, as a level 0.

The purpose of defining the system boundary is to assure further understanding of the subunits and maintainable items that are included within these limits. In general terms, for defining the boundary and consequently the limits of the analysis, following considerations have been followed:

(a) It is included only elements of the same class (in this case level 1), which are considered comparable.
(b) It is excluded those items specifically connected across boundaries. The failures occurred in a connection will not be attributed to the element connected inside the boundary.
(c) The monitoring and control instrumentation should not be included in this analysis (outside usually).

The RAMS expressions previously developed are easily applied in this case following a formulation in series. So if λ_1 is the failure rate of the electric system and λ_2 is the failure rate of the mechanical system, the ratio for the complete system will be therefore as follows:

- $\lambda_{\text{Complete Syst.}} = \lambda_1 + \lambda_2$

The rest variables will be then:

- $\text{MTBF}_{\text{Complete Syst.}} = \frac{\text{MTBF}_1 \cdot \text{MTBF}_2}{\text{MTBF}_1 + \text{MTBF}_2}$
- $A_{\textit{Complete Syst.}} = A_1 + A_2$
- $\text{UA}_{\textit{Complete Syst.}} = \text{UA}_1 + \text{UA}_2 - \text{UA}_1 \cdot \text{UA}_2$
- $\text{MDT}_{\text{Complete\_Syst.}} = \frac{\text{MTBF}_1 \cdot \text{MDT}_2 + \text{MTBF}_2 \cdot \text{MDT}_1}{\text{MTBF}_1 + \text{MTBF}_2}$

Deepening the level of subunit or component (level 2), we establish that the industrial asset does not have redundant equipment. That is, the complete operation of the system under study could be represented by a scheme which would contain only AND Boolean gates. A Boolean scheme would offer a fairly intuitive correct operation of industrial assets as a whole entity. However, it is not the purpose of this case to analyze the performance of a particular device, but to show a practical application of calculus in order to view the benefits of a RAMS analysis. For the purpose of representing a functional configuration of the equipment under analysis, it is rather intuitive the block diagram shown in Fig. 9.9.

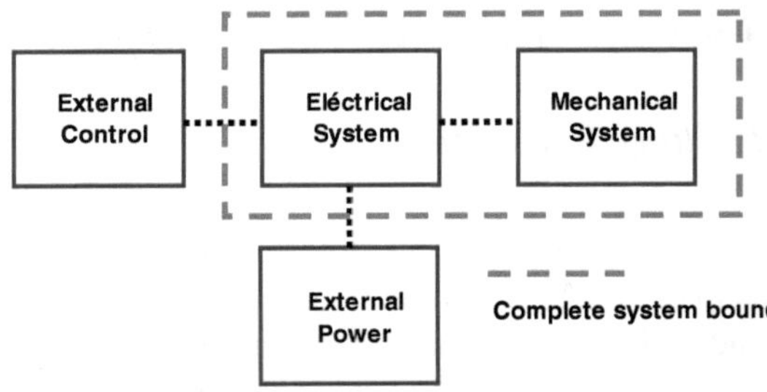

Fig. 9.8 Boundary definition of the industrial asset under study

Based on Fig. 9.9, let us consider the notation (Table 9.2) for the failure rates, in order to subsequently obtain the corresponding variables subunits level RAMS.

9.3.2 Application of RAMS Variables to the Practical Case

Next, we will develop the already discussed RAMS expressions for the electrical and mechanical system, in order to apply afterward as in-series formulation between these two systems. In the case of the electrical system (Fig. 9.10), and using the notation given above, the expressions development yields as follows:

- $\lambda_1 = \lambda_{11} + \lambda_{12} + \lambda_{13} + \lambda_{14}$
- $\mathrm{MTBF}_1 = [\mathrm{MTBF}_{11} \cdot \mathrm{MTBF}_{12}/(\mathrm{MTBF}_{11} + \mathrm{MTBF}_{12})] \cdot [\mathrm{MTBF}_{13} \cdot \mathrm{MTBF}_{14}/(\mathrm{MTBF}_{13} + \mathrm{MTBF}_{14})]/[[\mathrm{MTBF}_{11} \cdot \mathrm{MTBF}_{12}/(\mathrm{MTBF}_{11} + \mathrm{MTBF}_{12})] + [\mathrm{MTBF}_{13} \cdot \mathrm{MTBF}_{14}/(\mathrm{MTBF}_{13} + \mathrm{MTBF}_{14})]]$
- $A_1 = A_{11} \cdot A_{12} \cdot A_{13} \cdot A_{14}$
- $\mathrm{UA}_1 = (\mathrm{UA}_{11} + \mathrm{UA}_{12} - \mathrm{UA}_{11} \cdot \mathrm{UA}_{12}) + (\mathrm{UA}_{13} + \mathrm{UA}_{14} - \mathrm{UA}_{13} \cdot \mathrm{UA}_{14}) - (\mathrm{UA}_{11} + \mathrm{UA}_{12} - \mathrm{UA}_{11} \cdot \mathrm{UA}_{12}) \cdot (\mathrm{UA}_{13} + \mathrm{UA}_{14} - \mathrm{UA}_{13} \cdot \mathrm{UA}_{14})$
- $\mathrm{MDT}_1 = [[\mathrm{MTBF}_{11} \cdot \mathrm{MTBF}_{12}/(\mathrm{MTBF}_{11} + \mathrm{MTBF}_{12})] \cdot [(\mathrm{MTBF}_{13} \cdot \mathrm{MDT}_{14} + \mathrm{MTBF}_{13} \cdot \mathrm{MDT}_{14})/(\mathrm{MTBF}_{13} + \mathrm{MTB}_{14})] + [\mathrm{MTBF}_{13} \cdot \mathrm{MTBF}_{14}/(\mathrm{MTBF}_{13} + \mathrm{MTBF}_{14})] \cdot [(\mathrm{MTBF}_{11} \cdot \mathrm{MDT}_{12} + \mathrm{MTBF}_{12} \cdot \mathrm{MDT}_{11})/(\mathrm{MTBF}_{11} + \mathrm{MTB}_{12})]]/[[\mathrm{MTBF}_{11} \cdot \mathrm{MTBF}_{12}/(\mathrm{MTBF}_{11} + \mathrm{MTBF}_{12})] + [\mathrm{MTBF}_{13} \cdot \mathrm{MTBF}_{14}/(\mathrm{MTBF}_{13} + \mathrm{MTBF}_{14})]]$

In the same way, the expressions development for the mechanical system (Fig. 9.11), yields as follows:

- $\lambda_1 = \lambda_{21} + \lambda_{22} \cdot [\lambda_{23} \cdot \lambda_{24} \cdot (\mathrm{MDT}_{23} + \mathrm{MDT}_{24})] \cdot [\mathrm{MDT}_{22} + [(\mathrm{MDT}_{23} \cdot \mathrm{MDT}_{24})/(\mathrm{MDT}_{23} + \mathrm{MDT}_{24})]$
- $\mathrm{MTBF}_1 = \mathrm{MTBF}_{21} \cdot [\mathrm{MTBF}_{22} \cdot [(\mathrm{MTBF}_{23} \cdot \mathrm{MTBF}_{24})/(\mathrm{MDT}_{23} + \mathrm{MDT}_{24})]/[\mathrm{MDT}_{22} + [(\mathrm{MDT}_{23} \cdot \mathrm{MDT}_{24})/(\mathrm{MDT}_{23} + \mathrm{MDT}_{24})]]]/[\mathrm{MTBF}_{21} + [\mathrm{MTBF}_{22} \cdot [(\mathrm{MTBF}_{23} \cdot \mathrm{MTBF}_{24})/(\mathrm{MDT}_{23} + \mathrm{MDT}_{24})]/[\mathrm{MDT}_{22} + [(\mathrm{MDT}_{23} \cdot \mathrm{MDT}_{24})/(\mathrm{MDT}_{23} + \mathrm{MDT}_{24})]]]]$

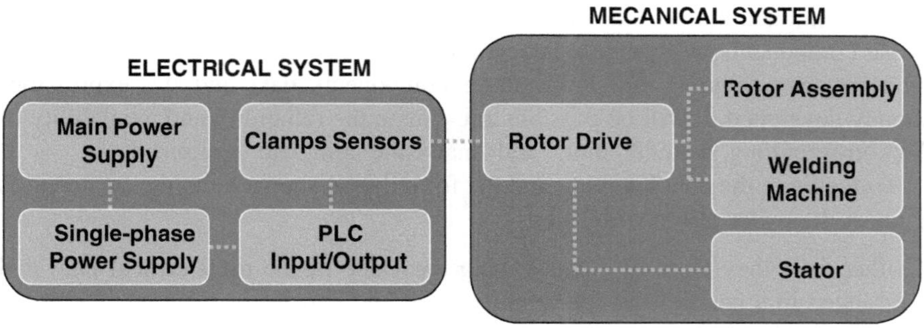

Fig. 9.9 Functional blocks according to subunits

Table 9.2 Notation for failure rates

Electrical system	Mechanical system
• General power supply: λ_{11}	• Rotor drive: λ_{21}
• Single-phase power supply: λ_{12}	• Rotor assembly: λ_{22}
• Clamps sensors: λ_{13}	• Estator: λ_{23}
• PLC input/output: λ_{14}	• Welding machine: λ_{24}

Fig. 9.10 Electrical system block diagram

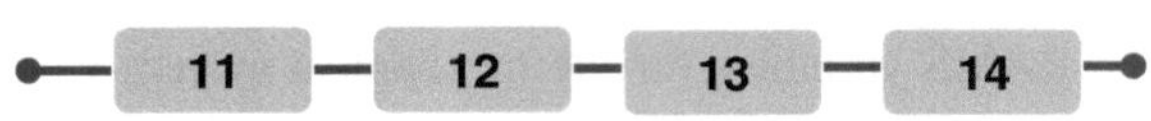

Fig. 9.11 Mechanical system block diagram

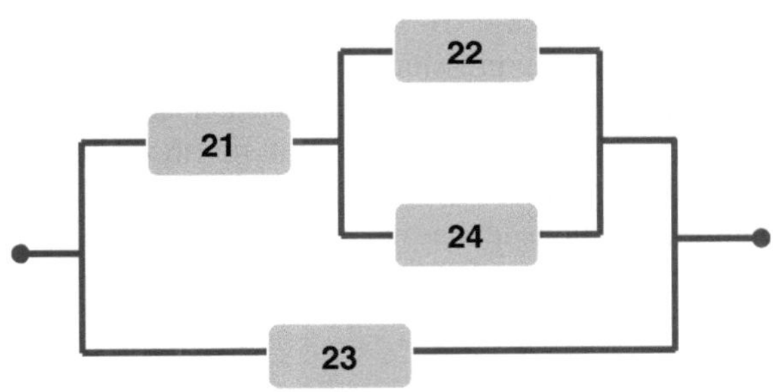

- $A_1 = A_{21} \cdot [A_{22} + (1 - A_{22}) \cdot (A_{23} + A_{24} - A_{23} \cdot A_{24})]$
- $UA_1 = UA_{21} + (1 - UA_{21}) \cdot UA_{22} \cdot UA_{23} \cdot UA_{24}$
- $MDT_1 = [MTBF_{21} \cdot [MDT_{22} \cdot [MDT_{23} \cdot MDT_{24}/(MDT_{23} + MDT_{24})]]/ [MDT_{22} + [MDT_{23} \cdot MDT_{24}/ (MDT_{23} + MDT_{24})]] + MDT_{21} \cdot [MTBF_{22} \cdot [MTBF_{23} \cdot MTBF_{24}/ (MDT_{23} + MDT_{24})]/[MDT_{22} + [MTBF_{23} \cdot MTBF_{24}/(MDT_{23} + MDT_{24})]]/ [MTBF_{21} + [MTBF_{22} \cdot [MTBF_{23} \cdot MTBF_{24}/(MDT_{23} + MDT_{24})]/ [MDT_{22} + [MTBF_{23} \cdot MTBF_{24}/(MDT_{23} + MDT_{24})]]$

As mentioned above, the subunits shown here are constituted by components. These components will provide later numerical data regarding time failure rates, time between failures, etc. From these components data and the expressions already developed, it will be possible to estimate the reliability and availability of the corresponding subunits, units, and finally the complete equipment.

To facilitate the data handling and the formulation shown here, the treatment is going to be under two possibilities:

- either as if they were identical (taking the value of the most critical one) and arranged in a parallel configuration (hereinafter, case -A-),
- or as if they were arranged in a series configuration (hereinafter, be case -B-).

Maintainable items for electrical and mechanical systems are indicated in Table 9.3.

Taking into account the above-mentioned considerations, the RAMS variables for the electrical system results as given in Table 9.4:

Here, the subindex *cxx* in case -A- refers to that component (level 3), critical to the proper function of the corresponding subunit. Also, the value (*) of case -B- is assumed under an estimation of 8 h for repairing any subsystem. Identical subunits would apply to 1.3. (sensors and clamps) and 1.4. (PLC input–output) which, being composed of a single item maintainable, their variables coincide with the one of the items. In the same way and also taking into account the above considerations, the RAMS variables for the mechanical system would be as given in Table 9.5:

As in the electrical system, the subindex *cxx* in case -A- refers to that component (level 3), critical for the proper function of the corresponding subunit. Also, the value (*) in case -B- is assumed under an estimation of 8 h for repairing any subsystem. The expressions obtained in terms of maintainable items (level 2) for both the electric system and mechanical system can be applied now to the expressions developed above and for those units (level 1) that constitute the complete system of robot welder (level 0). However, this substitution becomes so complex relations in these expressions that their manipulation is therefore difficult. In any case, below it is applied particular numerical data using spreadsheets to obtain useful results.

9.3.3 Application of Numerical Data and Results

Once we have correlated components RAMS variables with the assets functions ones, we will try to reach a quantitative assesssment. With this quantitative assessment, we will see what subunit or component is critical for the function of the entire system and will require more (or more frequent) preventive maintenance than other subunits. If we want these results to be useful, data for component parameters should be gathered and, in case this is not possible, we may relay on existing data bases. There are nowadays several databases that can provide interesting information. One of the largest databases is the EPRD-97—electronic

Table 9.3 Taxonomy of electrical and mechanical systems

Electrical system		Mechanical system	
Maintainable item	Part or component	Maintainable item	Part or component
General power supply	Cable section 6 mm^2	Estator	Estator structure
	Cable section 16 mm^2		Tube tighter assembly
	Cable section 10 mm^2 monophase		Start-up tighter
	Cable section 6 mm^2 monophase		Roller
	Protections 5SY6 215-8	Rotor assembly	Fixed rotor
	Protections 5SY6 206-8		Removable rotor
	Protections 5SY6 308-8		Edge preparation device
	Protections 5SY6 303-8		Welding group
	Protections 5SY6 332-8		Bolt
	Protections 5SY6 315-8		Torch hose
	Protections 5SY6 463-8		Grip hose
	Contactor CN20 220		Grip hose cover
	Contactor AL9-30-10-24		Torch
	Contactor AL12-30-10-24	Rotor driver	Auger
	Contactor VH300		Motor reductor
Single-phase power supply	Cable section 6 mm^2		Conical tighter
	Protections 5SY6 502-7		Interphase worm gear
	Protections 5SY6 214-7	Welding machine	Machine plug
	Protections 5SY6 503-7		Pressure reductor
Sensors	Clamps		Assembly support
PLC input/output	PLC		Others

Table 9.4 Formulation of cases A and B for the electrical system

Case -A-	Case -B-
General power supply	
• $\lambda_{11} = 15 \cdot \lambda_{c11}$	• $\lambda_{11} = \sum \lambda_{11k} \quad (k = 1, \ldots, 15)$
• $MTBF_{11} = MTBF_{c11}/15$	• $MTBF_{11} = 1/\lambda_{11}$
• $A_{11} = 1 - 15 \cdot (1 - A_{c11})$	• $A_{11} = MTBF_{11}/(MTBF_{11} + MDT_{11}) \times 100$
• $UA_{11} = 15 \cdot UA_{c11}$	• $UA_{11} = 1 - A_{11}$
• $MDT_{11} = MDT_{c11}$	• $MDT_{11} = (*)$
Single-phase power supply	
• $\lambda_{12} = 4 \cdot \lambda_{c12}$	• $\lambda_{12} = \sum \lambda_{12k} \quad (k = 1, \ldots, 4)$
• $MTBF_{12} = MTBF_{c12}/4$	• $MTBF_{12} = 1/\lambda_{12}$
• $A_{12} = 1 - 4 \cdot (1 - A_{c12})$	• $A_{12} = MTBF_{12}/(MTBF_{12} + MDT_{12}) \times 100$
• $UA_{12} = 4 \cdot UA_{c12}$	• $UA_{12} = 1 - A_{12}$
• $MDT_{12} = MDT_{c12}$	• $MDT_{12} = (*)$

Table 9.5 Formulation of cases A and B for the mechanical system

Case -A-	Case -B-
Rotor drive	
• $\lambda_{21} = 4 \cdot \lambda_{c21}$	• $\lambda_{21} = \sum \lambda_{21k}$ $(k = 1,\ldots,4)$
• $\mathrm{MTBF}_{21} = \mathrm{MTBF}_{c21}/4$	• $\mathrm{MTBF}_{21} = 1/\lambda_{21}$
• $\mathrm{A}_{21} = 1 - 4 \cdot (1 - A_{c21})$	• $A_{21} = \mathrm{MTBF}_{21}/(\mathrm{MTBF}_{21} + \mathrm{MDT}_{21}) \times 100$
• $\mathrm{UA}_{21} = 4 \cdot \mathrm{UA}_{c21}$	• $\mathrm{UA}_{21} = 1 - A_{21}$
• $\mathrm{MDT}_{21} = \mathrm{MDT}_{c21}$	• $\mathrm{MDT}_{21} = (*)$
Rotor assembly	
• $\lambda_{22} = 8 \cdot \lambda_{c22}$	• $\lambda_{22} = \sum \lambda_{22k}$ $(k = 1,\ldots,8)$
• $\mathrm{MTBF}_{22} = \mathrm{MTBF}_{c22}/8$	• $\mathrm{MTBF}_{22} = 1/\lambda_{22}$
• $\mathrm{A}_{11} = 1 - 8 \cdot (1 - \mathrm{A}_{c22})$	• $A_{22} = \mathrm{MTBF}_{22}/(\mathrm{MTBF}_{22} + \mathrm{MDT}_{22}) \times 100$
• $\mathrm{UA}_{22} = 8 \cdot \mathrm{UA}_{c22}$	• $\mathrm{UA}_{22} = 1 - A_{22}$
• $\mathrm{MDT}_{22} = \mathrm{MDT}_{c22}$	• $\mathrm{MDT}_{22} = (**)$
Estator	
• $\lambda_{23} = 4 \cdot \lambda_{c23}$	• $\lambda_{23} = \sum \lambda_{23k}$ $(k = 1,\ldots,4)$
• $\mathrm{MTBF}_{23} = \mathrm{MTBF}_{c23}/4$	• $\mathrm{MTBF}_{23} = 1/\lambda_{23}$
• $A_{11} = 1 - 4 \cdot (1 - \mathrm{A}_{c23})$	• $\mathrm{A}_{23} = \mathrm{MTBF}_{23}/(\mathrm{MTBF}_{23} + \mathrm{MDT}_{23}) \times 100$
• $\mathrm{UA}_{23} = 4 \cdot \mathrm{UA}_{c23}$	• $\mathrm{UA}_{23} = 1 - A_{23}$
• $\mathrm{MDT}_{23} = \mathrm{MDT}_{c23}$	• $\mathrm{MDT}_{23} = (**)$
Welding machine	
• $\lambda_{24} = 22 \cdot \lambda_{c24}$	• $\lambda_{24} = \sum \lambda_{24k}$ $(k = 1,\ldots,22)$
• $\mathrm{MTBF}_{24} = \mathrm{MTBF}_{c24}/22$	• $\mathrm{MTBF}_{24} = 1/\lambda_{24}$
• $A_{24} = 1 - 22 \cdot (1 - \mathrm{A}_{c24})$	• $A_{24} = \mathrm{MTBF}_{24}/(\mathrm{MTBF}_{24} + \mathrm{MDT}_{24}) \times 100$
• $\mathrm{UA}_{24} = 22 \cdot \mathrm{UA}_{c24}$	• $\mathrm{UA}_{24} = 1 - A_{24}$
• $\mathrm{MDT}_{24} = \mathrm{MDT}_{c24}$	• $\mathrm{MDT}_{24} = (**)$

parts reliability data, and NPRD-95—non-electronic parts reliability data. Both databases have been developed by the Reliability Information Analysis Center (RIAC), considered as a Center of Excellence by the Department of Defense United States [33]. Another database also fairly widespread is offshore reliability data (OREDA) [34]. This database is for projects sponsored by international oil companies, whose purpose is to gather and exchange data about reliability and maintenance of facilities and exploration devices between the organizations participating in the project, acting as a forum for management and reliability data collection within the oil and gas sector. For our case study, we used the database failure rate data in perspective (FARADIP) [1]. This is one of the largest databases on ratios and failure modes and is widely used as a baseline. It provides data ranges for failure rates in electrical, electronic, mechanical, pneumatic elements, instrumentation, etc. In any case, taking into account that:

- λ_{ijk} = Failure rate = No failures per million hours, and
- MTBF_{ijk} = Mean Time Between Failures = $1/\lambda_{ijk}$ $(\times 10^6$ h)

This database FARADIP provides values that can be applied as approaches for each component, given on one hand for case -A- (where the critical element is taken for each subunit); and similarly, considering the case -B- (which it is considered a series configuration for such parts or components). Finally, it is obtained the results shown in Table 9.6.<Query ID="Q7" Text="*Kindly check and confirm the usage of decimal system in 'Tables. 9.6–9.14' and Fig. '9.12–9.17' have been correctly identified and amend if necessary.*" –>

From the above results, it is possible to obtain RAMS parameters for higher-level units, applying the formulation resulted from the block diagrams. Considering the values of case -B- which corresponds to a more realistic situation and taking into account that:

- A_{ijk} = Availability = $MTBF_{ijk}/(MTBF_{ijk} + MDT_{ijk}) \times 100$
- UA_{ijk} = Unavailability = $100 - A_{ijk}$
- MDT_{ijk} = Mean Downtime (*)

(*) It has been assumed in the table an estimated time of 8 h for the repair of any subsystem [7]. With the above consideration, we obtain the values shown in Table 9.7.

Here, the parameters have been obtained for the whole system RAMS by applying an in-series formulation.

Based on the above information, it is going to obtain the reliability values $R(t)$. For that purpose, it is going to get $R(t)$ assuming a mixed behavior of the elements [35] (Table 9.8). That means, the mechanical components will follow a Weibull distribution with $\beta = 0.5$ (cells with white background) and electrical components will follow an exponential distribution $\beta = 1$ (cells with gray background). Representing these values graphically, it is possible to observe the evolution for items maintainable over a year (Fig. 9.12).

In the same way, the graphical evolution of mechanical and electrical systems, as well as the complete system, will be (Fig. 9.13).

Comparing the results of reliability after one year, we obtain the following order of priority and criticality matrix (Table 9.9). Focusing on the maintainable items, one can get a ranking of items from lower to higher reliability, which will define a preliminary list of recommended spares. This list can be formed first by taking those items whose reliability fell by 80 % after one year of operation and, within these items, by selecting those components or parts, whose failure rates (λ_{ijk}) account the 80 % of the total failure rate considered for the corresponding maintainable item, as a Pareto principle or rule 80-20.

9.4 Case Study on the Life Cycle Cost Analysis

9.4.1 Application Scenario

As it has been mentioned before, there are different types of warranties that are appropriate for different types of products [36, 37]. The case hereby explained deals with a commercial product whose warranty is negotiated with the customer as well as the possibility of extensions and other kinds of after-sales contracts.

Table 9.6 Results for cases A and B

ij	Maintainable item (Level 8 ISO14224)	-A- $l_{ij} = n \times \lambda_{ijk}$	-B- $l_{ij} = \Sigma k\ (\lambda_{ijk})$
1.1.	General power supply	90.00	47.00
1.2.	Single-phase power supply	12.000	6.500
1.3.	Clamps sensors	10.000	10.000
1.4.	PLC input/output	50.000	50.000
2.1.	Rotor driver	80.000	23.000
2.2	Rotor assembly	400.000	180.506
2.3.	Estator	200.000	155.000
2.4.	Welding machine	1,100.000	315.600

In this particular scenario, the customer is interested in the acquisition, use, and maintenance of one of the specific engines that have been developed and previously tested by the manufacturer. Some of these engines are already in use by the customer. This means there is plenty of information about the performance of these engines; therefore, the service company can offer periods of specific warranty to the customer after determined maintenance services, assuring the repaired engine will present 80 % availability during a specific period of time. Such period of time is to be determined following the guidelines previously detailed in this chapter. The availability is determined by the company, according to the risk that the organization is willing to take. The specific period of time is what here below is going to be determined, applying two different stochastic methods, the NHPP and the GRP. As mentioned, the first one is usually applied to simple repair cases, and the second one is usually implemented when repairs are complex. This case refers to a complex repair; therefore, results obtained by GRP will be more appropriate to determine a specific warranty period.

9.4.1.1 Exercise 1: NHPP Application

As above mentioned, the manufacturing company has at its disposal historical data about the engine behavior. As Table 9.10 can illustrate, failure and repair events are presented according to *ti*. With this t_i ($i = 1,\ldots, 6$), the values already defined are as follows:

$$T_i\,(\text{hours});\qquad T_n/T_i;\qquad \text{Ln}\ (T_n/T_i)$$

With these values and applying the above developed formulas, we will obtain: α_n, β_n, TNF_n, $t_{\text{MTW}n}$, Λ, $R(T_s)$, and $F(T_s)$. However, in order to see how impacts the NHPP in the calculation process, we will apply the mentioned formulas considering different numbers of known events. That means, we will analyze the system behavior under $n = 4$, $n = 6$, and $n = 8$.

Applying the process for $n = 6$ and $n = 4$, we obtain the results shown in Tables 9.11 and 9.12.

Table 9.7 Results according to critical values from case -B

ij	Maintainable item (Level 8 ISO14224)	λ_{ij}	$MTBF_{ij}$	A_{ij}	UA_{ij}	MDT_{ij}
1.1.	General power supply	47.00	0.02127660	99.96241413	0.03758587	0.000008
1.2.	Single-phase supply	6.500	0.15384615	99.99480027	0.00519973	0.000008
1.3.	Clamps sensors	10.000	0.10000000	99.99200064	0.00799936	0.000008
1.4.	PLC input/output	50.000	0.02000000	99.96001599	0.03998401	0.000008
2.1.	Rotor driver	23.000	0.04347826	99.98160338	0.01839662	0.000008
2.2	Rotor assembly	180.506	0.00553998	99.85580343	0.14419657	0.000008
2.3.	Estator	155.000	0.00645161	99.87615357	0.12384643	0.000008
2.4.	Welding machine	315.600	0.00316857	99.74815586	0.25184414	0.000008
i	Subunit (Level 7 ISO14224)	λ_i	$MTBF_i$	A_i	UA_i	MDT_i
1.	Electrical system	113.50	0.00881057	99.87735061	0.12264939	1.08194E−05
2.	Mechanical system	23.002	0.04347506	99.99765368	0.00234632	1.02009E−06
Complete system (Level 6 ISO14224)		λ	MTBF	A	UA	MDT
Industrial asset		136.502	0.00732592	99.87501005	0.12498995	9.16812E−06

Table 9.8 Reliability matrix for a mixed distribution

	Jan	Feb	Mar	Apr	May	Jun	Jul	Aug	Sep	Oct	Nov	Dec
days/month	31	28	31	30	31	30	31	31	30	31	30	31
T (month)	1	2	3	4	5	6	7	8	9	10	11	12
T (hours)	744	1416	2160	2880	3624	4344	5088	5832	6552	7296	8016	8760
1.1.	0.9656	0.9356	0.9035	0.8734	0.8434	0.8153	0.7873	0.7603	0.7350	0.7097	0.6861	0.6625
1.2.	0.9952	0.9908	0.9861	0.9815	0.9767	0.9722	0.9675	0.9628	0.9583	0.9537	0.9492	0.9447
1.3.	0.9926	0.9859	0.9786	0.9716	0.9644	0.9575	0.9504	0.9433	0.9366	0.9296	0.9230	0.9161
1.4.	0.9635	0.9316	0.8976	0.8659	0.8343	0.8048	0.7754	0.7471	0.7207	0.6943	0.6698	0.6453
2.1.	0.8774	0.8349	0.8002	0.7731	0.7492	0.7290	0.7103	0.6933	0.6783	0.6639	0.6509	0.6384
2.2	0.6932	0.6032	0.5356	0.4863	0.4454	0.4125	0.3835	0.3584	0.3371	0.3174	0.3003	0.2844
2.3.	0.7121	0.6259	0.5607	0.5127	0.4726	0.4402	0.4115	0.3864	0.3650	0.3453	0.3280	0.3118
2.4.	0.6160	0.5125	0.4380	0.3854	0.3432	0.3101	0.2816	0.2575	0.2374	0.2193	0.2038	0.1896
1.	0.9190	0.8515	0.7826	0.7212	0.6628	0.6108	0.5613	0.5159	0.4754	0.4369	0.4026	0.3700
2.	0.8774	0.8349	0.8002	0.7731	0.7492	0.7290	0.7103	0.6933	0.6783	0.6639	0.6509	0.6383
S. Comp.	0.7271	0.6443	0.5810	0.5342	0.4949	0.4630	0.4346	0.4097	0.3884	0.3686	0.3513	0.3350

AFTER-SALES SERVICE OF ENGINEERING INDUSTRIAL ASSETS
A REFERENCE FRAMEWORK FOR THE WARRANTY MANAGEMENT

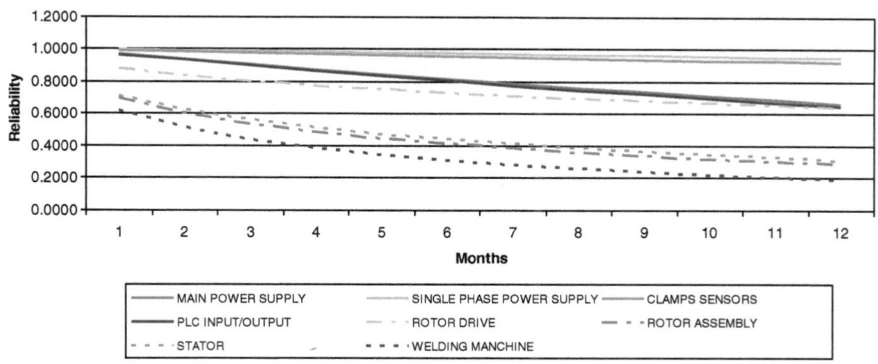

Fig. 9.12 Reliability evolution along a year for maintainable items

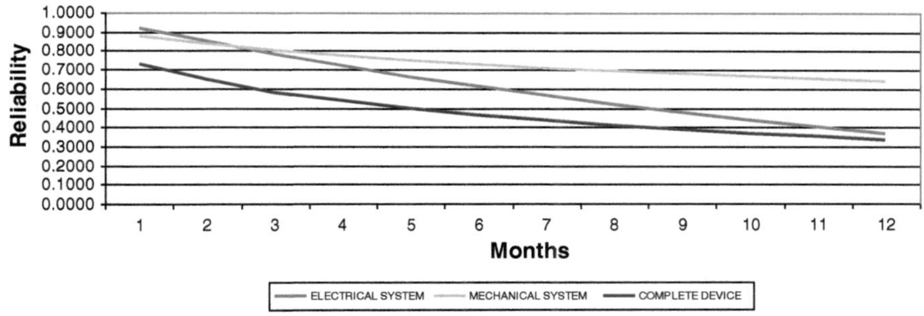

Fig. 9.13 Reliability evolution along a year for systems

Table 9.9 Priority order and criticality matrix

	ij	Element	R (8.760 h)
1	2.4.	Welding machine	0.189621467
2	2.2	Rotor assembly	0.284372381
3	2.3.	Estator	0.311845703
4	2.1.	Rotor driver	0.63835212
5	1.4.	PLC input/output	0.645325783
6	1.1.	General power supply	0.662509753
7	1.3.	Clamps sensors	0.916127254
8	1.2	Single-phase supply	0.944650747

Reliability	Severity		
	I	II	III
C	Estator	Rotor assembly	Welding machine
B	General power supply	Rotor driver	
A	Single-phase supply	Clamps sensors	PLC input/output

Table 9.10 NHPP engine historical data related to time per warranty event ($n = 8$)

Event (i)	t_i (h)	T_i (h)	T_n/T_i	Ln(T_n/T_i)
1	13,460.00	13,460.00	0.09	1.63
2	9,287.00	22,747.00	0.01	1.10
3	3,128.00	25,875.00	2.65	0.97
4	10,148.00	36,023.00	1.90	0.64
5	9,037.00	45,060.00	1.52	0.42
6	8,771.00	53,831.00	1.27	0.24
7	7,459.00	61,290.00	1.12	0.11
8	7,252.00	68,542.00	1.00	–

Applying different percentages (80, 60, and 40 %) for reliability, we can obtain different times as recommended warranty period based on the time till next failure. The use of one $t_{\mathrm{MTW}n}$ or another is a decision which can be more easily taken if we know the expected amount of failure Λ, as well as the system reliability $R(T_s)$. For that calculation, it has been here considered one year of function after the repair ($t_s = 8{,}760$ h).

With this exercise, we observe how decreases the time till next failure as well as the system reliability, when the amount of failures and repairs events decreases. If we analyze separately the development with the time of these cases, the mentioned differences in the system reliability are even more remarkable. The calculated parameters not only depend on the correct number of failures but also the order of them. A small number of divergent failures can distort the parameters, as in the first four events (see event 3, $t_i = 3.12800$ h), so it is desirable to consider convergent times of failures avoiding the mixture of different failure modes.

Consequently, the different risks assumed for warranty (in other words, the different choices for the percentage taken in the warranty time calculation) yield to different results in the values for the minimal time for the warranty period (Table 9.13). Figures 9.14 and 9.15 show the evolution of $R(T_s)$ along 2 years after the events $n = 4, \ldots, 8$, have taken place for $T_s = [2{,}190,\ 4{,}380,\ 6{,}570,\ 8{,}760,\ 10{,}950,\ 13{,}140,\ 15{,}330,\ 17{,}520]$.

This exercise contributes to illustrate the calculation of warranty periods. Nevertheless, it is also possible to explore how the time until next failure changes under the application of different warranty policies. In other words, it can be considered TNF as a variable. For this purpose, as an extension to this exercise, we can take into account the company's preference in maintaining a certain level of risk in the possibility of failure. That is, we will consider as an example that $F(T) \leq 0.20, 0.30$ and 0.40. This means that the company will not admit more than a 20, 30, or 40 % probability of failure during the warranty period. Therefore, if we maintain a fixed failure level, we obtain different values for the warranty periods that, the more probability of failure there is, the greater the slope of the warranty curve. In other words, if the company assumes greater risks in its warranty period, the decrease in the period of warranty time will be more significant when the probability of failure and repair increases.

Table 9.11 NHPP engine historical data related to time per warranty event ($n = 6$)

Event (i)	t_i (h)	T_i (h)	T_n/T_i	$\mathrm{Ln}(T_n/T_i)$
1	13,460.00	13,460.00	4.00	1.39
2	9,287.00	22,747.00	2.37	0.86
3	3,128.00	25,875.00	2.08	0.73
4	10,148.00	36,023.00	1.49	0.40
5	9,037.00	45,060.00	1.19	0.18
6	8,771.00	53,831.00	1.00	–

Table 9.12 NHPP engine historical data related to time per warranty event ($n = 4$)

Event (i)	t_i (h)	T_i (h)	T_n/T_i	$\mathrm{Ln}(T_n/T_i)$
1	13,460.00	13,460.00	2.68	0.98
2	9,287.00	22,747.00	1.58	0.46
3	3,128.00	25,875.00	1.39	0.33
4	10,148.00	36,023.00	1.00	–

Table 9.13 NHPP results obtained with $n = 8$, $n = 6$, and $n = 4$

	$n = 8$	$n = 6$	$n = 4$
αn	18,107.47	18,594.02	19,471.96
βn	1.562172	1.685553	2.253470
TNFn	5,367.67	5,155.20	3,749.66
$t_{\mathrm{MTW}n}$ (20 %)	1,217.77	1,178.91	878.32
$t_{\mathrm{MTW}n}$ (40 %)	2,770.34	2,673.73	1,973.40
$t_{\mathrm{MTW}n}$ (60 %)	4,926.88	4,735.69	3,452.79
Λ	1.653580	1.736069	2.532629
$R(T_s)$	0.191364	0.176212	0.079450
$F(T_s)$	0.808636	0.823788	0.920550

9.4.1.2 Exercise 2: GRP Application

In addition to the described exercise, which has helped to illustrate the calculation of warranty periods, it is possible also to explore now considering GRP model in order to obtain the different warranty parameters. For that purpose and as a second example, we are going now to consider the same repair events, obtaining the results of Table 9.14.

In addition, now for the GRP model, Figs. 9.16 and 9.17 show the evolution of $R(T_s)$ along 2 years after the events $n = 4, \ldots, 8$ have taken place for $T_s = [2{,}190, 4{,}380, 6{,}570, 8{,}760, 10{,}950, 13{,}140, 15{,}330, 17{,}520]$.

Now as with NHPP, through GRP model, the higher is the failure probability, the higher the slope of the warranty period curve. In other words, if the company assumes higher risks in its warranty policies, the increase in the warranty times shall be more significant. The adjustment of the reliability obtained from the GRP versus time and number of expected failures confirm its better fit thanks to the

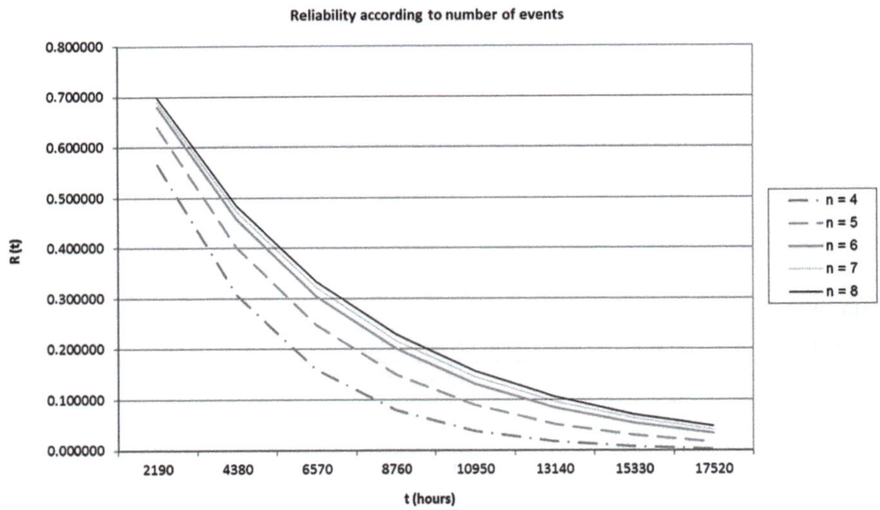

Fig. 9.14 NHPP reliability evolution depending on time and the amount of events

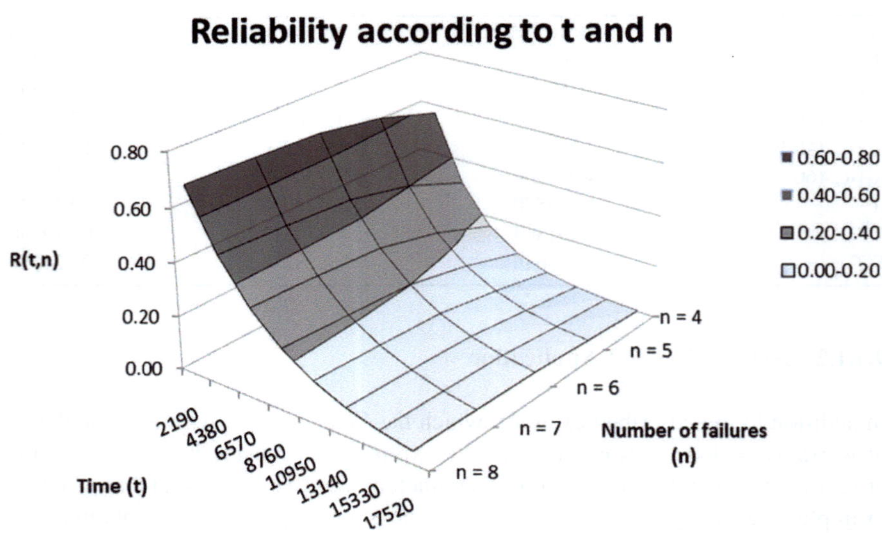

Fig. 9.15 NHPP surface representation of $R(t, n)$

inclusion of the parameter q, see Table 9.15 where a comparison of both models, NHPP and GRP, is realized using the likelihood value as a representation of the adjustment.

GRP is a general model, providing a way to describe the rate of occurrence of events over time understanding the effects of the repairs on the age of that system,

Table 9.14 GRP results obtained with $n = 8$, $n = 6$, and $n = 4$

	$n = 8$	$n = 6$	$n = 4$
αn	10,901.91	11,551.98	11,971.63
βn	3.464468	3.355038	2.663281
TNFn	9,016.30	8,850.22	8,087.37
tMTWn (20 %)	5,199.15	4,735.58	3,301.54
tMTWn (40 %)	7,099.28	6,770.40	5,548.63
tMTWn (60 %)	8,745.11	8,554.81	7,715.04
Λ	0.920753	0.973892	1.162304
$R(T_s)$	0.398219	0.377610	0.312765
$F(T_s)$	0.601781	0.622390	0.687235

and including the renewal process or the NHPP representation. For example, consider a system that is repaired after a failure, where the repair does not bring the system to an as good as new or an as bad as old condition. In other words, the system is partially rejuvenated after the repair.

9.4.2 Results and Conclusions

The obtained result as a warranty period in exercise 1 is, as it has been previously mentioned, a minimum value calculated with a NHPP model. It has considered different percentages (20, 40, and 60 %) of waiting TNF. Now, the company will have to decide whether this period is convenient or not to be extended with the consequent warranty cost increase. Therefore, the calculation of this percentage is one of the most important steps that need to be taken into account in this process. We also need to have some input on the risks the company is willing to assume for the warranty period. For this reason, two exercises have been here represented. In case the warranty period is a prefixed value included in the repair order or maintenance contract (or when imposed in the effective legislation [36]), the manufacturer (or subcontracted company offering the corresponding warranty maintenance services) can forecast with the previously described calculations the time until the next warranty service. This is in order to save part of the general budget for this warranty service and to have at its disposal a technical team ready for possible future events. To such effect, the NHPP model has a demonstrated ability to give good results in real situations, with minimum repairs [38]. Based on this, due to its conservative behavior and being manageable with respect to mathematical expressions, the NHPP model was chosen specifically for this task. However, when the repairs are not minimal, the best adjustment can be obtained with GRP model through its adaptability with three dimensions (α, β, and q), at the expense of more complexity. Of course, the previously described models have their advantages and limitations so that the more realistic the model is, the more complex the involved mathematical expressions are. The main strengths and weaknesses of these models are summarized as follows:

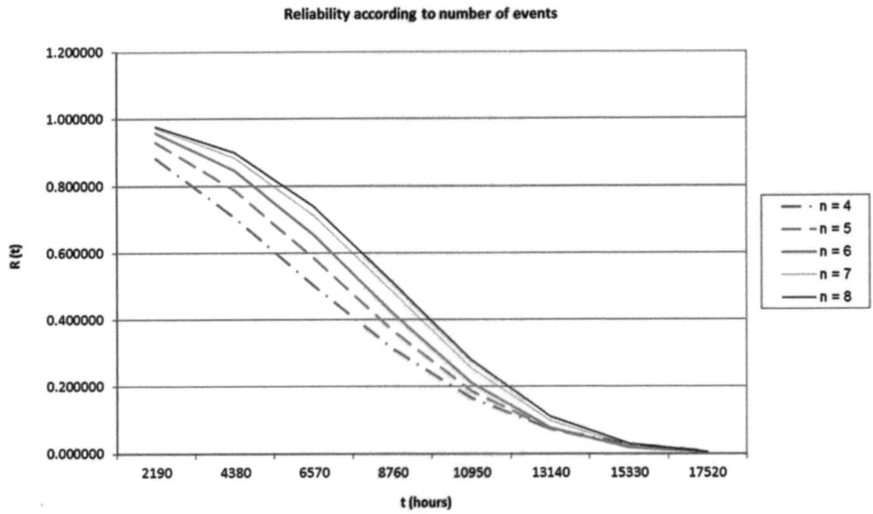

Fig. 9.16 GRP reliability evolution depending on time and the amount of events

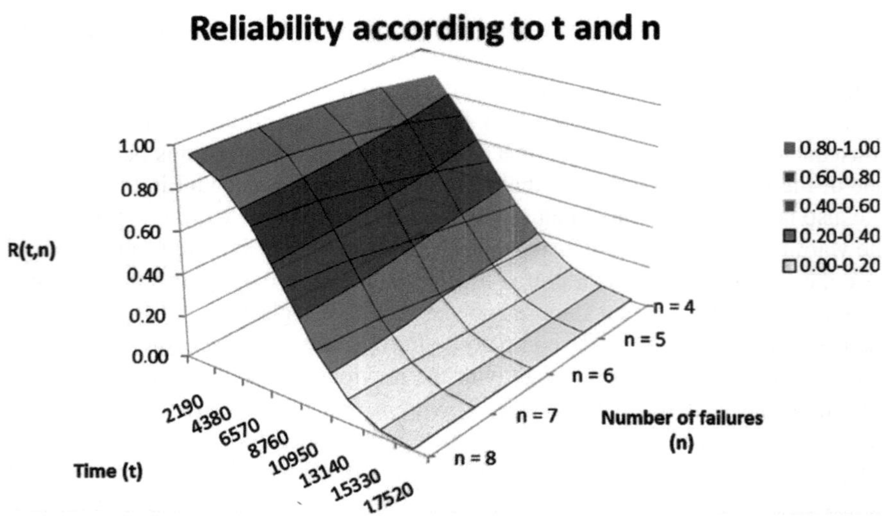

Fig. 9.17 GRP surface representation of $R(t, n)$

- Strong points:
  - Both are useful and simple models to represent the aging and deterioration of an industrial asset.
  - NHPP involves mathematical expressions that are relatively simple.

Table 9.15 Comparison of NHPP and GRP likelihood values obtained with $n = 8$, $n = 6$, and $n = 4$

LK	$n = 8$	$n = 6$	$n = 4$
NHPP	−79.756	−59.919	−39.398
GRP	−73.907	−55.900	−37.981

- NHPP is a conservative approach, but GRP is better for the evaluation of the warranty period regarding any given risk [39].

- Weak points:
  - NHPP is not appropriate to simulate repair tasks that restore the equipment to a different state of "as that of the old" [26].
  - GRP is not adequate to simulate repair actions when the number of events is small.
  - GRP involves relatively complex mathematical expressions.

References

1. Smith DJ. 2001. Reliability, maintainability, and risk: practical methods for engineers. Newnes. ISBN 0-75060-854-4.
2. Høyland A, Rausand M. 1994. System Reliability Theory, Models and Statistical Methods. J. Wiley. Universidad de Michigan. ISBN 0-471-59397-4.
3. Crespo Márquez A, Iung B. 2007. A Estructured Approach for the Assessment of System Availability and Reliability Using Montecarlo Simulation. Journal of Quality in Maintenance Engineering. 13(2):125-136.
4. Woodhouse J. 1993. Managing Industrial Risk. Chapman Hill Inc. London.
5. Markeset T, Kumar, U. 2001. R&M and risk analysis tools in product design to reduce life-cycle cost and improve product attractiveness. Proceedings of the Annual Reliability and Maintainability Symposium. Philadelphia. Pp. 116-122.
6. Blanchard BS. 2001. Maintenance and support: a critical element in the system life cycle. Proceedings of the International Conference of Maintenance Societies. Paper 003. Melbourne.
7. Blanchard BS, Fabrycky WJ. 1998. Systems Engineering and Analysis, 3rd ed., Prentice-Hall, Upper Saddle River, NJ.
8. Goffin K. 2000. Design for supportability: essential component of new product development. Research-Technology Management. 43(2):40-7.
9. Smith C, Knezevic J. 1996. Achieving quality through supportability: part 1: concepts and principles. Journal of Quality in Maintenance Engineering. 2(2):21-9.
10. Woodward DG. 1997. Life Cycle Costing - Theory, Information Acquisition and Application. International Journal of Project Management. 15(6):335-344.
11. Durairaj S, Ong S. 2002. Evaluation of Life Cycle Cost Analysis Methodologies. Corporate Environmental Strategy. 9(1):30-39.
12. Kirk S, Dellisola A. 1996. Life Cycle Costing for Design Professionals. Mc Graw Hill, New York. Pp. 6-57.
13. Woodhouse J. 1991. Turning engineers into businessmen, 14th National Maintenance Conference. London.

14. Fabrycky WJ, Blanchard BS. 1993. Life Cycle Costing and Economic Analysis. Prentice Hall, Inc, Englewood Cliff. New Jersey.
15. Willians D, Scott R. 2000. Reliability and Life Cycle Costs, RM-Reliability Group, Technical Paper. Texas.
16. Riddell H, Jennings A. 2001. Asset Investment & Life Cycle Costing. The Woodhouse Partnership, Technical paper, London.
17. Parra C, Omaña C. 2003. Análisis determinístico del Ciclo de Vida y evaluación del factor Confiabilidad en Motocompresores de gas para extracción de petróleo. Congreso Internacional de Mantenimiento, Colombia, Bogotá.
18. Ahmed NU. 1995. A design and implementation model for life cycle cost management system. Information and Management. 28:261-269.
19. Levy H, Sarnat M. 1990. Capital Investment and Financial Decisions. 4th Edition, Prentice Hall, New York.
20. Alsyouf I, Al-Aomar R, Al-Hamed H, Qiu X. 2011. A framework for assessing the cost effectiveness of lean tools. European J. of Industrial Engineering. 5(2):170-197.
21. González-Prida V, Crespo Márquez A, Moreu P, Gómez J, Parra C. 2009. Availability and reliability assessment of industrial complex systems: A practical view applied on a bioethanol plant simulation. Safety and Reliability for Managing Risk. London, UK. Taylor & Francis Group. ISBN 978-0-415-48513-5. Pp. 687-695.
22. Crespo Márquez A, Iung B. 2007. A Structured Approach for the Assessment of System Availability and Reliability Using Montecarlo Simulation. Journal of Quality in Maintenance Engineering. 13(2):125-136.
23. Parra C, Crespo Márquez A, Moreu P, Gómez J, González-Prida V. 2009. Non-homogeneous Poisson Process (NHPP), stochastic model applied to evaluate the economic impact of the failure in the Life Cycle Cost Analysis (LCCA). Taylor & Francis Group, London, ISBN 978-0-415-48513-5. Pp. 929-939.
24. Ascher H, Feingold H. 1984. Repairable System Reliability: Modeling, Inference, Misconceptions and their Causes. New York, Marcel Dekker.
25. Crow LH. 1974. Reliability analysis for complex repairable systems. Reliability and biometry, Proschan F, Serfling RJ, eds., SIAM, Philadelphia, pp. 379-410.
26. Yañez M, Joglar F, Mohammad M. 2002. Generalized renewal process for analysis of repairable systems with limited failure experience. Reliability Engineering & System Safety. 77:167-180.
27. Modarres M, Kaminskiy M, Krivtsov V. 1999. Reliability engineering and risk analysis. Marcel Dekker Inc., New York.
28. Yañez M, Gómez de la Vega H, Valbuena G. 2004. Ingeniería de Confiabilidad y Análisis Probabilístico de Riesgo. Editor: Reliability and Risk Management S.A., México. ISBN 980-12-12-0116-9.
29. Parra C. 2009. Desarrollo de modelos de cuantificación económica del factor "Fiabilidad" en el coste total del ciclo de vida de un sistema de producción. Tesis doctoral dirigida por Adolfo Crespo Márquez. Universidad de Sevilla.
30. Mettas A, Wenbiao Z. 2005. Modeling and analysis of repairable system with general repair. Proceedings of the Annual Reliability and Maintainabilty Symposium, Jan. 24-27, Alexandaria, Virginia, pp: 176-182.
31. Kijima M, Sumita N. 1986. A useful generalization of renewal theory: Counting process governed by non-negative markovian increments. J. Applied Prob., 23: 71-88.
32. International Standard. 2004. ISO/DIS 14224/2004. Petroleum, petrochemical and natural gas industries - Collection and exchange of reliability and maintenance data for equipment.
33. American Institute of Chemical Engineers Center for Chemical Process Safety. 1989. Guidelines for Process Equipment Reliability Data, with data tables. ISBN 0-8169-0422-7.
34. SINTEF Technology and Society. 2002. Offshore Reliability Data Handbook. 4th Edition, OREDA. ISBN 8-21400-438-1.
35. Parra C, Crespo Márquez A, Cortés P, Fygueroa S. 2006. On the Consideration of Reliability in the Life Cycle Cost Analysis (LCCA). A Review of Basic Models. Safety and Reliability

for Managing Risk. Londres, Reino Unido. Taylor & Francis. Pp. 2203-2214. ISBN: 0-415-41620-5
36. González-Prida V, Barberá L, Gómez J, Crespo Márquez A. 2011. Contractual and quality aspects on warranty: best practices for the warranty management and its maturity assessment. International Journal of Quality and Reliability Management, IJQRM.
37. Murthy DNP, Blischke WR. 2005. Warranty management and product manufacturing. Springer-Verlag London Limited. Pp 302 + xviii. ISBN 1852339330.
38. Asiedu Y. Gu P. 1998. Product lifecycle cost analysis: state of art review. International Journal of Production Research. 36(4):883-908.
39. Hurtado JL, Joglar F, Modarres M. 2005. Generalized Renewal Process: Models, Parameter Estimation and Applications to Maintenance Problems. International Journal on Performability Engineering. 1(1-3):37-50.

Chapter 10
Continuous Improvement

10.1 Application of ICTs in the Process of Warranty Management

10.1.1 The E-technology Concept in the Warranty Management

One of the most important parts inside the management of a warranty program is the relationship with the customer. The continuous improvement in the warranty management of a company or an organization requires the application of emerging techniques and technologies in those areas considered with more impact. The implementation of new technologies to the warranty management brings with itself the introduction of the concept "e-warranty" which is proposed as another component of "e-manufacturing" and parallel to the concept of "e-maintenance" (see Chap. 4), disposing the advantages of the new and emerging technologies on information and communications, in order to apply them inside a multiple user environment, by a cooperative and distributed way. Therefore, the e-warranty can be defined as the support to a warranty program which includes those resources, services, and management needed to enable proactive decision-making. Apart from the e-technologies, the aforementioned support also includes activities which are typically associated with warranty such as monitoring, diagnostics, and prognosis. Together with the application of new technologies for the warranty management, the involvement of the technicians from the after-sales service is essential in the continuous improvement process of the warranty in order to achieve a success goal which supposes higher levels of product quality and service effectiveness of the technical assistance. Consequently, it will be required higher levels of knowledge, experience, and training. Considering the proposed framework, the application of new information and communication technologies will have an effect on certain processes, which will be turned into "e-processes," while others will not be affected. The concept of e-warranty basically proposes a new way of managing a warranty program by which the products are controlled via

V. González-Prida Díaz and A. Crespo Márquez, *After-sales Service of Engineering Industrial Assets*, DOI: 10.1007/978-3-319-03710-3_10,

Internet by the manufacturer, e.g., information can be gathered or is possible to manage alarms. Thus, the manufacturer obtains real-time data from the equipment under warranty through digital technologies. The data obtained by sensors preinstalled on the equipments provide information about their status (temperature, run time, pressure, etc.) and allow a continuous diagnostic, which enables the failure prediction due to the malfunction of equipment or due to a bad use. The e-warranty incorporates the same principles already applied in the conventional process of warranty management, adding a Web service and the principles of collaboration and/or monitoring. This application may also be a useful strategic marketing tool when seeking new business, offering to the users the possibility to extend the warranty contract by the remote tracking of product performance, and, consequently, being able to foresee possible failures. In any case, this new advanced model of customer service management breaks the physical distances between manufacturer and customer through the use of ICT, transforming the company into a manufacturing services' business that provides support to its customers anywhere, anytime.

10.1.2 Technologies Applied to the Technical Assistance

Based on the ICT application to maintenance (see Chap. 4), in the case of warranty assistance, we also find the following options which can be applied on warranty management tools to generate new emerging processes and aiming a better performance of the technical assistance:

1. *Remote assistance* Using tools for remote access to information (Internet, Bluetooth, etc.), the warranty service technicians are able to access the product which presents a failure or where a complaint has been made, allowing them the remotely start-up, control, configuration, condition monitoring, data collection, and analysis of the product. One of the biggest advantages is to connect products distributed in different and distant geographic locations, to a warranty service center that allows a remote decision-making.
2. *Cooperative warranty service* The e-warranty would allow implementing an information infrastructure connecting clients (users) with suppliers (manufacturers) or connecting engineers and technicians with the products using Internet.

The resulting platform would allow then a strong cooperation between different areas of the company:

- Production (to implement changes in the manufacturing process).
- Purchasing (acquisition of a specific spare part).
- Engineering (change in the product design).
- Quality (to certify the compliance of the product itself and/or the repair works performed on it).

- Logistics (to control spare parts inventory and to calculate the minimum path to transport those spares to the place where the related device is located).
- After-sales or management team (to assign teams, tools, materials, planning, etc., when the failure cannot be repaired remotely).
- Etc.

In this manner, process errors are minimized and communication is optimized (it facilitates the bidirectional flow), feedback is generated, and to sum up, the quality and the perception of the service by the client are improved.

3. *Online assistance* The remote follow-up of sold units during the warranty period, together with the possibility of programming status alerts, allows the manufacturer or supplier to respond quickly to any situation and even before the user has been detected failure, preparing accordingly any intervention in an optimal way.
4. *Predictive warranty service* The e-warranty would allow a failure prediction in those units offered for sale, taking into account the product current state and the intended use on it, together with the life predictions on component performed by the different manufacturers, and obtained through a reliability analysis (RAMS analysis) [1].
5. *Failure diagnosis* The e-warranty allows technicians to carry out online diagnosis to those occurred failures, as well as to suggest solutions to the end user. Consequently, it reduces the communication time between supplier and customer, improving the quality of the shared information and, therefore, the resolution time.

10.1.3 Influence on the Proposed Management Framework

The proposed framework for the management of warranty assistance (see Fig. 5.9) consists in several tools and methodologies which involve the handling of a large amount of data relating to warranty. The stage we have defined as "e-warranty" is included in a fourth block of the management framework related to the improvement in the warranty program.

This improvement is carried out based on the performance of management according to the policy adopted by the company and it needed a computerized information system for the warranty program, which enables the cyclical processing of data management as well as the analysis of results. In this step is included the concept of e-warranty, which facilitates the decision-making, processing for that purpose large volumes of data and information from multiple sources and with a great variety of types. e-warranty is therefore the use of information technology in the context of after-sales service management as for instance may be the case for data storage and their analysis in relation to consumer complaints, among others. The e-warranty concept has a multidisciplinary approach (similar to the Six Sigma methodology), based on the monitoring, diagnosis, prediction, and real-time control of products and the detection of abnormal conditions. The diagnosis identifies possible causes of early degradation or failure in the sold unit (important when doing the RCA of the failure), and the prediction analyzes the impact of the failure on the product itself (essential for the criticality analysis). Thus, the manufacturer is able to obtain in real time whether a failure has occurred on equipment under warranty and, consequently, the failure (in terms of the causes that produced it) is included in the coverage of the warranty program, or it is a result of a misuse or an accident with the equipment. With this information, the manufacturer can anticipate the failure, warning the user [such as an added value in the customer relationship management (CRM)] about the existence of anomalies in the control parameters and recommending, for instance, the preventive shutdown of the equipment, reducing thereby the repair time and cost (avoiding further damage with the shutdown), and achieving at the same time a mutual benefit for both, manufacturer and customer. The collected information also allows the manufacturer to make decisions and to carry out maintenance actions if necessary, generating automatically work orders and speeding up the administrative procedures. Sometimes the historical information from the equipment may be sufficient to verify the "natural" cause of the failure, or an abnormal situation; thus, it is not required a previous physical assessment to start planning the repair. The influence of ICT on the steps proposed as reference framework for the warranty management can be summarized very succinctly in the following guidelines, according to the numbering shown in Fig. 9.17:

1. Analysis of customers and complaints.
2. Multidisciplinary approach.
3. Prioritization in the decision-making.
4. Analysis of the failure impact on the product.
5. Identification of possible degradation causes or anticipated failure.
6. Modernization of technical assistance plans.
7. Processing of large volumes of information.
8. Update of reliability data and failure rates.
9. Data feedback on the product life cycle cost.

From the steps proposed as a framework, the processes that can be considered as emerging processes are those ones innovative in the warranty management and susceptible to be implemented with new information and communication technology. They are rounded in Fig. 10.1: risk–cost–profit analysis, life cycle cost, and RAMS analysis. Although it is not indicated in the figure, obviously, the customer communications are made usually using ICTs (e-mail, mobile telephone, Internet…), which cannot be considered as an emergent process. In few words, following a cyclical approach of continuous improvement to the warranty management, the fact of applying new information and communication technologies to the customer service, effects in the management framework itself a feedback at every stage with information and digital data directly from the units launched to the market and sold.

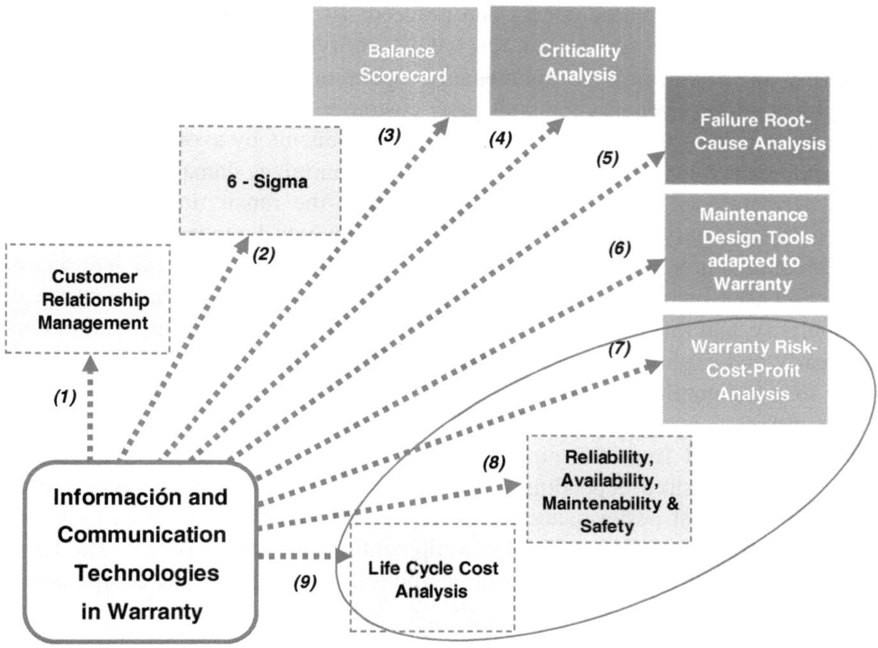

Fig. 10.1 ICTs influence on the proposed warranty framework

10.1.4 Advantages and Future Utilizations

The e-warranty concept allows a full knowledge on how products or devices behave by receiving signals from sensors on remote. Thus, during the specified warranty period, possible failures can be predicted reducing substantially the administrative procedures, increasing functioning time, and optimizing the equipment performance and productivity. A possible future use of the e-warranty system might be the optional hire by the customer of an e-warranty extension on the product; thus, the customer would take advantage of the sensor network pre-installed on the equipments in order to predict failures which may affect the equipment or may have an impact on the production line where this device develops its functions. The e-warranty system or platform enables to warn manufacturers via Internet when:

- There is a failure in equipment.
- Before the failure occurs (alarms), and
- When the equipment reaches certain conditions in its operation.

Moreover, it also provides the customer and the manufacturer with the necessary information from all equipments in real time, using for that purpose any Internet browser. The manufacturer receives the message (alarm) with detailed information when a failure occurs and proceeds (if possible) to send those technical and material resources to resolve it, notifying the customer about the necessity of a warranty or maintenance operation. The failure prediction in equipments brings benefits for both parts customer and manufacturer during the warranty period of the equipment. The customer benefits by avoiding or reducing the equipment downtime. Anticipating failures minimizes damage on the equipment as mentioned above, and this also reduces the repair time and therefore reduces significantly the labor and material (components damaged) costs in charge of the manufacturer. All this is translated into a fast, effective, and efficient service to the end user, increasing the operating time of the equipment under warranty, its productivity, and finally the customer satisfaction. Obviously, the application must be clearly and explicitly agreed in the acquisition contract of industrial assets, in order to avoid conflicts or problems regarding the privacy of customer activities. The e-warranty is a proactive system, where information is automatically and instantly received from the equipment, being able to configure what kind of information is desired depending on the equipment under analysis, for example, failures, equipment performance, anomalies in measurable parameters (temperature, pressure, etc.) workload peaks, equipment status, among others. The transmitted data can only provide information about the operation of the equipment. In addition to the above mentioned, the application of this system has other advantages:

- It eliminates spatial barriers between customer and manufacturer.
- It eliminates the time barriers.
- It reduces the cost per failure to the customer and manufacturer.
- It collaborates in the organizational familiarity with ICT.
- It enables permanent access and control in real time to all that information on the equipment under warranty.
- It facilitates fast communication between customer and manufacturer.
- It eliminates paperwork for the manufacturer in case of equipment failure (cause verification, repair record, etc.).

10.2 Management of Customer Relations in After-sales Service

10.2.1 Importance of Customer Relations

As it has been mentioned in several occasions, nowadays we acknowledge that companies, in the practice, pay more attention to their warranty management processes when there are repercussions at high levels of responsibility and costs, putting at stake the product's demand on the market or the recuperation of lost customers. As a consequence, there are losses of opportunities of significant cost savings as well as the possibility of increasing the business' value through better-quality products and higher levels of customer satisfaction. It is why it is mandatory to take into consideration the interactions between the engineering, the marketing, and the after-sales support elements from an operational and strategic level [2]. Therefore, warranty management should be handled, nowadays, not as a cost focus, but as an active that can create more value in the performance of the business. For instance:

- Improving the reclamation response time (speeding up their process).
- Decreasing reclamation costs (giving them an automatized treatment).
- Reducing personnel in charge of the reclamation process.
- Reducing the loss of customers due to the poor quality of the product.
- Improving distribution costs (speeding up the information flow with suppliers).

The result will be to reduce the costs of the warranty, improve the product and services, and gain consequently higher levels of customer retention. Therefore, the after-sales service requires improving the strategies with the use of more efficient tools and integrated systems, as well as through an organizational structure. The application of new technologies represents higher levels of knowledge, experience, and personnel training in the technical service. Logically, one of the most important parts within a warranty management program is the relation with the customer. The techniques used in these relations are generally considered as a part of an enterprise resource planning (ERP) system or as a complementary part of them. The customer relation management (CRM) is exclusively used not only to define a business strategy centered on the customer, but also to include a group of information technology applications useful to sort and manage data related to customer and reclamations, and in general to the commercial activity of the company. As it has been mentioned before, the CRM is often used as a module for customer management including ERP software. Originally, the CRM was once focused mainly on marketing and commercial aspects. However, the CRM module can be adapted to warranty management, allowing the company to:

- Identify products and services requiring assistance.
- Reduce the time of the assistance and optimize channels of information.
- Identify groups of customers in order to develop common strategies.
- Be conscious of customer's real needs after the sale.
- Increase sales on the same level as customer satisfaction, etc.

Customer relation management (as part of a ERP or as a complementary system) includes operative areas dealing with tasks directly related to the customer (front office) and other more analytical areas closely associated to different internal parts of the company (back office).

This representation is shown in Fig. 5.6. The main part of any CRM strategy will be a database that can provide information about claims, repetitive failures, hidden defects [2], quantitative and qualitative analysis, statistical results [3], etc.

10.2.2 Contractual Aspects on Warranty

10.2.2.1 Strategic Issues

A warranty program management, viewed from a strategic perspective, begins with a strategy linked to technical and commercial planning from the very start of the product development process [4]. However, an after-sales support is frequently offered, while the production line is still open. That means, that the information obtained from this after-sales service can be very helpful for the production lines and, of course, very useful to improve the design and development of the product. In addition to this inference on the product life cycle, there are currently different international standards (that will be forward treated) to be considered when defining any after-sales strategy, which approach the whole business activity [5]. The management process defined in such standards and quality management systems usually fulfills certain features already defined in Chap. 4. Considering these characteristics was able to develop the reference framework for the warranty management in Chap. 5. However, many of the information required to this close loop is not possible to obtain from the very beginning of the process. Therefore, trying to adapt the different steps of any project (see Fig. 4.3) to the phases of the framework will have an evolution as shown in Fig. 10.2. We can observe here how during design and development of the product, it is the time to get a first approach of a strategical formulation for the warranty program through an incipient balance scorecard as well as an approximated evaluation of the product behavior. As a result of such analysis, one can obtain a deep knowledge about the product capacity to perform properly the functions expected from it, under established conditions and during a period of specific time [6], as well as the cost of spare parts, reparation times, etc. [7].

As a result of the previous steps, this prelaunch stage is also the perfect moment to start considering the implementation of new e-technologies and activities such as e-monitoring, e-diagnosis, and e-prognosis using remote devices. After this preliminary stage, comes the launch stage, which is highly influenced by the production and the marketing phases. From the proposed framework point of view, it is the moment to turn on some new steps with new iterations of our close loop. For instance, with the results obtained from the RAMS analysis, an initial maintenance plan, applied to the warranty time horizon, can suppose a good first approach for the warranty capacity planning, the spare parts' provisioning, the warranty task schedule, skill level of the technicians, and so on. In the same way, with former studies in the product life cycle cost during the prelaunch stage, it is possible now to carry out an incipient analysis on Warranty Policy Risk-Cost-Benefit. In this launch stage is also of importance to start implementing a CRM, which will be used, in general, for the definition of a business strategy centered in the client. Afterward, it comes the post-launch stage where every steps of our proposed framework will be finally in action. Here, we can see how a criticality

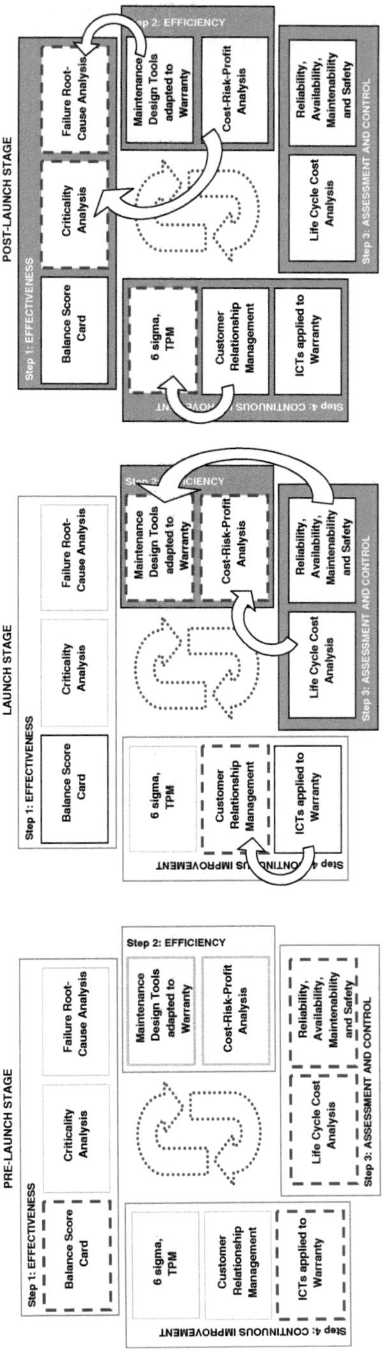

Fig. 10.2 Framework evolution, according to the industrial asset life cycle

analysis based on risk–cost combines the probability of an occurring event, with the impact that this event would cause. Or how useful can be the application of a new management philosophy like Six Sigma (which will be later discussed), as a methodology which integrates the human factor with improvement tools, mainly statistical. For posterior iterations of the suggested warranty management reference framework, it will deal mainly with aspects more related to the warranty logistics and the achievement of a balance between costs and customer satisfaction.

10.2.2.2 Legal Aspects on the Warranty Management

Once analyzed the strategic management of the product life cycle from a warranty perspective, it is time to briefly discuss about possible warranty extensions and how they can be expressed by a contract which encompasses the conditions for such extended assistance. In the literature, we observe that warranty is sometimes called as a "service contract." However, there are some differences between both concepts [8]. Mainly, a service contract is purchased separately, while warranty is part of the product purchase and integrated to the sale. Therefore, a warranty extension should be considered as a "service contract," whose scope can be the same as the one for warranty, or on the contrary, it can include and/or exclude some different topics. Afterward, during the warranty time (and normally close to the end of this period), the warranty extension will be agreed between buyer and seller and finally committed into effect, if both parts are interested to prolong or to contract this service. In legal terms (Fig. 10.3), warranty and warranty extensions are concepts expressed through a contract between the parties for specific assurances and applied to customers during a determined period of time ($W + L$, where W and L are the warranty and the extended warranty period, respectively).

Such a warranty can be given for purchased products or for services with a specific workmanlike quality. In the case of warranty extensions applied to a product, this contract can give different conditions of the technical service to be provided by the manufacturer, which are described in the fundamental terms of the contract. These agreement clauses give consequently a legal obligation to the manufacturer in front of the buyer. Usually, there is an inherent understanding by the buyer about the warranty concept itself, which is independent of any written expression. In any case, the sold products and the products whose warranty has been extended must reasonably conform (or keep conforming) to an ordinary user's expectations, and the seller must be sure that the item is fit for that particular purpose. Those countries that have ratified the United Nations Convention on Contracts for the International Sale of Goods (CISG) find the obligation that, excepting when the parties have agreed otherwise, a seller must provide goods fit for their ordinary purpose [8]. Comparing for instance the American law, with the European law, in the USA, when the seller at the time of contracting has reason to know any particular purpose for which the goods are required and that the buyer trusts in the seller's skill or judgment to select and provide suitable goods, there is

Fig. 10.3 Manufacturing, warranty, and legislation (adapted from [4])

then an implied warranty that the goods shall be fit for such purpose [9]. In general terms, the products must comply the trade standards applicable to the contract for sale and for the warranty extension, being uniform in quality. They must fit those purposes which such goods are ordinarily used to, even if the buyer ordered them for use otherwise. The Uniform Commercial Code (UCC) allows sellers to disclaim the implied warranty of products. Therefore, in the USA, a disclaimer must be marked or in evidence in the contract. Other codes and laws (as the Magnuson-Moss Act which covers warranties on consumer products used for commercial purposes), together with interesting references on this field, can be found in [4]. In Europe, countries usually have a general law for the defense of consumers and users [10] and other complementary laws for warranty and after-sales services, where all those seller's responsibilities together with the consumer's or user's rights are depicted. Particularly, the repair and the substitution according to this law will be adjusted mainly to the following rules:

(a) They will be free of charges for the user. This free repair considers the necessary expenses to correct the non-conformity of the product with the contract, including the shipping and handling, as well as the costs related to the manpower and the materials.
(b) They will be carried out in a reasonable term and without more inconveniences for the user, depending of course on the products sort and the purpose that they have for the user.
(c) The repair stops the computation of warranty terms. This suspension period will begin since the user puts the product at the seller's disposal, and it will conclude with the delivery to the user of the product repaired.
(d) Once concluded the repair and given the product to the customer, if the item continues being not according to the contract, the user will be able to demand the substitution of the product (unless this option would be disproportionate), a price discount, or the resolution of the contract.
(e) The user will not be able to demand the substitution of the failed item in the case of secondhand products or fungibles.

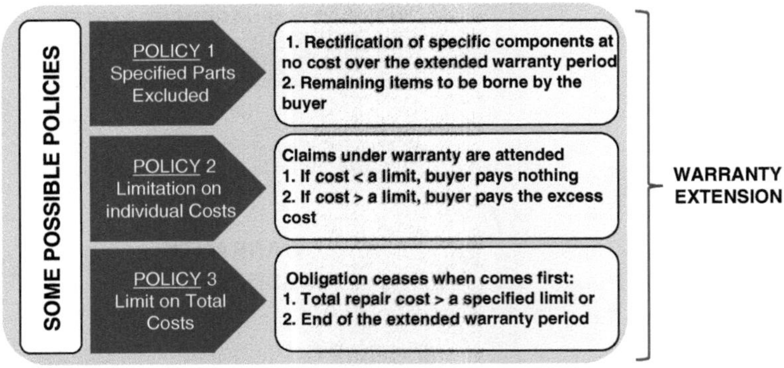

Fig. 10.4 Possibilities for the warranty extensions (adapted from [10])

It is interesting to add here that in connection with suspension periods, the warranty term as well as the extended period is sometimes suspended for all the affected elements from the moment when the appearance of the defect is communicated to the contractor, until the repair or substitution is performed. It is not considered to be affected by this suspension those elements which stay operative in spite of the failure. This issue is frequently not clear in contracts and supposes usually a confrontation topic between user and seller. Further research on this field can be for instance about how warranty is seen by other markets, mainly emerging countries, and the interaction between such markets with the developed countries, since the first ones usually are manufacturers and the second ones usually are users or consumers. Some possible policies on service contract extensions can be extracted from references [4, 11, 12]. Figure 10.4 tries to illustrate some possible policies to take into account at the moment of offering a contract for warranty extensions. Apart from deciding what kind of policy is more interesting for the parts, it is important to include in the service contract the following paragraphs (see Table 10.1). As already commented, due to the fact that there is usually an inherent understanding by the buyer on the warranty concept itself, it is important to define as clear as possible all those requirements, conditions, work scope, etc., by written in the service contract for the mentioned warranty extension. Further researches on this topic (mainly on possible policies) would be very interesting and welcome, in order to contribute to the development of the after-sales service area.

10.3 The Methodology Six Sigma in After-sales Warranty

In the proposed warranty management reference framework (Chap. 5), we can observe that the forth step refers to the continuous improvement which will be carried out based on indicators that are generally accepted [13]. The obtained information from the described steps in previous chapters can represent a great

Table 10.1 Example guideline for a service contract on warranty extensions

• Extended period	• Remedying of warranty defects	• Damaged products
• Predelivery inspection form	• Costs for replacement of components, repair, etc.	• Expiry of warranty
• Geographic area	• Non-refundable costs	• Changes in design
• Warranty defect	• Replaced components	• Measurements
• Service and maintenance	• Products sold by other vendors	• Limitation of liability
• Payment	• Refund, payment of compensation	• Force majeure
• Notice of warranty claim	• Normal wear and tear	• Representations

amount of data as well as a wide variety in their typology. Therefore, the challenge will be the treatment of such information and also the knowledge to discriminate between useful data from which are not. With the aim of assuring a more continuous cycle of improvement, it was been included in the forth step the application of new technologies, to process great data volumes with multiple sources, and the incorporation of management applications that makes proper decision-making and contractual customer relations easier. Before all that variety of tools, it is necessary to implement a quality methodology that, in a certain way, lubricates all these management and organization machinery. Whit this in mind, we now continue to describe in a succinct manner the Six Sigma method [14, 15]. Also, there will be a brief description of its interactions with other tools and management and quality methodologies for its possible application in the after-sales service.

10.3.1 Brief Description of the Method

We have analyzed that in order to engage all this tool and methodology mechanisms, it is necessary to apply programs or methods that improve the efficiency of the organization, specifically in our case, that improve the management and organization of warranty services. With that aim, it is possible and interesting to adapt concept, for instance coming from the total productive maintenance (TPM) philosophy, which attempts to achieve a continuous improvement in the maintenance application [16, 17]. This objective resides also in other methodologies such as total quality management (TQM) [18, 19] or even the ISO-9000 [5, 20] family that has been used in our case as a base for the reference framework. Definitely, and with the attempt to go beyond, a lot of organizations are recently applying a new management philosophy that can also have positive results and utility for the after-sales service. The Six Sigma is a methodology that integrates the human factor with improvement tools [21, 22], mainly statistic, so that it is possible to engage, as mentioned, complex mechanisms within the company [23]. The sigma is better known in statistics as a dispersion measure (variability) [24]. This variability is considered as the main negative factor of quality; thus, the Six Sigma proposes the following measures [25]:

- To select those indicators or variables that are critical for quality.
- To measure these variables so that real data are gathered from the current situation.
- To detect the variation source of the variables of interest and their admissible tolerances.
- To implement solutions.
- To apply statistic controls that allow the medium- and long-term improvements.

The application of the Six Sigma philosophy requires real data and the use of a great amount of variables [26]. Therefore, and as it has been highlighted previously, this methodology combined with new electronic information and communication technologies can throw some great results in the warranty management program and service field. The application of the Six Sigma requires a full-time dedicated team in such matters [27]. In the case of technical assistance to the customer, this team must consist of medium-level managers, independent from the departments of not just after-sales but engineering, manufacturing, or quality. That is to say, they must be constituted as an autonomous team that assures the resolution of issues related to customer reclamations. In short, the aim of the Six Sigma is directed through three main areas: customer satisfaction improvement [28], time reduction in the assistance cycle [23], and defect reduction [29]. To implement the Six Sigma [30, 31], a group of people is needed and should be organized as follows (refer to Fig. 10.5):

- *BLACK BELT* is the person 100 % in charge that deals with critical opportunities of change and getting the right results. It is almost an expert in solving problems through Six Sigma tools.
- *MASTER BLACK BELT* is the trainer, consultant, and monitor of the BLACK BELT. It is an expert in Six Sigma tools. It is also the responsible for an efficient performance on behalf of the groups or teams.
- *GREEN BELT* is trained in the Six Sigma methods. The difference between them and the BLACK BELT is that the GREEN BELT usually has another real job within the organization; thus, they only serve part-time as members or leaders of a team.
- *CHAMPION* This position is very common in the Six Sigma method, and normally, it is a high-rank executive the one who sponsors a GREEN BELT or a group. Generally, they are key managers in changing the direction of a company.
- *IMPLEMENTATION LEADER* The implementation leader is a professional in the field of business improvement.
- *DMAIC TEAMS* are in charge of quality processes as well as of solving problems of the company that are reflected in the customers and equally have repercussions in the company's capital.

DMAIC is the acronym in English that stands for "define, measure, analyze, improve, and control." It is designed so that teams are formed with specific tasks supervised by a BLACK BELT or a GREEN BELT. At the same time, these tasks

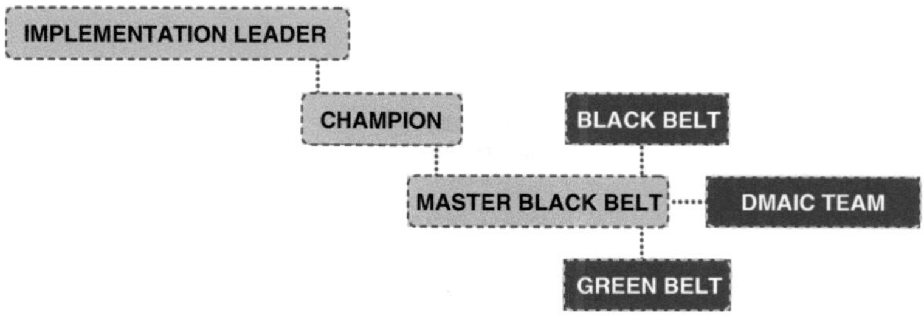

Fig. 10.5 Organization for the implementation of Six Sigma

are reviewed by a MASTER BLACK BELT, who is a CHAMPION and finally the IMPLEMENTATION LEADER. Each part of the acronym DMAIC entails the following actions [13, 14]:

- *DEFINE* There is a need to consider the program framework as a whole. Groups must define what the considerations of a problem are and how to visualize it.
- *MEASURE* It has two main objectives:
  - Gathering data to validate and quantify the problem.
  - Obtaining key indicators for problem resolution.
- *ANALYZE* Once the information is gathered, it must be analyzed, and the criteria for problem resolution must be set out.
- *IMPROVE* This is the stage where problems are identified and solved, adapting the necessary measures to prevent them from occurring again.
- *CONTROL* This is the part that is in charge of localizing defects and prevents them from occurring again.

10.3.2 Interaction of the Six Sigma with Other Tools and Methodologies

In terms of corporate management, Six Sigma is already a mature and solid methodology and, at the same time, versatile and applicable, designed to reach, maintain, and optimize results in the company. This methodology uses a set of statistical and management tools (Fig. 10.6), which when they are synergistically combined allows to reach significant short-term results [32]. The application of this methodology proves that the company adopting it improves incredibly the satisfaction of its customers and company value, eliminating all that is done but that does not increment value [28, 33]. In short, the implementation of the Six Sigma methodology in the corporate culture reduces costs and increases productivity and benefits, and the satisfaction of its customers [34, 35]. In general, all corporations that implement it deal with the same basic question: What is the flow of the process? Who are the actors? What are the objectives of the process? What

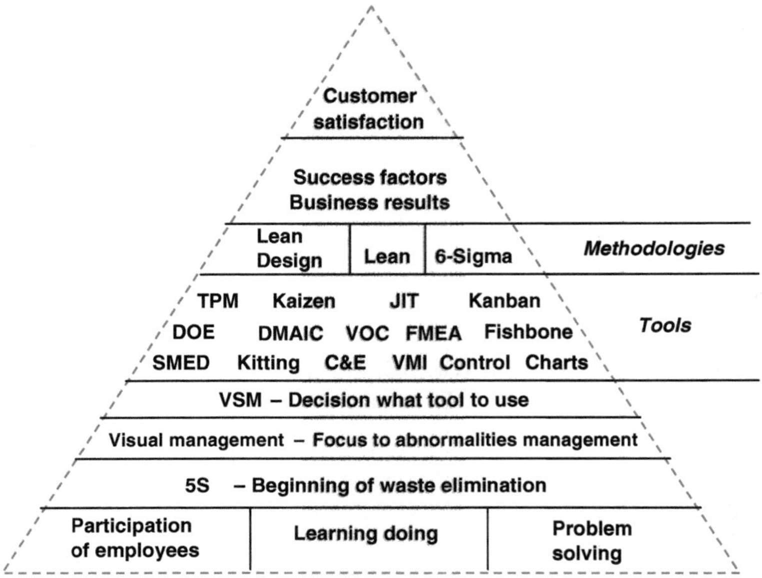

Fig. 10.6 Methodologies to improve customer satisfaction (adapted from [32])

does the customer expect from it? What are parameters used by the managers and the people responsible for controlling it?

The cultural aspect undoubtedly affects the way in which this methodology is implemented. Taking into account context and type of company implementing this methodology, certain tendencies and interaction combining the Six Sigma with other quality tools emerge. For instance, usually these combinations are to be found:

(a) Lean Six Sigma

The application of integrated *Lean* techniques to statistic tools is a solid fact. And it is gaining popularity. The combination properties of this technique allow an easier simulation of processes with the aid of computer information technologies [36, 37]. As it is shown in Fig. 10.7, Lean methodology makes it possible to eliminate activities that do not provide an added value, creating at the same time, a flow of processes. Basically, this technique is focused on reducing all sorts of waste such as

- Overproduction
- Waiting time
- Transport
- Excess of processing
- Inventory
- Movement
- Defects.

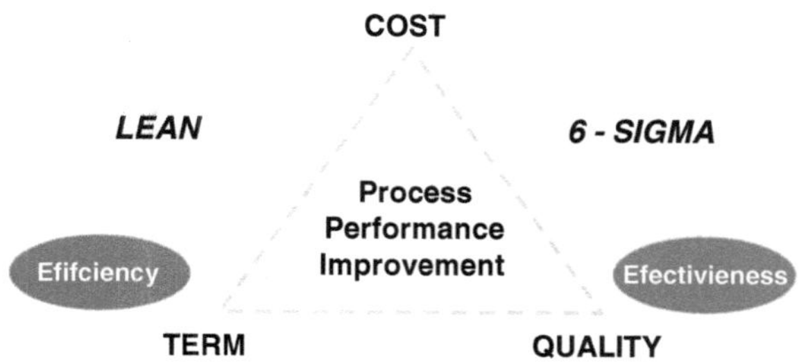

Fig. 10.7 Complementation between the objectives of Lean methodology and Six Sigma

These Lean tools include continuous processes of analysis (Kaizen), production "pull" (Kanban), and elements and failure-proof processes (Poka-Yoke).

(b) DFSS: Design for Six Sigma

Also, gaining popularity is the system known as define–measure–analyze–design–validate (DMADV). Techniques are incorporated to reach a higher creativity and innovation level, integrating as well TRIZ (Russian acronym for theory of inventive problem-solving) tools [38]. Basically, DFSS, or design for Six Sigma, it is usually defined as a rigorous focus for product design and services with the aim of reaching (and exceeding) customer expectations [39, 40]. Companies that apply Six Sigma usually find many defects that in reality are originated from the design process. This is the reason why DFSS makes it possible to redesign processes of manufacturing as well as initial capacities and assures that final products are produced with fewer defects through the existing technology. That is to say, DFSS integrates engineering and designing processes so that defects are eliminated before they can even occur. The integration of DMAIC principles and designing tools such as "initiate, design, execute, and sustain process" (IDEaS) produces as a result product designs that comply, always and from the beginning, with the Six Sigma norms. In short, the DFSS design stems from the Six Sigma technique and other quality methodologies related to DMAIC that were developed originally by Motorola in order to systematically improve processes through defect elimination [41, 42]. As opposed to Six Sigma/DMAIC predecessors, which are normally allocated to solve existing manufacturing problems, DFSS pretends to avoid such manufacture problems through the practice of a more proactive focus for problem-solving and to compromise the company's effort in an early stage so that potential problems can be reduced. The main objective of the DFSS is to achieve a significant reduction in the number of faulty units and production variations. It stems from the comprehension of customer expectations, carrying out critical evaluations of quality (CTQs), before the design is completed. Normally, only a small part of the CTQs are related to reliability in a DFSS program; therefore, reliability does not get the proper attention, considering that the DFSS program does not usually set its objectives to analyze long-term problems (after manufacture) that can arise from the product.

(c) DFR: Design for Reliability

The good practice philosophy can be summarized in three important principles that can be applied nowadays to successful companies [43, 44]:

1. Reliability must be designed in products and processes using the best scientific methods available.
2. It is important to know how to calculate reliability, but more importantly, to know how to achieve and maintain it.
3. The practice reliability must begin at a very early stage, in the design process, and must be integrated with the development cycle of the product in general.

Comparing it to the DFSS, the design for reliability is indeed a process specifically orientated toward getting high levels of reliability in the long term [45, 46]. This process tries to identify and prevent designing problems at a development stage, instead of finding these problems when the product is later in customer hands. For that, a wide variety of tools are used [47, 48]. These tools are different from the ones used in DFSS, although there is certain superposition.

Figure 10.8 illustrates the tools used in DFSS and DFR, as well as the overlap between them. As it can be seen in the graphic, the types of tools used in DFR are based on models related to the useful life of the product, the physical property of the failure, warranty predictions, etc. The common zone between DFSS and DFR includes such tools as voice of the customer (VOC), design of experiment (DOE), and the failure modes and effects analysis (FMEA) that are essential to any type of improvement program of a product. As a closing statement of this section, it is worth mentioning that besides the interactions of the Six Sigma with other tools and methods in both the quality field and continuous improvement (many of them referred to in this chapter), it can also be observed that some tendencies are affecting, for instance, the proper training of the staff responsible for the Six Sigma implementation. That is to say, nowadays there is a certain change in the Black Belt and Green Belt profiles, where each time there is a higher requirement of knowledge and training, as well as a minor individualist consideration of the projects. In another respect, there is a current search for certain specialization. This way, Black Belt and Green Belt specialists come forward according to the implantation sector. The following chapter suggests consulting next section where aspects of Six Sigma methodology in warranty management are dealt with, together with the focus on what was here elaborated.

10.4 Case Study on Six Sigma Methodology

This chapter briefly described the theoretical foundation upon which Six Sigma methodology is based. In this next section, this theoretical base is going to be applied in a succinct manner to a generic manufacturing plant in the automobile sector. It has been decided, for that purpose, to modify the starting data in order to

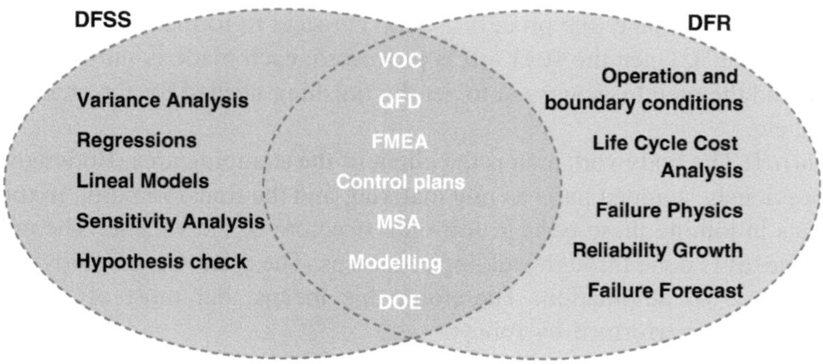

Fig. 10.8 Tools and methodologies used in DFSS and DFR

develop this application from an academic point of view, with an easy extrapolation to other contexts and with the sole purpose to emphasize the potential of the Six Sigma methodology. As it has been seen throughout this document, when a customer notices a fault in a product (in this case his/her vehicle), the situation can be considered as abnormal since it is expected that the customer's product (a vehicle) works perfectly. Nevertheless, these abnormal situations, which should be treated as warranty incidences, represent highly important percentages in the expenses incurred by automobile companies. Therefore, their study, analysis, and improvement belong to a justified field, as it has been observed in Fig. 2.1. We have seen that the improvement in reliability and quality of the product does not just have a favorable impact on the customer but also considerably reduce expected costs of warranty services [49]. However, in the automobile industry for instance, to consider a small ratio of customer claims might be merely a simplification. That ratio could be related to the fact that some companies reject those claims more effectively, encouraging and redirecting customers to the suppliers. This might result in an interesting line of research for future researches related to this particular field.

10.4.1 Application Scenario

The study scenario is based on a real automobile manufacturing plant, from where we have extracted data regarding the number of vehicles sold a month, the failure ratios these sales represent, the time it takes the technical service staff to react and repair under warranty failures, etc. In short, a typical assembly automobile plant has the following different areas or productive departments: stamping, bodywork, painting, and assembly. Such areas form the general sequence of the vehicle manufacturing. A brief description of each of them and their existing interactions is listed as follows:

- *Stamping*: This is the area in which all the metallic parts (of the bodywork or structure of the vehicle) are formed. The stamping area uses steel rolls as raw material, and these are cut in rectangular forms of different sizes depending on the use they will have (the process used to cut steel to form rectangles is known as “blanking”). Once the steel roll is processed, each blade is introduced into a press, and the metal is squeezed to get the finishing needed for the chassis, roof, doors, etc.

- *Bodywork*: The bodywork area is the client of the stamping area. Bodyworks use the previously stamped parts as raw material, and the transformation in this area consists in joining these parts to form the structure of the vehicle. The union of these pieces is done through welding processes. The plant we are studying has a high degree of automation. Therefore, this means that most of its welding operations are performed by robots.
- *Painting*: Once the bodywork of the vehicle is built, it goes to the area of painting where the unit is submerged in inks so anticorrosive materials can be applied. Then, the unit is painted and undergoes a dry-heating process. The finished product at this stage of the process consists in a painted shell which is later sent to the assembly area.
- *Assembly*: The final assembly area is the biggest and with the largest number of operators. In this area, the painted unit is received without any parts installed. The process in the assembly area normally consists of several hundreds of work stations (our study plant has 284 work stations in the assembly area) at which the operators take part in an assembly-line manufacturing process. The vehicle is first assembled with the installation of wires, sound insulators, modules, and sensors. After that, the carpets, instrument panels, and interior fabrics are installed. Finally, it is completed with the installation of the engine, transmission, suspension, and wheels.

Once the vehicle is assembled, it is delivered to the area of quality control where some functional tests are run. One of the most important functions, which come together with the transformation of the vehicle throughout all the areas or departments, is related to logistics of material. The modalities of material gathering to the production process vary depending on the characteristics of the operation and on the material itself, even though, generally, there is a need to minimize the inventories.

10.4.2 Set Out and Application of the Methodology

One of the biggest problems, in short, will be to reduce the number of incidences in/under warranty. To achieve this, there should be an improvement in the quality control during the productive process as well as in the efficiency and technical servicing performance in warranty, in the way that staff can deal quickly and effectively with an abnormal situation of systematic failures on sold vehicles. The application of the Six Sigma methodology in the after-sales service allows us to analyze the reliability of a product once is put on sale, improve it, and also boost the performance of the warranty service when dealing with failures. With this purpose, the Six Sigma methodology will be applied in our particular case, in order to reach these objectives using multidiscipline equipment, statistic tools, recorded data, etc., just as other previously mentioned criteria.

10.4.2.1 Roles and Processes of the Six Sigma Application

This section defines the strategy based on roles and responsibilities (functions) in which the personnel in charge of dealing with problems regarding warranty will be focused on. This is a key activity within a quality system to all those companies that, nowadays, pretend to be competitive in the market. Staff focused on this area of work should follow certain guidelines. The following is a description of the different guidelines or roles that are considered in our generic plant:

- *Role 1*: It is related to the total comprehension of customer problems, prioritizing them and adjusting quality inspections in the manufacturing process so that those problems can be prevented if possible (Fig. 10.9).

That is to say, the purpose of this role consists of understanding, in a very clear way, the difference between customer expectations and the current quality level of the vehicles, so that these concerns can be later translated into specific requirements to be adjusted in the manufacture process and/or the product design. This role includes two critical processes: first, definition of defects and priority dissatisfactions, which refer to these key activities: classifying of defects, internal indicator analysis, benchmarking, communication, contact with the client, etc. Secondly, the other critical process involved in this role is the early detection of failures, which consist of adjusting quality inspections in the factory, providing feedback to the specific process generating the problem, etc.

- *Role 2*: This role is focused on solving those problems that were previously prioritized (role 1). The Six Sigma methodology is of great help at this stage as it is a very efficient tool for solving problems in a systematic way. This role has as a concept to bring abnormal situations to an earlier stage of a product's life cycle. An abnormal situation would be when a customer detects a failure in his/her vehicle, whereas a normal situation would be to expect a perfect performance from the product. The real problem-solving issue lies in the ability to control these abnormal situations at the earliest stage possible and correct them as quickly as possible (see Fig. 10.10).

In other words, these abnormal situations should be stopped, for instance, at the assembly-line plant, or at the facilities of the specific supplier of the defective part; another option would be that the risk of the defect being foreseen during the design process, and from that moment, some actions should be taken in order to eliminate its possible occurrence. Therefore, this role plays a critical part in problem-solving, and for that purpose, the activities of this role will be both to prevent critical problems and the methodology to solve them.

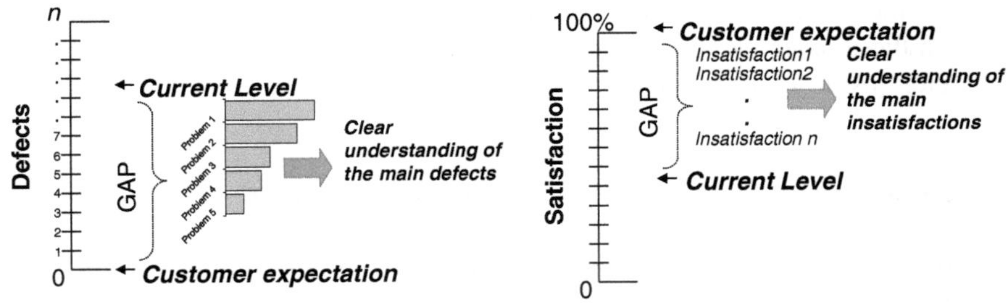

Fig. 10.9 Objectives of Role 1: to comprehend customer concerns and define priorities

Fig. 10.10 Objectives of Role 2: solution to non-evident problems

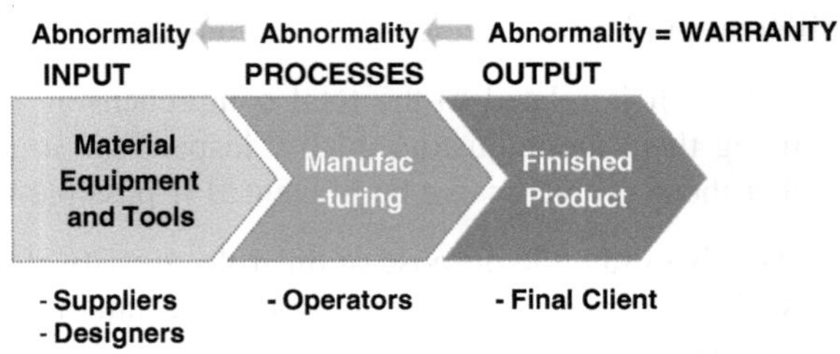

- *Role 3*: As problems are being solved (Role 2), causes generating these situations must be controlled in order to prevent the recurrence of the defect. This is a very valuable activity as it allows the organization to focus on the control of critical variables of quality and how to react when such critical variables overstep the parameters of control, avoiding the need to react when the effect of the uncontrolled variables reaches the customer. This is a role that prevents the generation of problems regarding quality (Fig. 10.11).

Therefore, the purpose of this role refers to the improvement in systems and processes as well as the control of critical quality characteristics (CTQ or critical to quality) and the selection of criteria (SC).

- *Role 4*: Ideally, the system should be able to solve problems regarding warranty (Role 2) and will have control over its critical quality variables and thus prevent the recurrence of such defects (Role 3). Role 4 is related to the improvements made from product design and the manufacturing process, so that the product is not likely to present any defects. This means that the product should be solid on its own design in a way that the probabilities of experiencing failures are minimized. At the same time, that product depends less on critical quality variables. The activities of this role could be considered as proactive as they are implemented back from the product designing stages, the prevention of defects based on the acquired experience of roles 1, 2, and 3 (Fig. 10.12). Therefore, the purpose would be the change in the product or process so a certain capacity is achieved, greater than the allowed, by the current design conditions. In this role,

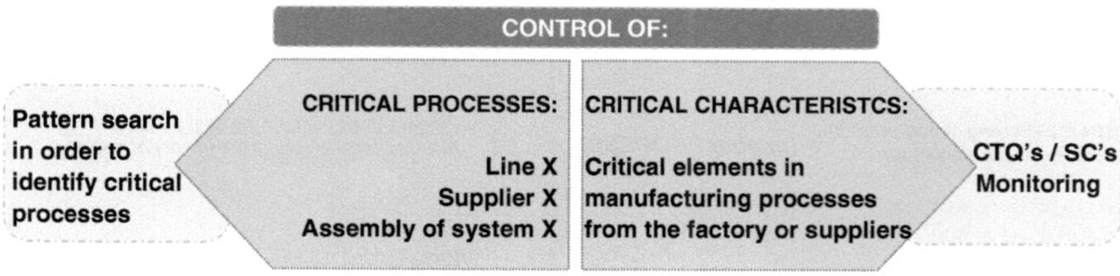

Fig. 10.11 Objectives of Role 3: process assurance and critical characteristics

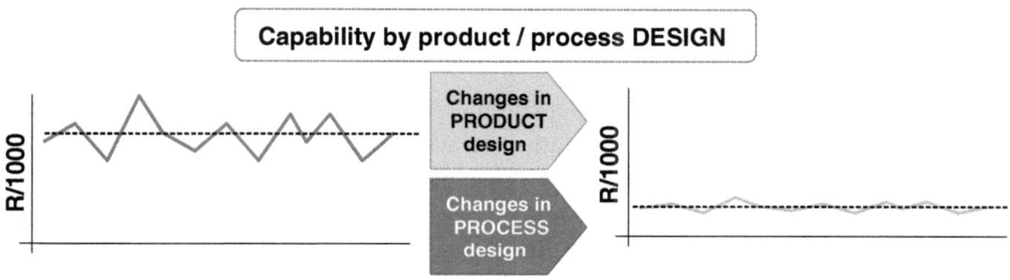

Fig. 10.12 Objectives of Role 4: improve the capacity for product/process design

we will have as critical processes, on the one hand, the alignment of the product characteristics with customer expectations. Thus, key activities would involve to analyze what aspects result in customer satisfaction, to detect problems at an initial stage of prototype testing, to encourage interaction between the responsible staff for warranty analysis and the designers of the product, to validate cost reductions, to participate in preproduction stages, etc.

On the other hand, the purpose of this role will be also to eliminate non-solid processes, so that its key activities would be both the follow-up and the validation of changes made to the product.

10.4.2.2 Variability Reduction Team (VRT) Activities

In the previous chapter, it was indicated that the personnel focused on the application of the Six Sigma methodology must work following determined guidelines or roles. Within those work, teams are those usually known as "variability reduction team leaders" or VRT leaders. As it will be later seen, the role of VRT leaders is used in the Six Sigma application mainly in Role 2, even though it also applies to the additional roles. This term could be interpreted as those responsible for such teams in charge of the variability reduction. Figure 10.13 shows the focus and responsibility that those VRT teams have within the roles described in the previous section.

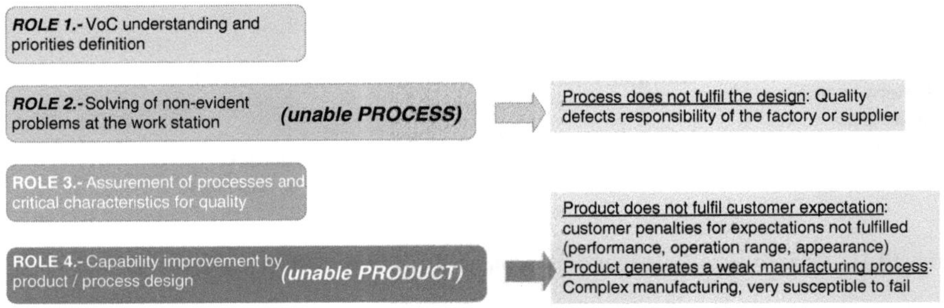

Fig. 10.13 Focus and responsibility of the VRT in respect to distinct roles

In short, these people have the primary responsibility of solving problems which can affect the company in relation to warranties reported by the final customer or by the automobile distributor. The definition of this role is a starting point in the Six Sigma methodology. Among these TRV activities, we can find the follow-up of customer requirements, the participation in the preproduction/planning phases, warranty result projections, product change validation, and improving weak processes derived from a deficient design.

10.4.3 Example and Results of the Application

Let us take into account a quality problem in which the customer starts to claim, through the after-sales service, that the front fender is detaching itself from the bumper of the vehicle. In respect of being a repetitive claim on behalf of diverse customers, the incidence acquires an urgent resolution. The reclamations from VOC customers regarding this situation indicate that the bumper does not remain in its place, specifically in the front part of the driver's side; it comes loose from its position to the left. Given the importance and the repetition of the incident, the same system shown in Fig. 10.10 is applied as a process map of resolution. This system is related to solving non-evident problems in the work station. With the aim of getting 0 defects, it is necessary to start analyzing the internal indicators which are reported, observing that evidently the component presents poor quality. This first study can facilitate to determine the moment when the problem first started (Fig. 10.14).

In this first analysis, it could also be observed that the assembly of this piece required a great deal of effort in contrast to other parts in the process. This elevated effort was confirmed in the work stations through studies that measured effort. Once these first studies were carried out, a study of each of the affected component parts was also performed, as well as a follow-up to the corresponding suppliers of these elements. Regarding the metrological testing of the pieces, it is important to do a dimensional evaluation, with a certain frequency if possible and mainly in

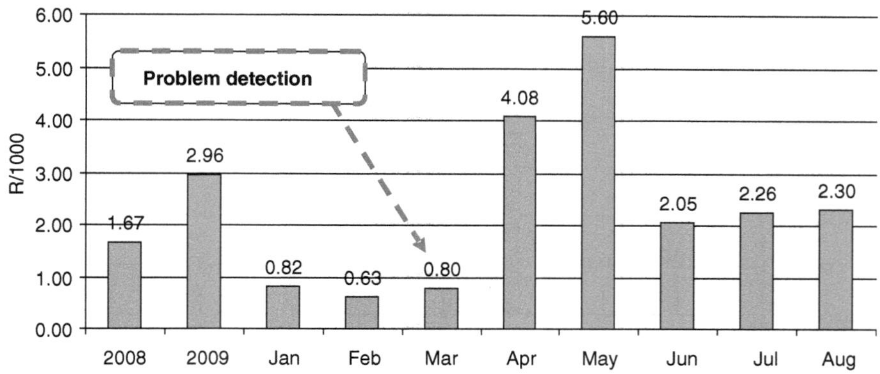

Fig. 10.14 Internal indicators related to the problem

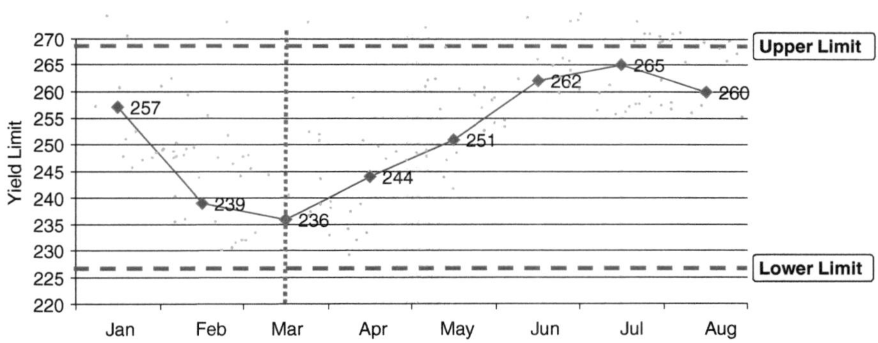

Fig. 10.15 Mean average of the limit of tensile strength of steel used

respect to the anchorage of the piece to the vehicle, so that the suitability of the design is verified. A critical point was observed throughout all this analysis during the stamping process relative to the tensile strength limit of steel used: reaching levels that surpassed the superior limit (Fig. 10.15). In particular, it was determined that the tensile strength of the steel was the cause for the displacement of 2.45 mm in the anchorage zone. An engineering change in the zone was proposed as a corrective action to solve the dimensional problem.

With regard to all this, it was detected that the bad anchorage of the piece was taking place during the assembly operations. Moreover, it was also found that the cut of the panel edge presented different conditions. The solution to this issue consisted in making an instruction sheet to correct the faults, concluding that the steel was in fact the critical input for the follow-up stamping process.

In the same way, special means of measurement were added and integrated to achieve the improvement in the current state of the component, thus consequently avoiding future customer reclamations. After implementing the improvements, we observed how important it is to carry out an abnormality analysis to detect the

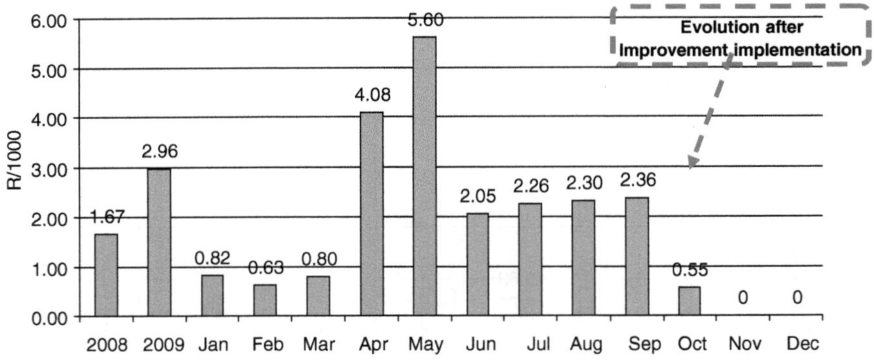

Fig. 10.16 Internal indicators after the improvement

anomaly [50]. In this specific case, an improvement of 70 % was obtained in relation to the effort measurements so that the operators did not need to beat the block to anchor it. Likewise, such improvement is also reflected when taking into account the internal indicators to that respect (Fig. 10.16), in comparison with those originally shown in the graphic in Fig. 10.13. Interesting references in this respect are [41, 42, 51, 52].

References

1. González-Prida V, Barberá L, Crespo Márquez A. 2010. Practical application of a RAMS analysis for the improvement of the warranty management. 1st IFAC Workshop on Advanced Maintenance Engineering, Services and Technology. Lisbon, Portugal.
2. Wang H. 2006. Warranty Cost Models Considering Imperfect Repair and Preventive Maintenance. Bell Labs Technical Journal. Lucent Technologies Inc. Published by Wiley Periodicals. 11(3):147-159.
3. Lawless JF. 1998. Statistical analysis of product warranty data, International Statistical Review. 66:41-60.
4. Murthy DNP, Blischke WR. 2006. Warranty management and product manufacture. Springer-Verlag London Limited. ISBN 1852339330.
5. Mihok L. 2004. Quality management systems – New approach by ISO 9000/2000. Metalurgija 43(3):193-198. ISSN 0543-5846.
6. International Std. 2004. Petroleum, petrochemical and natural gas industries - Collection and exchange of reliability and maintenance data for equipment. ISO/DIS 14224, ISO TC 67/SC/WG 4.
7. Parra C, Crespo Márquez A, Moreu P, Gómez J, González-Prida V. 2009. Non-homogeneous Poisson Process (NHPP), stochastic model applied to evaluate the economic impact of the failure in the Life Cycle Cost Analysis (LCCA). Taylor & Francis Group, London, ISBN 978-0-415-48513-5. Pp. 929-939.
8. United Nations Convention.1980. Contracts for the International Sale of Goods. Chapter II, Obligation of the seller.
9. Uniform Commercial Code. General obligation and construction of contract. Art. 2, Sales, Part 3.

10. Real Decreto Legislativo 1/2007, de 16 de noviembre. Ley General para la Defensa de los Consumidores y Usuarios y otras leyes complementarias. BOE núm. 287, de 30-11-2007, pp. 49181-49215.
11. Padmanabhan V. 1996. Extended Warranties. Chapter 18 in Product Warranty Handbook, W. R. Blischke and D. N. P. Murthy (eds), Marcel Dekker, New York, 439-452.
12. Padmanabhan V. 1995- Usage heterogeneity and extended warranties. Journal of Economics and Management Strategy. 4:33–53.
13. González-Prida V, Parra C, Gómez J, Crespo Márquez A. 2010. Reference framework proposal for the management of a warranty program. Euromaintenance. 20th International Industrial Maintenance Congress, from the 12th to the 14th of May in the Verona Conference Centre, Italy, pp 100-105.
14. Pyzdek T. 2003. The Six-Sigma Handbook. McGraw-Hill, 0-07-141596-3.
15. Pyzdek T. 2003. The Six Sigma Project Planner. A Step-by-Step Guide to Leading a Six Sigma Project Through DMAIC. McGraw-Hill, 0-07-142555-1.
16. Ahuja IPS, Khamba JS. 2008. Assessment of contributions of successful TPM initiatives towards competitive manufacturing. Journal of Quality in Maintenance Engineering. 14(4). Emerald.
17. Manzini R, Regattieri A, Pham H, Ferrari E. 2010. Maintenance for Industrial Systems. Springer Series in Reliability Engineering. ISBN: 978-1-84882-574-1.
18. Johnson RS. 1993. TQM: The Mechanics of Quality Processes. ASQ Quality Press.
19. Sheridan BM. 1993. Policy Deployment: The TQM Approach to Long-Range Planning. ASQ Quality Press.
20. European Committee for Standardization, Brussels. 2005. Quality management systems - Fundamentals and vocabulary. EN-ISO 9000: standards.
21. Kirkpatrick DL. 1996. The ASTD Training and Development Handbook: A Guide to Human Resources Development. Craig RL., Editor-in-chief, New York: McGraw-Hill.,Pp. 294-312.
22. Phillips JJ. 1996. Measuring the Results of Training. The ASTD Training and Development Handbook: A Guide to Human Resources Development. Craig RL. Editor-in-chief, New York: McGraw-Hill. Pp. 313-341.
23. Alwan LC, Roberts HV. 1989. Time Series Modeling for Statistical Process Control. Statistical Process Control in Automated Manufacturing, New York, Marcel Dekker.
24. Harrington HJ. 1992. Probability and Statistics. Quality Engineering Handbook. Quality Press, pp. 513-577.
25. Shainin D, Shainin PD. 1988. Statistical Process Control. Juran's Quality Control Handbook. 4th Edition. New York: McGraw-Hill.
26. Berry MJA, Linoff G. 1997. Data Mining Techniques for Marketing, Sales, and Customer Support. New York: John Wiley & Sons. Pp. 369-370.
27. Scholtes PR. 1988. The Team Handbook: How to Use Teams to Improve Quality. Madison, WI: Joiner Associates, Inc.
28. Hayes BE. 1992. Measuring Customer Satisfaction: Development and Use of Questionnaires. Milwaukee. ASQ Quality Press.
29. Filippini R, Forza C, Vinelli A. 1998. Sequences of operational improvements: some empirical evidence. International Journal of Operations & Production Management. Emerald. Pp. 195.
30. Johnson DW, Johnson FP. 1999. Joining Together: Group Theory and Group Skills. Prentice-Hall.
31. Hutton DW. 1994. The Change Agent's Handbook: A Survival Guide for Quality Improvement Champions. ASQ Quality Press.
32. Mizuno S. 1988. Management for Quality Improvement: The 7 New QC Tools. Cambridge, MA: Productivity Press.
33. Reichheld FF. 1996. The Loyalty Effect: The Hidden Force Behind Growth, Profits, and Lasting Value. New York: McGraw-Hill.
34. Ohno T. 1988. Toyota Production System: Beyond Large-Scale Production. Portland, OR: Productivity Press.

35. Pyzdek T. 1976. The Impact of Quality Cost Reduction on Profits". Quality Progress. Pp. 14-15.
36. Womack JP, Jones DT. 1996. Lean Thinking: Banish Waste and Create Wealth in Your Corporation. New York: Simon & Shuster.
37. Burcher P, Dupernex S, Relph G. 1996. The road to lean repetitive batch manufacturing: Modelling planning system performance. International Journal of Operations and Production Management. 16:210-220.
38. Stegmaier R, Viveros P, Nikulin C, González-Prida V, Barberá L, Crespo Márquez A. 2011. Aplicación e integración de procedimientos para la generación de soluciones innovadoras y sustentables, mediante la metodología rca y teoría inventiva TRIZ. Congreso MAPLA, Chile.
39. Kackar R. 1985. Off-line Quality Control, Parameter Design, and the Taguchi Method. Journal of Quality Technology. 17:176-188.
40. Meyers RH, Montgomery DC. 1995. Response Surface Methodology: Process and Product Optimization Using Designed Experiments. New York: John Wiley & Sons.
41. Taguchi G. 1986. Introduction to Quality Engineering: Designing Quality into Products and Processes. White Plains, NY: Quality Resources.
42. Akao Y. 1990. Quality Function Deployment: Integrating Customer Requirements into Product Design. Cambridge, MA: Productivity Press.
43. González-Prida V, Gómez JF, López M, Crespo Márquez A, Moreu P. 2009. Availability and reliability assessment of industrial complex systems: A practical view applied on a bioethanol plant simulation. Safety, Reliability and Risk Analysis: Theory, Methods and Applications – Martorell et al. (eds). Pp 687-695. Taylor & Francis Group, London, ISBN 978-0-415-48513-5.
44. González-Prida V, Barberá L, Crespo Márquez A. 2010. Practical application of a RAMS analysis for the improvement of the warranty management. 1st IFAC workshop on Advanced Maintenance Engineering Services and Technology, Lisbon.
45. Rose KH. 1995. A Performance Measurement Model. Quality Progress, February. Pp. 63-66.
46. Kristjanpoller F, Arata A, González-Prida V. 2010. Caso de estudio sobre el análisis de la fiabilidad en una planta de triturado de mineral en Chile. Revista Ingeniería y Gestión de Mantenimiento. Vol. Abril/Mayo/Junio 2010 España.
47. Parra C, Crespo Márquez A, Gómez J, González-Prida V, Arata A, Kristjanpoller F. 2010. Reliability stochastic model applied to evaluate the economic impact of the failure in the life cycle cost analysis (LCCA). Case of study in the oil industry. ESREL, Rhodes, Greece.
48. Barberá L, González-Prida V, Moreu P, Crespo Márquez A. 2010. Revisión de herramientas software para el análisis de la fiabilidad, disponibilidad, mantenibilidad y seguridad (RAMS) de equipos industriales. Revista Ingeniería y Gestión de Mantenimiento. Vol. Abril/Mayo/Junio 2010 España.
49. Chukova S, Hayakawa Y. 2004. Warranty cost analysis: non-renewing warranty with repair time. John Wiley & Sons, Ltd. Appl. Stochastic Models Bus. Ind. 20:59–71.
50. Liker J. 2004. The Toyota Way. New York: McGraw Hill. ISBN 0-07-139231-9.
51. Pyzdek T. 2003. The Six-Sigma Handbook. McGraw-Hill, 0-07-141596-3.
52. Pyzdek T. 2003. The Six Sigma Project Planner. A Step-by-Step Guide to Leading a Six Sigma Project Through DMAIC. McGraw-Hill, 0-07-142555-1.

Part V
Extensions to the Framework

Chapter 11
Intellectual Capital in the Technical Assistance

11.1 Background

Warranty assistance is characterized by a complex field that involves different disciplines: Management, Organization, Human Resources, Business Economics, Safety, and Environmental Management and, of course, a good knowledge of the product helpful for the customer. In another respect, it is an activity that can only be seen if it responds properly when there is a failure in the product, and therefore a reclamation issued by the user, which leads to a certain pressure when such claims take place. In these situations, it is necessary to make decisions rather quickly, making it often difficult to identify the best choice (cost reduction, low impact on production, urgency, and priority customer service quality, appropriate human resources, etc.). Eventually, the decision made is not always liked by the entire organization so, in case of any doubts, we must act with diligence and high professional ethics in order to face the responsibility and social impact that may arise from the decision adopted [1].

The customer service requires a special management (Fig. 11.1) due to the fact that, as a difference from other departments within the company, it deals directly with the customer and must be able to meet user's needs [2]. In customer-oriented organizations [3, 4], the customer service department is a key [5] for its contribution not only for customer satisfaction [6] through the assistance quality, but also enriches the entire organization in terms of service [7]. In this sense, knowledge management is critical because the significant amount of operation links with intensive exchange of information internally and externally (Fig. 11.2).

Basically, the need to update and share knowledge of technical assistance, whether for maintenance or warranty incidents [8–10], becomes a key issue. In order to extend this knowledge, activities such as coaching, mentoring, or on-job training [11, 12] are used. Sometimes these activities are not so easy to perform as many procedures and technical instructions refer to equipment under "laboratory conditions" and not under actual operating conditions. Obviously, when the product is launched into the market and distributed by different users, the effects on the physical assets are very diverse. Human capital is crucial for continuous

V. González-Prida Díaz and A. Crespo Márquez, *After-sales Service of Engineering Industrial Assets*, DOI: 10.1007/978-3-319-03710-3_11,

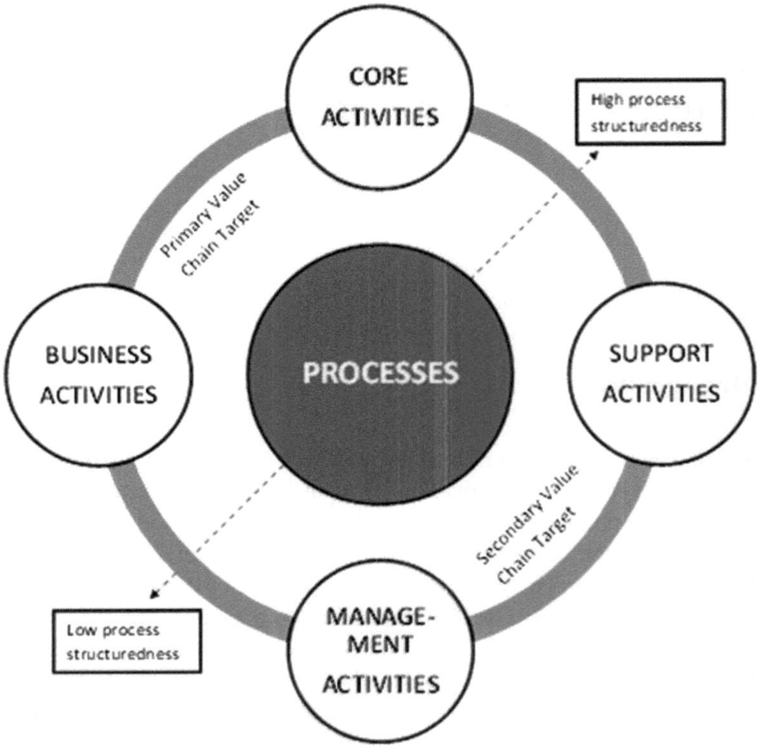

Fig. 11.1 Process topology

Fig. 11.2 Information flow among after-sales service and other entities

improvement and decision-making [13–15], since the main source for creating competitive advantage relies mainly on knowledge, how to use what is known and the ability to do new things [16]. For this reason, in order to improve customer service, human capital is one of the best forms of management, through knowledge, skills, attitudes, interests, and competence [17]. This must be assessed by technical aspects, staff flexibility to changes, adaptability, learning, teamwork, and so on, as well as the ability to work in diverse and interrelated tasks. Consequently, the assessment of the value generated by customer service is not an easy task and takes different perspectives into account. In order to quantify such an assessment, this section proposes a methodology to measure the advantages of after-sales service management in companies that have distributed their products in the market and considering not only the benefits, but also the reduction of risks in the reliability scope.

11.2 After-sales Service Contribution in a Company

Now, we are going to assess the impact of different areas in the company's customer service. In order to achieve this, this assessment will be based on a comparison of the maintenance department, where different authors [18–22] have summarized their contributions to that area. Basically, customer service is influenced as follows (Fig. 11.3):

- *Management and Organization*. Management is a mandatory subject, which consists of the ability to lead the involved activities and resources (skills, knowledge, responsibilities, etc.) to defined objectives, in order to facilitate their control and knowledge management.
- *Finance—Economy*. From the company's practical point of view, the customer service can incur significant company savings by reducing the consequences of failures that could lead to more serious incidents.
- *Business—Functionality*. Customer service is aimed at maintaining the product functions, once it is already in the hands of the client, in order to fulfill the expected functions satisfactorily.
- *Quality and customer relationships*. Customer service can increase its contribution to gaining customers and customer satisfaction and retention.
- *Impact on safety*. After-sales service plays an active and important role in protecting the internal and external resources. Now, there is also a growing interest in society for the protection of the environment. Thus, the objectives of customer service must also be the conservation of industrial assets in safety conditions, to the environment and people, in accordance with the general legal regulations on that matter.
- *Improvement and development*. Experience in technical assistance can improve products and services through continuous improvement of processes, always looking for the creation of value, and the ability to quickly react to changes.

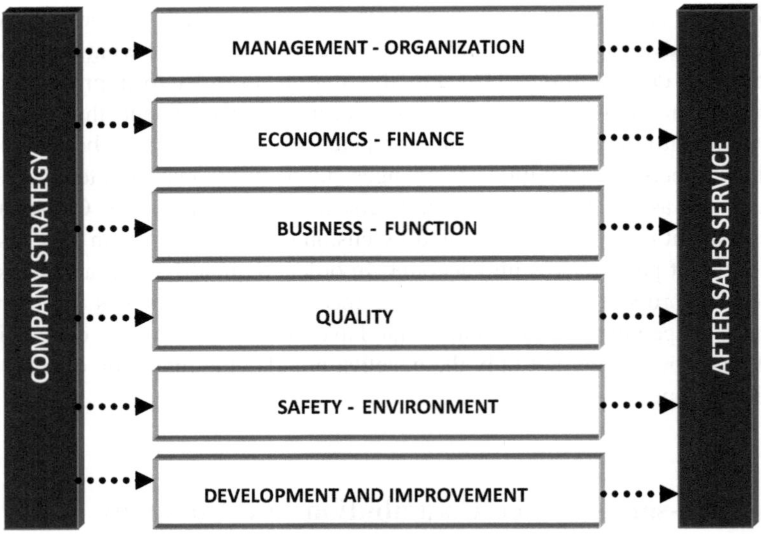

Fig. 11.3 Contribution between after-sales service and business strategy

In order to evaluate the contributions in all these areas, we need historical data and a consistent application of indicators [23]. In this way, indicators can be used as a support for decision-making and as a guide for the development and improvement of technical assistance. In after-sales services (mainly due to its high interaction with customers and other dispersed elements), the contribution of such a service will consider the measurement of performance on warranty activities, plus the impact analysis of claims. The contribution of customer service in a context of failure analysis cannot be considered from a deterministic point of view, due to the uncertainty in the occurrence of a failure, thus, it must be managed from a probabilistic approach [1, 24], based on the study of random variables as the MTBF and MTTR which are characterized by probability distribution functions [25]. The importance of the claims is not only relevant in terms of operational performance, but also in terms of costs [26] and the life cycle phases of industrial assets (Fig. 11.4).

The risk–cost model [26] makes the decision-making easier for a higher benefit (i.e., "lower cost," as defined in Chap. 7) with less impact. In general terms, post-sales costs (CP) are the sum of costs from planned activities (if any), corrective (of the claim itself), and other risk–costs arising in production, safety, reliability, and product quality that may affect customer service:

$$\begin{aligned} \mathrm{CP} = {} & C_{\mathrm{Planned}} + C_{\mathrm{Corrective}} + C_{\mathrm{Risk\_production}} \\ & + C_{\mathrm{Risk\_safety}} + C_{\mathrm{Risk\_reliability}} + C_{\mathrm{Risk\_quality}}. \end{aligned} \tag{11.1}$$

Summarizing, we could classify after-sales service costs based on their impact into two categories [20, 27, 28]:

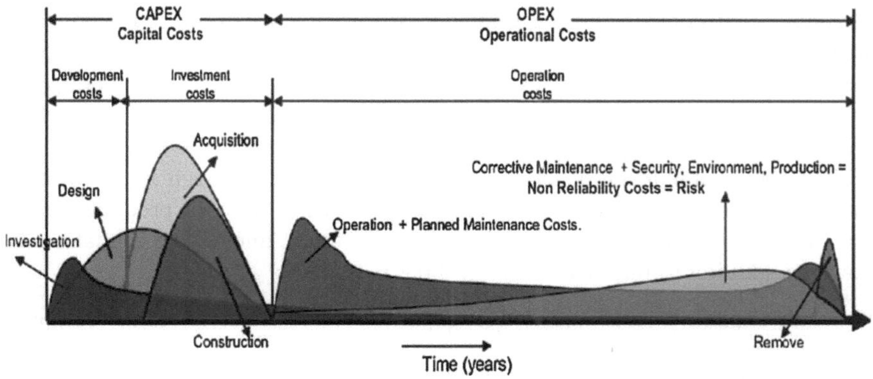

Fig. 11.4 Costs generated along the product life cycle

1. Direct costs such as planned and corrective costs.
2. Costs associated with risks in order to quantify the consequences due to the lack of reliability, environmental impact, production, and quality.

Risk costs can be expressed as:

$$\text{Risk } (R) = \sum \left(P_r \times C_{\text{Consequence}}\right) \quad \text{with} \quad i = 1, \ldots, n \tag{11.2}$$

where

- P_r is the probability that a complaint can be issued
- $C_{\text{Consequence}}$ is the resulting cost of a specific complaint
- i is the type of possible failures (1, 2,..., n).

In order to know all possible after-sales service costs, one should measure direct costs and the quantification of risks due to the lack of reliability. These types of costs can represent the actual and potential capacity for the customer service organization to produce benefits (i.e., lower costs) and savings for the company. In other words, we use these measurements to assess the intellectual capital of customer service.

11.3 The Technical Assistance as Intellectual Capital

As previously mentioned, a knowledgeable and experienced technical support staff is a strategic asset for the company and can be considered as a competitive advantage against other companies in the same sector. These aspects should be evaluated and for this purpose, the American Accounting Association (AAA) has defined a series of nonfinancial indicators to facilitate decision-making regarding the evaluation of human capital [29]. There are many studies on the valuation of knowledge that demonstrate the difficulty of this practice when it is applied to intangible issues that are complicated to measure and reproduce. Referring to the measurement of intellectual capital [30, 31], it is considered that knowledge generates a competitive advantage for the company [32] with the potential to produce future benefits. Intellectual capital has been defined in 1998 by Euroforum as all those intangible assets in an organization which, although have not been considered in traditional accounting, certainly have the effect of producing company value or to potentially produce it in the near future. Intellectual capital can be shown in addition to physical assets in three ways [33]:

- Human Capital, such as knowledge, experience, and creativity of the company employees.
- Structural Capital that considers the organization's ability to perform its function. That is, information resources, technological systems, industrial properties, franchises, corporate culture and the process efficiency, service quality, innovation capacity, etc.
- Relational Capital of the company with agents in its environment (customers, suppliers, government agencies, etc.), as well as the brand valuation, customer portfolio, relationships with them, etc.

According to this classification, the majority of valuation methods for intellectual capital [30–34], the sum of different types of capital applies as a reference equation, weighed by efficiency indexes. Accordingly and consistently with the calculation method developed by [32], the assessment of an organization's intellectual capital can be expressed as the sum of human capital, structural capital, and relational capital multiplied by efficiency ratios specificied for each type of capital:

$$\mathrm{CI} = i_H \cdot \mathrm{CH} + i_S \cdot \mathrm{CS} + i_R \cdot \mathrm{CR} \tag{11.3}$$

where

- CH, CS, and CR represent the economical value of human capital (CH), structural capital (CS), and relational capital (CR).
- i_H, i_S, and i_R, correspond to efficiency indexes for the intellectual capital. In case of more than one index for each type of capital, the arithmetic mean or the average of all of them is used: $[(i_1 + i_2 + \cdots + i_n)/n]$

11.4 Proposed Methodology

This section attempts to show, in a quantitative value, how the establishment of an after-sales service management can create value for the organization. In order to achieve this, we will quantitatively assess the effectiveness of technical assistance in the area of reliability.

11.4.1 Generated Value Regarding Staff

In relation to people and intellectual capital, the two main investments in human capital, from the accounting point of view, are investment in salaries and investment in education and training. Thus, the effectiveness of these investments can be represented by the following indicators:

- Investment in salaries ($\mathrm{CH}_{\mathrm{salary}}$):
  - Staff motivation ($i_{H\_\mathrm{motivation}}$). This indicator is the rate on personnel motivation.
  - Staff performance ($i_{H\_\mathrm{performance}}$). This indicator represents the performance of the technicians.
  - Task productivity ($i_{H\_\mathrm{productivity}}$). That means: [Task hours]/[total of hours]
- Investment in training ($\mathrm{CH}_{\mathrm{training}}$):
  - Staff percentage covered by training plans (i_{H_training}).
  - Staff percentage with skills to develop the assigned tasks (i_{H_skills}).
  - Relationship between staff that perform work orders according to the standard number of hours and the total work orders ($i_{H\_\mathrm{efficiency}}$).

Apart from this, from the point of view of reliability, people dealing with stressful circumstances, time limits, concerns, etc. can cause errors and failures with significant consequences for the business. In a few words, there is always a human error probability (HEP) [35, 36]. The contribution of human capital in the context of reliability must be quantified in order to know the real potential of the staff. As an example of the importance of human errors, there are studies such as the one developed by [37] in different sectors whose results concluded that at least between 20 and 30 % of failures in industrial assets were caused by a human error. Most studies [38, 39] on human reliability analysis (HRA), apply a probabilistic risk assessment. Summarizing, HEP can be used as an indicator, which will affect the amount of technical assistance so that the higher the HEP, the higher the number of errors that can take place and, consequently, the higher the costs of technical assistance. Multiplying $i_{H\_\mathrm{reliability}}$ by the average cost of a technical assistance (cost_a) and the number of all types of assistance (n_a) in the evaluation period, the contribution of this negative impact on intellectual capital is:

$$-i_{H\_\mathrm{reliability}} \cdot \cos t_a \cdot n_a = -\mathrm{HEP}_H \cdot \cos t_a \cdot n_a. \tag{11.4}$$

Therefore, human capital must be reduced in relation to human errors, because it causes extra corrective activities.

$$\begin{aligned} &\frac{\left(i_{H\_\mathrm{motivation}} + i_{H\_\mathrm{performace}} + i_{H\_\mathrm{producvity}}\right)}{3} \cdot \mathrm{CH}_{\mathrm{salary}} \\ &+ \frac{\left(i_{H\_\mathrm{training}} + i_{H\_\mathrm{skills}} + i_{H\_\mathrm{efficiency}}\right)}{3} \cdot \mathrm{CH}_{\mathrm{training}} \\ &- i_{H\_\mathrm{reliability}} \cdot \cos t_a \cdot n_a. \end{aligned} \tag{11.5}$$

11.4.2 Generated Value Regarding Quality

Regarding customers and intellectual relational capital, companies invest in customer service (CR_{client}) so that the efficiency of the investment can be represented by the following indicators:

- The rate of customer satisfaction due to the received assistance service ($i_{R\_satisfaction}$).
- The percentage of clients who are not affected by warranty incidents ($i_{R\_clients}$). That means: (1 − [sum of clients affected by failures/total sum of clients]).

From the reliability point of view, customer service contributes to customer perception in the way that, the more reliable a product is, the fewer claims from the customer there will be and, therefore, less need for technical assistance and a higher customer satisfaction. There are many methods for measuring quality based on assistance attributes, considering their importance and the company's contribution to providing such a value [40]. With this in mind, the measurement of assistance quality has to be carried out by a viable assessment with certain requirements, such as those related to the technical assistance itself, quick response time, assistance reliability and safety, and so forth. However, it is difficult to measure the quality perceived by the customer. The assistance influences on customer perception through the service quality, i.e., satisfaction increases when there is a high quality in the resolution of an incident and its quick treatment. We will limit the assistance effects on the quality of the service by two criteria:

- A poor assistance causes more damage than a good assistance in the customer's perception. Client's perception is diminished with a bad service quality and transmits negative product advertising to the market.
- It is essential to determine the level of incidents that require a certain level of service, according to the market as a standard and regarding to the agreements established with the clients.

Furthermore, the client's behavior may differentiate greatly from one client to another. For example, the customer can expect or demand either a monetary refund, or they can wait or take immediate legal action. In each situation, the client can stop buying our products and/or give negative publicity. In most cases, the client keeps a bad experience in mind to remind us when it is more convenient for him. Basically, each customer perceives service quality differently. It is possible to estimate the quality of assistance using statistical methods such as "Survival Data Analysis." This type of analysis examines a group of individuals and their reaction to an incident after a certain period of time [41–44]. Besides, the assistance quality can be measured by the compliance degree with those agreements on warranty service, delivery date, response time, recovery time, or accuracy on the level of service, etc. To achieve this, we consider that quality is accepted by the customer when it is within a tolerance level. This tolerance level can be defined by $P_a(t_R)$, which is the probability of customer loss or abandonment, depending on the repair time (T_R). Therefore, one can use the loss average probability in a particular sector, which can be estimated based on historical data of those industrial assets launched to the market for sale. There are different techniques to solve this type of analysis [45, 46]. In certain sectors using a variable called "net present value of the customer" (NVC), which correspond to the updated profit per customer. This would be equivalent to the total income obtained from a client during his relationship with the company, minus the direct costs from sales, acquisition, and retention (discounts). Therefore, the risk due to customer loss is calculated multiplying $i_{R\_\text{reliability}}$ by NVC, the average number of customers affected by incidents (n_c), and the average number of failures (n_f).

$$-i_{R\_\text{reliability}} \cdot \text{NVC} \cdot n_c \cdot n_f = P_a \cdot \text{NVC} \cdot n_c \cdot n_f. \tag{11.6}$$

Therefore, the relational capital must be reduced in accordance with the probability of clients that can abandon the service, as shown in the following formula:

$$\frac{\left(i_{R\_\text{satisfaction}} + i_{R\_\text{income}}\right)}{2} \cdot \text{CR}_{\text{clients}} - i_{R\_\text{reliability}} \cdot \text{NVC} \cdot n_c \cdot n_f. \tag{11.7}$$

11.4.3 Generated Value Regarding the Company Structure

Next, the main structural capital investments are from the accounting point of view: investment in standardization, prevention and performance assessment ($\text{CS}_{\text{process}}$); and investment in information and communication technologies (ICT) and innovation (CS_{ICT}). The effectiveness of these investments can then be represented by the following indicators:

- Investment in standardization, prevention, and performance assessment ($\text{CS}_{\text{process}}$).
  - Change in the planned activities (i_{S_planned}). That is, the relationship between planned assistance and total assistance.
  - The capacity rate used in relationship to the maximum capacity in support services (i_{S_capacity}).
- Investment in ICT and innovation (CS_{ICT}).
  - Percentage of staff that apply TICs ($i_{S\_\text{personnel\_ICT}}$).
  - Percentage of industrial assets covered by ICT as tools for monitoring or forecasting ($i_{S\_\text{equipment\_ICT}}$).

Once more, we focus on the value generated by the reliability, in this case, related to the structure of the company. The indicators here shown can be understood as a sign of the assets conservation. As in previous sections, we must assess the impact that reliability has on the company structure. For this purpose, a good indicator is the asset availability by the MTBF average in the critical components and the mean time to repair average (MTTR). Depending on the industrial sector, the indicator of equipment availability (A_e) affects the total revenue by the loss of customers. Therefore, the indicator on asset unavailability is multiplied by the NVC and the total number of clients (n_c). The contribution of this impact will be negative in the overall intellectual capital.

$$-i_{S\_\text{equipment\_availability}} \cdot \text{NVC} \cdot N_c = -(1 - A_e) \cdot \text{NVC} \cdot n_c. \tag{11.8}$$

Therefore, the structural capital must be reduced in accordance with the product's availability as can be seen in the following equation:

$$\begin{aligned} &\frac{\left(i_{S\_\text{planned}} + i_{S\_\text{capacity}}\right)}{2} \cdot \text{CS}_{\text{process}} \\ &+ \frac{\left(i_{S\_\text{personnel\_ICT}} + i_{S\_\text{equipment\_ICT}}\right)}{2} \cdot \text{CS}_{\text{ICT}} \\ &- i_{S\_\text{equipment\_availability}} \cdot \text{NVC} \cdot n_c. \end{aligned} \tag{11.9}$$

Finally, we should underline that this quantification of intellectual capital may be different depending on the department or sector of our concern. The methodology here proposed is based on the doctoral thesis indicated in Ref. [47]. Apart from this, as a continuation to Chap. 11, we suggest to consult the next section that deals with aspects regarding quantification of intellectual capital; as a complement to what was here developed and focused on.

11.5 Case Study on the Intellectual Capital in Technical Services

11.5.1 Scenario and Initial Considerations

This section looks at a case study whose context or scenario is similar to the one in Chap. 8 (that is to say, to a water distribution company). In this case, we intend to compare the service contribution in technical assistance in terms of intellectual capital of two companies that work on areas with similar characteristics. To that end, the described formula in Chap. 11 will be applied. It should be considered that the after-sales service needs a quantification of its maturity that would reflect the organization and their capacities. This situation, although it has been already mentioned in a previous case in Chap. 6, not only requires the assessment of its management but also the measurement of its efficiency in its applicability. In other words, the advantages of after-sales management should be evaluated as a quantitative value that reflects the produced improvements in reliability (in this case, a water distribution network) as well as the quality of the product (in this case, the water distribution service). In accordance with the obtained results, managers will

Table 11.1 International indicators indicadores internacionales

	Category	Benchmark
Yearly maintenance costs:	Total maintenance cost/Total manufacturing cost	<10–15 %
	Maintenance cost/Replacement asset value of the plant	<3 %
	Hourly maintenance workers as a % of total	15 %
Planned maintenance:	Planned maintenance/Total maintenance	>85 %
	Scheduled maintenance as a % of hours worked	~85–95 %
	Unplanned down time	~0 %
	Reactive maintenance	<15 %
	Run-to-fail (emergency + non-emergency)	<10 %
Maintenance overtime:	Maintenance overtime/Total company overtime	<5 %
Monthly maintenance rework:	Work orders reworked/Total work orders	~0 %
Inventory turns:	Turns ration of spare parts	>2–3
Training	For at least 90 % of workers, hours/Year	>80 h/year
	Spending on worker training (% of Payroll)	~4 %
Safety performance:	OSHA Recordable Injuries per 200,000 labor hours	<2
	Housekeeping	~96 %
Monthly maintenance strategies:	Total hours PM/Total maintenance hours available	~20 %
	Total hours PDM/Total maintenance hours available	~50 %
	Total hours PRM/Total maintenance hours available	~20 %
	Total hours REM/Total maintenance hours available	~2 %
	Total hours RNEM/Total maintenance hours	~8 %
	Available time/Maximum available time	>97 %
Contractors	Contractors cost/Total maintenance cost	35–64 %

be able to know a real situation and make decisions in an easier way, given that the information regarding quality costs helps to justify the investment in quality improvement, helping to control at the same time the efficiency of the efforts made [48]. In short, the aim is to show using quantitative values, that the investment in intellectual capital creates value for the organization. Moreover, it is important to highlight that human reliability is the most important factor when explaining and predicting the company's economic growth [49]. As it has been previously mentioned, the proposed methodology will allow us to measure the evolution of the norms and regulations as well as the comparisons with other competitors in the field [50].

External references are used as objectives in order to determine maturity [51, 52]. Table 11.1 shows some of the indicators evaluated in the study regarding different companies, detailed in Ref. [53]. The different evaluation methods of intellectual capital employ reference indicators as personnel motivation, customer satisfaction, personnel performance, rate of training, degree of TICs establishments, non-planned activities, efficiency in the use of resources, etc.

In this direction, there is a need to measure not only the consequences of negative impacts carried out by a technical service, but also the consequences of positive impacts that result from good service, particularly in the most relevant categories of intellectual capital:

- The human capital or related to personnel through personnel and human reliability investments.
- The relational capital referring to quality through investments in quality and consequences of customer satisfaction.
- The structural capital or company structure, through structural investments and equipment preservation.

As a consequence, the indicators of distribution companies must be linked to network reliability, availability, re-establishment time, etc., and its application should be through a quality service that increases customer satisfaction.

11.5.2 Results of the Practical Application

The contributions to after-sales services and their impacts will be analyzed, as it has been previously mentioned, in two water distribution companies that operate in the same distribution area, and where it is attempted to compare both of them in a determined period of time.

Table 11.2 shows as a starting point the applied investments and the main key indicators of performance that both companies present. With this data, the intellectual capital is calculated in order to compare companies A and B, as we can see the investments of both companies are the same 900,000 €/year. However, we now are going to apply the formulas developed in Chap. 11, obtaining the results that are shown in Table 11.3. Here, if we only take into account the capital formulas, the intellectual capital value is less than the investments of both companies, 695.417 in company A and 777.417 in company B.

Likewise, the following details can be distinguished:

- There is more structural capital in company B than in A due to the ICTs inversion.
- There is more relational capital in company B than in A due to the fact that customer satisfaction is higher given that fewer customers are affected by the failures.
- The human capital is the same for both although company A is more efficient due to the motivation factor.

In addition, the impact of reliability is also assessed; these values will reduce the intellectual capital in terms of possible risks in the network to the results of Table 11.4.

Table 11.2 Applied inversions and main KPI's

Investments	*Company A*	*Company B*
(a) Salaries	300,000	250,000
(b) Investment in learning and training	50,000	100,000
(c) Investment in customer attention	100,000	100,000
(d) Investment in quality and processes	300,000	250,000
(e) Investment in ICT and innovation	150,000	200,000
Total	900,000	900,000
KPI	*Company A*	*Company B*
1. Personnel motivation	90 %	80 %
2. Personnel performance	75 %	85 %
3. Productivity of labor	90 %	90 %
4. Personnel with learning	85 %	90 %
5. Fulfillment of personnel competences	95 %	90 %
6. Fulfillment of efficiency versus standards	85 %	80 %
7. Customer satisfaction	75 %	85 %
8. Customer incomes affected by failures	80 %	95 %
9. Planning	60 %	80 %
10. Capacity	75 %	85 %
11. Personnel with ICT	75 %	95 %
12. Equipments with ICT and optimization	80 %	80 %
13. HEP	0.25	0.28
14. Mean cost of activity (€)	150	150
15. Number of activities (n_a)	589	589
16. P_a (Re-establishment time)	0.0013	0.0005
17. CNPV	3,000	3,000
18. Number of affected customers (n_c)	200	200
19. Number of failures (n_f)	30	30
20. Availability	85 %	95 %
21. Unavailability	15 %	5 %

Table 11.3 Application of the formulas to assess intellectual capital

Capitals equations	Company A	Company B
a × average(1 + 2 + 3)	255,000	212,500
b × average(4 + 5 + 6)	44,167	86,667
c × average(7 + 8)	77,500	90,000
d × average(9 + 10)	202,500	206,250
e × average(11 + 12)	116,250	175,000
CI_human (a × (1 + 2 + 3)/3 + b × (4 + 5 + 6)/3)	299,167	299,167
CI_relational (c × (7 + 8)/2)	77,500	90,000
CI_structural (d × (9 + 10)/2 + e × (11 + 12)/2)	318,750	381,250
Total	695,417	770,417

Table 11.4 Negative impact when applying reliability

Reliability impacts	Company A	Company B
Impact of human reliability (13 × 14 × 15)	−22,088	−24,738
Impact on customers (16 × 17 × 18 × 19)	−23,400	−9,000
Impact on production (21 × 17 × 18)	−90,000	−30,000
Total	−135,488	−63,738

Table 11.5 Result of intellectual capital and reliability impact

Summary	Company A	Company B
CI_human-impact of human reliability	277,079	274,429
CI_relational-impact on customers	54,100	81,000
CI_structural-impact on production	228,750	351,250
Total	559,929	706,679

In other words:

- Taking into account the impact of human errors and, supposing that the average cost of an activity and the number of activities is the same for both, the negative contribution by the impact of human reliability is more in company B, due to the fact that the HEP is higher in B than in A.
- If the re-establishment time is different in both companies but the CNVP is the same, the impact in terms of costs is more in company A, considering that the probability for abandonment is higher.
- Finally, the impact on equipment reliability is also reduced. If the type of equipment supports 200 customers and the CNVP is of 3,000 €/year the negative contribution is higher in A than in B.

In short, as it is observed in Table 11.5, the difference now between both companies' increases. The value of company B is greater than A, mainly due to a lower level of risks (human error and network failures) in spite of, as we have seen, starting from an equal budget for both companies. This practical case is supported also in this doctoral thesis [47].

References

1. Crespo Márquez A, Moreu P, Sanchez A. 2004. Ingeniería de Mantenimiento. Técnicas y Métodos de Aplicación a la Fase Operativa de los Equipos. Aenor, España.
2. Earl MJ. 1994. The New and the Old of Business Process Redesign. Journal of Strategic Information Systems. Vol. 3.
3. Porter M. 1980. Competitive Strategy: Techniques for Analyzing Industries and Competitors. Free Press.
4. Ishikawa K. 1985. What is Total Quality Control? Prentice-Hall Inc. Englewood Cliffs, NJ.

5. Murthy DNP, Atrens A, Eccleston JA. 2002. Strategic maintenance management. Journal of Quality in Maintenance Engineering. 8(4):287-305.
6. Zhu G, Gelders L, Pintelon L. 2002. Object/objective-oriented maintenance management. Journal of quality in maintenance engineering.
7. Küssel R, Liestmann V, Spiess M, Stich V. 2000. ''Teleservice'' a customer oriented and efficient service. Journal of Material Processing Technology. 107:363–371.
8. Wireman T. 1991. Total Productive Maintenance. Industrial Press.
9. Nakajima S. 1992. Introduccion al TPM (Mantenimiento Productivo Total). Productivity Press.
10. Nonaka y Takeuchi. 1995. The Knowledge -Creating Company. USA: Oxford University Press.
11. Swanson L. 1997. An empirical study of the relationship between production technology and maintenance management. Elselvier Int. J. Production Economics 53:191-207.
12. Pintelon LM, Gelders LF. 1992. Maintenance Management Decision Making. European Journal of Operational Research. 58(3):301-317.
13. Fowler A. 1987. When chief executives Discovery HRM. Personnel Management.
14. Edwards JD. 1987. Improved utilization of maintenance manpower. Maintenance Management International. 6:285 -293.
15. Deming WE. 1989. Calidad, productividad y competitividad: la salida de la crisis. Madrid, Ediciones Díaz de Santos.
16. Prusak L. 1996. The Knowledge Advantage. Strategy & Leadership.
17. Smith AF, Kelly T. 1997. El capital humano en la economía digital. Ediciones Deusto.
18. Campbell JD, Jardine A. 2001. Maintenance excellence. New York: Marcel Dekker.
19. Pintelon L, Van Wassenhove L. 1990. A Maintenance Management Tool. Omega, 18(1):59-70.
20. Moubray J. 1997. Reliability-centered Maintenance. Industrial Press.
21. Deardeen J. Lilien G, Yoon E. 1999. Marketing and Production Capacity Strategy for Non-Differentiated Products: Winning And Losing At The Capacity Cycle Game". International Journal Of Research In Marketing.
22. Weule H. 1993. Life cycle analysis a strategic element for future products and manufacturing technologies. Annals of the CIRP. 42(1):181-184.
23. Kaplan RS, David PN. 1996. The Balanced Scorecard: Translating Strategy Into Action. Boston, MA: Harvard Business School Press.
24. Avizienis A, Laprie JC, Randell B. 2001. Fundamental Concepts of Dependability. Research Report No 1145, LAAS-CNRS, April.
25. Greves D, Schreiber B. 1993. Engineering costing techniques in ESA. http://esa.com/pub.esrin.esa.it.point/bullet/greves.html.
26. Woodhouse J. 1993. Managing Industrial Risk. Chapman Hill Inc, London.
27. Yañez M, Joglar F, Modarres M. 2002. Generalized renewal process for analysis of repairable systems with limited failure experience. Reliability Engineering and System Safety. 77:167–180.
28. Nowlan SF, Heap HF. 1978. Reliability-centred Maintenance. National Technical Information Service, US Department of Commerce, and Springfield Virginia.
29. American Accounting Association. 1977. Report of the Committee on Human Resource Accounting.
30. Stewart T. 1999. Intellectual Capital. Doubleday, N. York.
31. Stewart T. 2001. The Wealth of Knowledge: Intellectual Capital and the Twenty-First Century Organization. Nicholas Brealey.
32. Edvinsson L, Malone M. 1997. Intellectual Capital - Realizing your company's true value by finding its hidden brainpower. Harper Collins New York.
33. INTELECT. 1998. Medici6n del Capital Intelectual: Modelo Intelect. Euroforum. Madrid.
34. Riesco GM. 2007. El negocio es el conocimiento. Ediciones Díaz de Santos.
35. Bley D. 2002. New methods for Human Reliability Analysis. Environmental Management & Health. 13(3):277-289.

36. Hollnagel E. 2006. Human Reliability Analysis. International encyclopaedia of ergonomics and human factors. Karwowski Waldeman. Taylor and Francis Group. Pp. 753-757.
37. Dhillon BS. 1989. Human errors: a review. Microelectron. Reliability. 29(3):299-304.
38. Dougherty EM. 1990. Human reliability analysis - Where shouldst thou turn? Reliability Engineering and System Safety. 29(3):283-299.
39. Swain AD, Guttman HE. 1983. Handbook of Human Reliability Analysis with Emphasis on Nuclear Power Plant Applications. NUREG/CR-1278, USREG, Washington, DC.
40. Johnson RL, Tsiros M, Lancioni RA. 1995. Measuring Service Quality: A System Approach. Journal of Services Marketing. 9(5):6-19.
41. Parmar MKB, Machin D. 1996. Survival analysis: a practical approach. Chichester: Wiley.
42. Cox DR, Oakes D. 1984. Analysis of Survival Data. London; Chapman and Hall.
43. Kleinbaum DG. 1996. Statistics in the health sciences: Survival analysis. New York: Springer-Verlag.
44. Klein J, Moeschberguer M. 1997. Survival Analysis. Techniques for Censored and Truncated Data. Springer–Verlag. New York Inc.
45. Blischke WR, Murthy DNP. 2000. Reliability Modelling, Prediction and Optimization. Wiley.
46. Harrell FE. 2001. Regression Modelling strategies. Springer.
47. Gómez JF. 2010. Marco de referencia para la gestión y evaluación del mantenimiento de redes de distribución de servicios. Tesis doctoral dirigida por Adolfo Crespo Márquez. Universidad de Sevilla.
48. Dale y Plunkett. 1991. Quality Costing. Chapman Hall.
49. Becker GS. 1964. Human Capital. Columbia University Press, National Bureau of Economic Research N° 80.
50. Svantesson T. 2008. EFNMS-SMRP Maintenance and Reliability Indicator Harmonisation Project. Euromaintenance. Brussels.
51. Campbell JD. 1995. Outsourcing in maintenance management: A valid alternative to self-provision. Journal of Quality in Maintenance Engineering. 1(3):18-24.
52. Tsang A. 2002. Strategic Dimensions of Maintenance Management. JQME. 8(1):7.
53. Mitchell JS. 2007. Physical Asset Management Handbook. Fourth Edition. Clarion Technical Publishers, Houston, 440 pages, ISBN 0-9717945-4-5.

Chapter 12
Maintainability and the After-sales Service

The purpose of this chapter is to describe a process to obtain a maintainability index for industrial assets that, particularly, have been launched to the market and need to be technically assisted by the customer service. The concept of maintainability considered here is the one defined in the European Standard UNE-EN 13306:2010, which indicates that the item (element or device) maintainability is its ability to be maintained or restored to a state where it can perform a required function, considering that the provided technical assistance is carried out under specific conditions and using established procedures and means [1]. Maintainability is a key contributor to the dependability of an industrial device. The term dependability expresses a general concept, not quantitative, which encompasses all those properties used to describe the device availability and its conditioning factors: reliability, maintainability, and technical assistance support. Maintainability is therefore an aptitude for industrial devices related to their maintenance, which will be evaluated under certain conditions of use. This is a key aspect in this section, which requires previously to the evaluation process, the description of the device conditions of use for an average end user. These conditions not only have to do with the way or manner in which the equipment is operated, but also its disposition where it is located (a process plant, a defense system, an aerospace craft, etc.) and the environmental conditions where it performs the required function. In order to complete the proposed analysis, it is also necessary to know in detail the structural characteristics, the device assembly and installation, as well as the maintenance routines. This last section will clearly identify the different levels of technical assistance, the actions in each one, the procedures for carrying out these actions, etc. As a consequence of the foregoing paragraph, and in order to obtain comparable measures for the maintainability levels, training of the evaluators will be another key consideration and will be required for a successful implementation. Then, it will be established as a requirement to fulfill by the evaluators of maintainability indexes, the accreditation of certain levels of technical training (in general terms), and technical assistance and maintainability training (in particular terms). For one device, the proposed maintainability indexes will be then assessed under different conditions of use, even at different stages of its life cycle, always with different purposes. Thus, during the preparation phase of

V. González-Prida Díaz and A. Crespo Márquez, *After-sales Service of Engineering Industrial Assets*, DOI: 10.1007/978-3-319-03710-3_12,

the device, the evaluation of maintainability index can help and improve the design in order to facilitate subsequent warranty and maintenance services, regardless of environmental conditions in which the device operates finally. Consequently, in its operational phase (and therefore during the period specified under warranty assistance), in case of extensions to the technical assistance, the evaluation of maintainability index may be made periodically in order to measure how easy is technical assistance under specific operating conditions and environment, which may often be changing over time. When the maintainability index assessment is made during the operation phase of the device, it is important to remark that the thing to evaluate corresponds to the skills of the device, not the organization responsible for the technical assistance. For example, when the required training for the staff of technical support is mentioned as maintainability attribute, it refers to the qualification that must have the staff that is responsible for performing the actions at a certain technical level. It does not assess the training level of the existing staff. This nuance is essential for a proper interpretation and use of this chapter.

12.1 Brief Review of the State of the Art

12.1.1 Introduction

It is not the aim of this section to present a comprehensive and exhaustive study of the methods proposed for measuring the technical assistance performance, but to offer a quick overview of the main approaches found in the technical literature dealing with this issue. The different contributions can be grouped in two main basic approaches, according to the source of the input information and the nature of it:

1. *Expert methods*: Those ones based on the use of the experience and knowledge of maintenance experts combined with historical data and objective assessments.
2. *Statistical methods*: Those ones based on the application of historical data recorded from trials.

12.1.2 Summary of the Reviewed Bibliography

In our review to the literature, the most referred contributions that we could find in this area have been the followings:

- Eldon [2] uses MTTR as maintainability indicator (involving, in that MTTR, all the times related with the repair, from the fault location, to the final checking after repair and setup again). This MTTR is calculated throughout statistical inference, using historical data. The MTTR of a complex device can be calculated by combining the MTTR, individually determined for each of its components.

- Houshyar [3] presents a software tool for the operators of a device to record data from the device in order to later calculate the performance and the maintainability, based on that information.
- Sols [4] presents some indicators that measure different aspects of technical assistance operations relating to maintainability, but not giving an overview. All these indicators are derived through statistical calculations on historical data. For example, the probability that the repair time is below a certain time for certain operations (lognormal distribution), the average repair time to get up to 90 % of the functions of equipment, man hours used to relief efforts, the average cost of technical assistance, etc. All statistical parameters calculated are related to maintainability and can give an idea about this index.
- Wani [5], in Tribology (for mechanical devices), assesses the tribo-maintainability; that is, the maintainability from the tribology point of view. The tribology features are assessed by a table in which qualitative assessment corresponds to a numerical value given by experts. Using those values, the tribo-maintainability is obtained throughout an algorithm.
- Chen [6] the vector projection method (VPM) is based on a multi-objective evaluation and applies a qualitative assessment with weighted method in order to calculate the value, depending on several factors on which depends maintainability. The factors are grouped according to the physical design (simplicity, accessibility, assembly, etc.), logistical support (tools, documentation, etc.), and ergonomics (failure and operating indicators, skills of technical staff, etc.). A weight is assigned to each factor, depending on their importance.
- Smith [7], discusses the empirical U.S. Military Handbook 472, whose proposal presents the mean time to repair as an important indicator for maintainability. There are two main methods to calculate the MTTR. In the first one, the MTTR is calculated from the average repair time for each failure. In the second method, we measure a number of maintainability factors that result in MTTR. To perform this second procedure, each failure is studied with a checklist composed of 32 factors, 15 of them correspond to design factors, 7 to logistical support, 10 to the personnel qualification, and other related topics. An expert assesses the state of the requirements for each factor and, finally, with the points obtained in each group of factors, it is possible a measurement of mean time to repair based on a regression of repair time.
- Coulibaly [8] defines two main and general indicators for maintainability (one taking into account the components criticality of the device under study, and the other one without this consideration). These indicators include access and assembly time for all components. There are other important indicators for assessing diverse aspects that affect maintainability (MTTR and its component of time, etc.). All calculations are based on statistical data or simulation (for each new item without historical data).
- Guo [9] calculates MTTR from historical data and validate statistically the result. Although MTTR is related to maintainability, the author makes no considerations about that.

- Ben Daya [10] considers maintainability from a statistical point of view. As in other researches, he determined certain factors that influence the maintainability of industrial assets. These factors may relate to the skills requirements of staff working conditions as well as working environment. Maintainability indicators are calculated statistically, as the maintainability function, technical assistance durability, etc.
- Liu [11] mainly considers the maintainability dependency on the human factors required for maintaining the device, in order to be taken into account in the device designing process. The author groups those factors in four main categories: maintenance operations safety, accessibility, comfort, and practicability.
- Lu [12] presents an expert-based assessment procedure. The expert assesses a set of factors, from which, the author considers, the maintainability is depending on (such as identification, ergonomics, testability, simplicity, etc., and other experience evaluated factors). The evaluation of each factor is made by verbal scale which is translated into numbers by fuzzy logic.
- Abas [13] determines repair time using statistical methods (applying *Software Relex*) in a case study. The calculation also considers some environmental and operational conditions together with repair time data, with the aim of describing in a better way maintainability with such indicators based on statistics.

12.1.3 Conclusions of the Literature Review

Basically, we can consider that there are two main streams to measure the maintainability or the technical assistance performance regarding a system or equipment: either from the internal maintenance or from the customer service point of view. On one side, there are statistical approaches based on the probability that a technical or maintenance task can be performed at any given time. On the other hand, the second group of methods is the expert ones, based on observation to relief efforts and other factors related to the actual performance of after-sales service or maintenance department. These factors can be, for example, the environment, logistics, personnel training, etc. Technical assistance can be evaluated by assessing all these aspects and the convenience of combining them. In other cases, some authors apply graph theory or mathematical models, although they are not too extended, due to the complexity of these methods. The method presented in this section can be classified within the expert-based approach. By its use, a comprehensive assessment and updated maintainability can be achieved, in a way that is very practical, easy to understand, and to apply.

12.2 Definitions and Requirements for the Assessment

Just as it has been previously mentioned, maintainability is the aptitude of an element under certain given conditions of use, to keep in, or returned to, a state in which it can perform all required functions when the technical assistance is executed under determined conditions and using preestablished resources and procedures. The maintainability index is used to measure the maintainability of an item. This technique evaluates the general maintainability of a component or the maintainability of a component at a determined level of technical assistance. Such level of technical assistance will be therefore a set og actions (in our case) of after-sales service to carry out a specified intervention, which will correspond to the subdivision of an element from the warranty and maintenance point of view. Levels of intervention of an element could be system, subsystem, and component and will depend on the complexity of an element construction, of the subsystems accessibility, the level of competence of after-sales service staff, the availability of test equipment, levels of security and safety, etc. The term unavailability is understood as the time interval in which a product is in an unavailable state, once in customer's hands. Such state will be that characterized by a failure or a possible incapacity to develop the required function during preventice maintenance.

For more details, it is suggested to consult the UNE-EN 13306:2010 "Terminology on maintenance" [1], and the UNE 20654:1992 "Introduction guide for maintainability, requirements, and maintainability program" [14]. The assessment of maintainability indexes will be made according to a number of requirements regarding the evaluator's training, the operating conditions and environment where the assessment is carried out, and the frequency of such evaluations. These requirements are summarized here below:

- *Requirement for the evaluators*:
  The person in charge of the assessment must demonstrate:
  - A university degree in engineering;
  - An academic background in the maintenance field of at least 100 h;
  - A 25 h academic training in the maintainability field and after-sales area.
- *Requirements for the assessment conditions*
  For the evaluation of the maintainability index in a device that is during its operational phase, the operator and the technicians who may be involved in the technical assistance will always be available to the evaluator. So that they can always clarify, if necessary, the form and manner in which it conducts and coordinates any of the technical assistance on the device. The device operating conditions and environment (at the moment of the maintainability levels evaluation) will always be described in sufficient detail and recorded in documents showing the result of the evaluation.
- *Requirements for the assessment frequency*
  In order to achieve continuous improvement of maintainability index, it would be desirable periodical evaluations, depending on the level of activity to be

registered on the facilities where the device is located. The evaluation will also be recommended when it is important to analyze the implications for the technical assistance when significant changes in their environment and/or their conditions of use are expected.

12.3 Maintainability Indicators and Maintenance Levels Description

The maintainability indicators are assessed according to two kinds of attributes or characteristics of the device:

- *General attributes*: Those affecting any device maintenance level. That is to say, they are maintenance level independent.
- *Specific attributes*: Those depending on the maintenance level. That means, they are functions of all the maintenance actions to be performed on a specified maintenance indenture level.

The assessment of the maintainability specific attributes for a device requires, therefore, the classification of the maintenance actions within specific maintenance levels. In this paper, we propose five maintenance levels. For each level, the device maintainability performance will be assessed, checking whether the maintainability specific attributes have been taken into account for the development of maintenance actions at that level. The attributes ranking is done using a five values scale from 0 to 4 points, trying to reduce subjectivity and to obtain fair values easily to manipulate. The maintenance levels are here established based on the complexity of the tasks to be performed and the complexity of the human and material resources needed for its performance. The required unavailability time on the device is here also taken into account. In any case, these levels are due to common criteria accepted worldwide and are the following ones;

- *Level* 1: *Simple maintenance actions performed in the up state of the device*. At this level, the operator performs preventive or corrective activities which do not require setting the device into a down state. They are, for instance, simple adjustments foreseen by the manufacturer, without assembling or disassembling the device. Another example is the simple replacement of easily accessible components.
- *Level* 2: *Maintenance actions with replacement offunctional components*. These actions set the device into a down state. In this period, the operator performs preventive and/or corrective maintenance actions, usually considering functional components of the device for their replacement.
- *Level* 3: *Failures identification and diagnosis*. In these maintenance actions the operator, after setting the device into a down state, identifies and locates the causes of the failures.

- *Level* 4: *Inspections*. The inspections refer to an extensive amount of tests and preventive/corrective maintenance actions that may require full or partial disassembly of the device. The purpose of the inspection is to maintain, on a device, the required availability and safety level over the time. Revisions are usually performed at prescribed time intervals or after determined amount of operations.
- *Level* 5: *Updating, reconstruction and/or overhaul*. These operations are under the maintenance service responsibility of the plant or manufacturer. They are very important operations which may include modifications and/or improvements. It is therefore possible that these operations increase the lifetime of the original device.

12.4 Classification of the Maintainability Attributes

The maintainability indicators' assessment shall be performed taking into account attributes from the device. These attributes are classified into three groups according to their nature: attributes related to the design, related to work conditions and staff requirements, and related to the needs of logistic support. It is important to observe here how these attributes are connected to capabilities of the device under assessment, and not to the actual capabilities of the maintenance organization already existing, where the device is being used.

12.4.1 Attributes Related to the Device Design

Inside this group, we consider eight attributes:

- *Simplicity*: Existing reduction in the amount of elements and unnecessary assemblies.
- *Identification*: Clear location and signs from those components that are going to be mainly maintained, and from those existing points for inspection and test.
- *Modularity*: Device design in separated parts, functional assembly units, thus, it is not necessary to disassemble de whole equip in case of failure, but only the part(s) where the problem is located.
- *Tribology*: Right use of materials with appropriated quality in order to increase the use of life of fragile elements and to improve the lubrication of pieces wearing out quickly.
- *Standardization*: Spare parts compatibility to other similar materials when it is necessary to replace a device component. This attribute is conditioned by the standards choice during the design of elements such as bearings, gaskets, etc. and their dimensional and functional tolerances.

- *Failure watch*: Indications in the device about critical parameters and alarms to foresee failures.
- *Accessibility*: Schedule for accesses to all those elements to be maintained through gates, sliding doors, etc.
- *Assembly/disassembly*: Easiness to remove or replace elements in the different subsystems. This easiness shall be emphasized by gaskets, joints, welding etc. influencing also the elements size and volume.

12.4.2 Attributes Related to the Maintenance Staff and Work Conditions

Inside this group, we consider three attributes:

- *Ergonomics*: Space requirements in order to set up the proper working conditions for the development of maintenance activities. This attribute also assesses requirements in locations and spaces where materials to manipulate can be placed when it is necessary to perform interventions on the physical system.
- *Training*: Skill required to the maintenance staff for the kind of work to perform.
- *Environment*: Requirements on the environmental conditions are referred in order to enable the maintenance under proper conditions and complete safety.

12.4.3 Attributes Related to the Necessity of Logistics Support

Inside this group, we consider five attributes:

- *Relation with the manufacturer*: Requirements related to the coordination among the people responsible for the plant maintenance and the manufacturer: common language, same system for machines management, geographical remoteness, same working hours, same jurisdiction, etc.
- *Personnel organization*: Amount of people required to carry out the maintenance operation and possibilities of dividing the work into parallel tasks.
- *Spare parts*: Requirements in terms of spare parts for their use in the maintenance activity where the acquisition easiness shall be observed.
- *Maintenance tools and equipments*: Requirements in terms of tools and instruments for performing the maintenance activity. Functionality, ergonomics as well as the acquisition easiness will be observed.
- *Interdepartmental coordination*: Complexity in the task environment: Requirements for handling of hazardous parts or elements, for the permit application, for the communication among different departments, etc.

- *Documentation*: Indications given by the manufacturer for the device maintenance or prepared for the maintenance service, explaining how to perform the maintenance action.

12.5 Maintainability Attributes and Maintenance Level

The 17 attributes, commented in the paragraph above, are grouped in two sets, according to Sect. 12.4. The first group concerns those attributes that make influence on the device maintainability at any maintenance levels (General attributes); that is, a type of maintainability that is independent of the maintenance level. The second group of attributes could influence the device maintainability in a different way and degree, deepening on the level of maintenance that is being considered (Specific attributes).

12.5.1 General Attributes and Their Assessment

They are considered as eight general attributes that are maintenance level independent. For their assessment, the expert will consider each one of them independently. The result of their assessment is an integer number within the range 0–4. Hereafter, some examples are presented:

G1. Simplicity: It will be checked for the use of a minimal amount of components and assemblies in the devices, even those components that are redundant.

- 0: Very high number of components with redundant elements, easily visible.
- 4: Optimized, reduced, and without redundancy number of components.

G2. Identification: The identification of elements to be maintained and the points for testing will be checked, considering whether they are clearly indicated or not. It will be also observed that connectors are identified as well as danger areas, places where technicians have to be positioned themselves for working, etc.

- 0: No identification.
- 4: Complete identification, everything can be seen in front of the device.

G6. Standardization: For those components which could be replaced at any moment during their life cycle, it will be checked out of their compatibility with other ones out of the shelves of the market. It will result in a minimum storage of components and minimum amount of adjustments.

- 0: Very bad standardization. High difficulty to find a spare part in the market. High need of spare parts storage.
- 4: Good standardization. High easiness to find spare parts in the market and with competitive prices. There is no need to store spare parts.

Table 12.1 Accessibility

Level	Watching this attribute will have as a target	Assessment scale
1	Checking the proper access for first maintenance level tasks, for instance: basic inspections, consumable changes (lubricants…), etc.	
2	Checking the proper access for second maintenance level tasks, easy repairs by replacement of functional elements, etc.	In the range [0–4] where
3	Checking the proper access for third maintenance level tasks, for instance on the diagnosis and finding out of failures causes in the device, or the replacement or repair of minor components	0: Very difficult access, it is needed to move things and the technician himself, etc.
4	Checking the proper access for forth maintenance level tasks, which involve important revisions on the device, test performances, etc.	4: Very good access
5	Checking the proper access in order to perform reconstructions, updating or overhauls in the device	

Table 12.2 Assembly/Disassembly

Level	Watching this attribute will have as a target	Assessment scale
1 2 3 4 5	Checking the easiness for the assembly/disassembly (open, close, connect, disconnect, adjust, etc.) of those subsystems or elements which are involved in the device of first maintenance level tasks	In the range [0–4], where: 0: Many difficulties: Many tools are needed. Material weight, volume, and size are too important. 4: Very easy to assemble and disassemble

12.5.2 Specific Attributes (Maintenance Level Dependent) and Their Assessment

These attributes will be applied in the place where the device maintenance is performed (which can be different from its usual location), especially in the case of the last two maintenance levels. We consider nine maintainability attributes, which should be assessed by an expert, at each maintenance level, by assigning an integer number, ranged from 0 to 4. Here after, there are two of those attributes, as examples (see Tables 12.1 and 12.2).

There are similar tables for other seven specific attributes (they are nine in total).

12.6 Maintainability Indicators Assessment

In order to apply this methodology, it is required to calculate six indicators, based on the value of the attributes assessed at general level and at each maintenance level:

- One general maintainability indicator (GMI), which results from the device general attributes assessment.
- Five specific maintainability indicators, one for each maintenance level, named "maintainability indicator of maintenance level i" (where "i" takes the integer value 1–5) or LMI_i, as a result of assessing the specific attributes of the device.

The maintainability degree of an industrial asset can be understood by studying the scores obtained for each one of the eight general attributes and for those of the nine specific attributes (per maintenance level, i.e., 45 values considering five possible levels). However, it is obvious that such an amount of information is not practical to be managed (especially for comparison purposes). Moreover, the relevance of the different attributes could not be the same, for a device in particular. Additionally, an attribute could be very important for a divide in a certain operating conditions while the same attribute could be negligible for the asset in a different operating condition, or for a different asset in the same conditions. For the first reasons (the big amount of heterogeneous information concerning the maintainability of an asset), it would be convenient to aggregate all this information in a few indicators, leaving the whole attributes information for in deep analysis. For the second consideration (the different importance of the attributes, depending on the kind of asset and/or its operating conditions), it is recommended to assess also the relevance of each attribute. That means that we have to consider two figures for each attribute (the score for each attribute and the attribute relative importance within the set of attributes), instead of only one (the score of the attribute), as it has been considered until now. But this consideration drives to a new problem: to assess the relative importance for each attribute. Notice that the complexity related to the big amount of information to deal with the maintainability performance results now even increased when considering the relative importance of each attribute. For all previous reasons, the aggregation of the information becomes an absolute need, and in order to carry out such aggregation, in each case, the maintainability indicators are obtained by the weighted average of the values achieved in the attributes assessment.

Let us consider that

$$p_{A_i} = \{0,1\} \quad \text{with} \quad \sum_{i=1}^{n} p_{A_i} = 1$$

where

p_{A_i} Relative weight of attribute A_i. The treatment is similar for general attributes (G_i), as well as for specific ones (S_i) concerning a maintenance level;

i Ordinal number of each attribute;

n Number of attributes that configure the indicator under calculation ($n = 8$ for the general maintainability indicator, based on the eight general attributes; and $n = 9$ for any of the maintainability indicators of any maintenance level, based on the nine specific attributes).

To determine the contribution of each attribute to the maintainability indicator, we suggest the following procedure:

- To estimate $p_{A_i} = \{0, 4\}$: the importance of each attribute for the maintainability calculation. The expert who is assessing the maintainability should answer how important each attribute (general or specific for a maintenance level) is, concerning the maintenance of the device under evaluation, in its operation conditions.
- The weight of each attribute (relative importance of each attribute within the set of them), will be:

$$p_{A_i} = \frac{P_{A_i}}{\sum_{i=1}^{n} P_{A_i}} \tag{12.1}$$

The general maintainability indicator is defined as follows:

$$\text{GMI} = \sum_{i=1}^{8} G_i p_{G_i} \tag{12.2}$$

where

i Ordinal number of each of the eight maintainability general attributes;

G_i Integer value in the range 0–4, for each of the eight maintainability general attributes;

p_{G_i} Decimal value in the range 0–1, weight of each of the eight maintainability general attributes. These weights have been previously calculated [Eq. (12.1)] from the importance of the attributes for the maintainability performance.

The maintainability indicator of maintenance level j is defined as follows:

$$\text{LMI}_j = \sum_{i=1}^{5} S_{ij} p_{S_{ij}} \quad \text{with} \quad j = 1, \ldots 5 \tag{12.3}$$

where

j Ordinal number of each of the five maintenance levels;

i Ordinal number of each of the nine maintainability specific attributes;

S_{ij} Integer value in the range 0–4, for each of the nine maintainability specific attributes, for each of the five maintenance levels;

$p_{S_{ij}}$ Decimal value within the range 0–1, weight of each of the nine maintainability specific attributes, for each of the five maintenance levels.

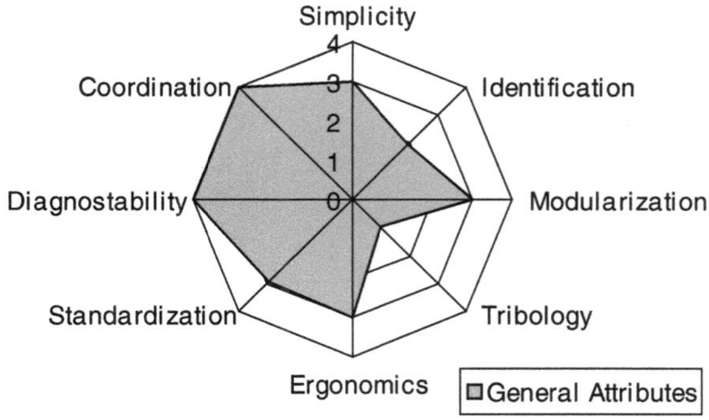

Fig. 12.1 Graphical representation of general maintainability indicator

These weights have been previously calculated [Eq. (12.1)], in a similar way to what we did for the case of GMI (Fig. 12.1).

Finally, it should be underlined here that many of the considerations made in this section have inspired the Spanish Norm Project PNE 151001 of AEN/CTN 151 "Maintenance" of AENOR [15]. Apart from this, as a continuation of Chap. 12, it is suggested to consult next section that deals with aspects of quantification of maintainability focused on what was here developed.

12.7 Case Study on Maintainability

12.7.1 Scenario and Previous Considerations

This chapter includes a case study with a practical application for the description of the maintainability index, the scenario of which is shown in Fig. 12.1. On this occasion, the industrial asset is a bridge-crane [16]. Guidelines described in Chap. 12 will be applied to this industrial asset in order to assess through evaluation sheets its maintainability index. For that, some global attributes as well as variables have been taken into account (level 0, 1, 2, 3 and 4).

As it has been indicated, the application scenario is an industrial asset that is specified as a bridge-crane. Following the indications in Fig. 12.2, it will be constituted by: (1) a bridge-crane, which type is of those that have a main car (2) carrying an engine (22) for its longitudinal displacement on the primary cranes, and a secondary car (3) carrying an engine (31) for its displacement on transverse cranes (21) installed in the main car (2) and of an engine (42) for the vertical displacement of a hoist (4) carrier of a transportation hook (41); being the engines (22, 31, 42) controlled by means of activation (24, 32, 43) that are characterized by having two radio receptors (51a, 51b) associated to a remote control (operated by radio) (52) to control the means of activation (24, 32, 43) of the engines (22, 31, 42) in a wireless fashion; this is because the radio receptor (51a) is located in the

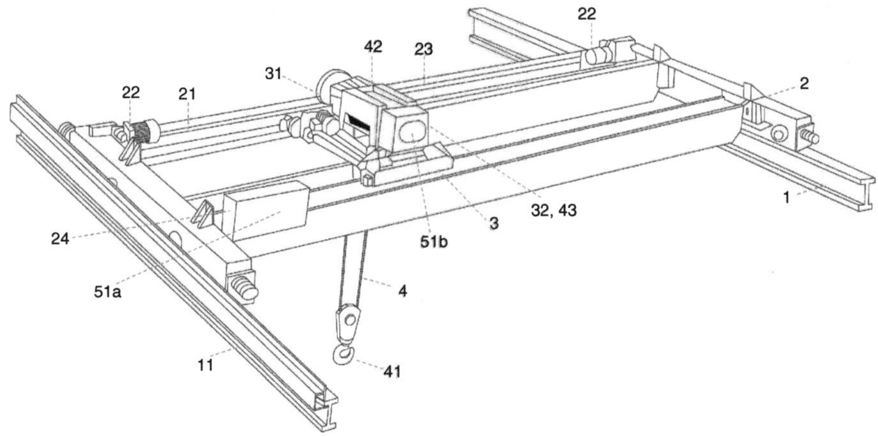

Fig. 12.2 Components of the bridge-crane

main car (2) to operate the means of activation (24) of the engine (22) for displacement of the related main car and because the radio receptor (51b) is located in the secondary car (3) to operate the means of activation (32, 43) of the engine (31) for the displacement of the secondary car (3) and the engine (42) of the hoist (4). This bridge-crane is a particular example because the longitudinal cranes (1) and/ or the main car (2) include electrified trails (11, 23) interconnected by power supply to the means of activation (24, 32, 43) of the engines (22, 31, 42) associated with radio receptors (51a, 51b).

12.7.2 Application of the Evaluation Sheets

Once the scenario has been described, where the methodology previously elaborated in Chap. 12 will be applied, we will then proceed in analyzing the evaluation sheets (Tables 12.3, 12.4, 12.5, 12.6, 12.7 and 12.8) together with the related graphics resulting from them (Figs. 12.3, 12.4, 12.5, 12.6 and 12.7).

12.7.2.1 Global Attributes

Table 12.3 Evaluation sheet of general attributes

Code	Attribute	Criteria	Scale [0–4] where	Value
G1	Simplicity	It will check the utilization of a minimum number of components and groups in the devices, even those that are redundant	0: High amount of duplicated components, easily visible 4: Optimal amount of components	3
G2	Identification	It will check that the elements to be maintained and the related places where the tests will be run are properly identified; together with that all the connectors, danger areas, specific working places, etc.	0: No identification 4: Perfect identification of equipment	2
G3	Modularization	It will check that there are different functional units of assembly in the device, which would allow minimizing handling in case of maintenance intervention	0: Complex change of units and requires the movement of other units 4: Excellent modularization	3
G4	Tribology	It will check that the correct material is selected for components which are subjected to friction, lubrication and wear, with the aim of maximizing their useful life	0: 0–10 % of the items are properly selected 4: 80–100 % of the items are properly selected	1
G5	Ergonomy	It will check the simplicity of maintenance tasks, analyzing the weight, size and shape of the components that are going to be handled. Also revised, in an equal manner, will be all those places where the tasks will be performed, making sure that are the proper ones (in relation to light, volume, etc.) for changes to be made easily	0: Tasks with complex implementation. Elements are not comfortably handled because of their weight, size and shape, making the operator become tired. Inadequate work space 4: Excellently ergonomic device, service is comfortable and agile	3

(continued)

Table 12.3 (continued)

Code	Attribute	Criteria	Scale [0–4] where	Value
G6	Standardization	It will check the compatibility of the components to be replaced with others that are available in the market. This will equally benefit minimizing the storage of components and the minimum number of adjustments especially in replaceable elements in low levels of maintenance	0: Poor standardization. High difficulty in obtaining spare parts in the market. High necessity of storage for spare parts 4: Good standardization. Easy to find replacements in the market at competitive prices. No need for spare part storage	3
G7	Diagnosticability	It will prove the existence of failure indicators in the device and the possibility to monitor those useful parameters for technical service	0: Poor diagnostic ability. There are no indicators of the condition of equipment and it is impossible to make a diagnosis of its condition 4: Good diagnostic ability. There are indicators that supply data of the condition of the equipment and the indicators are easily monitored	4
G8	Coordination	It will prove the necessity of coordination when it comes to develop the related maintenance tasks, the requirements of ways of communication and management common with other parts	0: Poor coordination. The technician is far away, speaks another language, etc. 4: Good coordination. The technician is nearby, speaks the same language, there are no communication problems, uses standard technologies, etc.	4
			Global index	2,875

12.7.2.2 Variables for a Level 0 Technical Assistance

Table 12.4 Evaluation sheet for variables of Level 0 for technical assistance

Code	Attributes	Criteria	Scale	Value
V1	Accessibility	-Hinges -Doors -Removable shelves	0: Very difficult access 1: Incredibly difficult access 2: Difficult access 3: Normal access 4: Very good access	5
V2	Assembly/ disassembly	-Lock -Welding -Splices -Connections -Edge connectors	0: A lot of difficulties 1: Too many difficulties 2: Some difficulties 3: Good manipulation 4: Very good manipulation	8
V3	Personnel	-Trained people according to the type of work -Training for operators	0: Very bad 1: Bad 2: Too bad 3: Regular 4: Good	2
V4	Technical equipment	-Number of people for technical operation -Preventing active maintenance time -Logistical cost	0: 5/6 people, no coordination 1: 4/5 people 2: 3/4 people 3: 2/3 people 4: 1 person	6
V5	Environment	-Isolation -Exhaust detection system -Presence of high tension cables	0: Very dangerous 1: Dangerous 2: Incredibly dangerous 3: Safe 4: Very safe	5
V6	Tools and equipment for technical assistance	-Weight and volume components for an easy handling -Standard components and tools	0: Too many rare things 1: Many things 2: Too many things 3: Few things 4: Too few things	3

(continued)

Table 12.4 (continued)

Code	Attributes	Criteria	Scale	Value
V7	Complexity of the technical assistance	-Possibility of simultaneous maintenance work -Permits to carry out operations	0: Very complicated 1: Complicated 2: Little complicated 3: Regular 4: Very well organized	5
V8	Documentation	-Convenient instruction and procedure manual for technical assistance	0: No documentation 1: Incomplete 2: Complete but difficult 3: Regular 4: Very good	4
			Level index 0	4,75

12.7.2.3 Variables for a Level 1 Technical Assistance

Table 12.5 Evaluation sheet of variables of Level 1 for technical assistance

Code	Attribute	Criteria	Scale	Value
V1	Accessibility	-Hinges -Doors -Removable shelves	0: Very difficult access 1: Incredibly difficult access 2: Difficult access 3: Normal access 4: Very good access	3
V2	Assembly/ disassembly	-Lock -Welding -Splices -Connections -Edge connectors	0: A lot of difficulties 1: Too many difficulties 2: Some difficulties 3: Good manipulation 4: Very good manipulation	2
V3	Personnel	-Trained people according to the type of work -Training of operators	0: Very bad 1: Bad 2: Too bad 3: Regular 4: Good	4
V4	Technical equipment	-Number of people for technical assistance -Preventing active maintenance time -Logistical cost	0: 5/6 people, no coordination 1: 4/5 people 2: 3/4 people 3: 2/3 people 4: 1 person	3

(continued)

Table 12.5 (continued)

Code	Attribute	Criteria	Scale	Value
V5	Environment	-Isolation -Exhaust detection system -Presence of high tension cables	0: Very dangerous 1: Dangerous 2: Incredibly dangerous 3: Safe 4: Very safe	3
V6	Tools and equipment for technical assistance	-Weight and volume components for an easy handling -Standard components and tools	0: Too many rare things 1: Many things 2: Too many things 3: Few things 4: Too few things	4
V7	Complexity of the technical assistance	-Possibility of simultaneous maintenance work -Permits to carry out operations	0: Very complicated 1: Complicated 2: Little complicated 3: Regular 4: Very well organized	3
V8	Documentation	-Convenient instruction and procedure manual for technical assistance	0: No documentation 1: Incomplete 2: Complete but difficult 3: Regular 4: Very good	2
			Level index 1	3

12.7.2.4 Variables for a Level 2 Technical Assistance

Table 12.6 Evaluation sheet of variables of Level 2 for technical assistance

Code	Attribute	Criteria	Scale	Value
V1	Accessibility	-Hinges -Doors -Removable shelves	0: Very difficult access 1: Incredibly difficult access 2: Difficult access 3: Normal access 4: Very good access	2
V2	Assembly/disassembly	-Lock -Welding -Splices -Connections -Edge connectors	0: A lot of difficulties 1: Too many difficulties 2: Some difficulties 3: Good manipulation 4: Very good manipulation	2

(continued)

Table 12.6 (continued)

Code	Attribute	Criteria	Scale	Value
V3	Personnel	-Trained people according to the type of work -Training of operators	0: Very bad 1: Bad 2: Too bad 3: Regular 4: Good	4
V4	Technical equipment assistance	-Number of people for technical assistance -Preventing active maintenance time Logistical cost	0: 5/6 people, no coordination 1: 4/5 people 2: 3/4 people 3: 2/3 people 4: 1 person	3
V5	Environment	-Isolation -Exhaust detection system -Presence of high tension cables	0: Very dangerous 1: Dangerous 2: Incredibly dangerous 3: Safe 4: Very safe	3
V6	Tools and equipment for technical assistance	-Weight and volume components for an easy handling -Standard components and tools	0: Too many rare things 1: Many things 2: Too many things 3: Few things 4: Too few things	3
V7	Complexity of the technical assistance	-Possibility of simultaneous maintenance work -Permits to carry out operations	0: Very complicated 1: Complicated 2: Little complicated 3: Regular 4: Very well organized	3
V8	Documentation	-Convenient instruction and procedure manual for technical assistance	0: No documentation 1: Incomplete 2: Complete but difficult 3: Regular 4: Very good	2
			Level index 2	2,75

12.7.2.5 Variables for a Level 3 Technical Assistance

Table 12.7 Evaluation sheet of variables of Level 3 for technical assistance

Code	Attribute	Criteria	Scale	Value
V1	Accessibility	-Hinges -Doors -Removable shelves	0: Very difficult access 1: Incredibly difficult access 2: Difficult access 3: Normal access 4: Very good access	2
V2	Assembly/ disassembly	-Lock -Welding -Splices -Connections -Edge connectors	0: A lot of difficulties 1: Too many difficulties 2: Some difficulties 3: Good manipulation 4: Very good manipulation	2
V3	Personnel	-Trained people according to the type of work -Training of operators	0: Very bad 1: Bad 2: Too bad 3: Regular 4: Good	3
V4	Technical equipment assistance	-Number of people for technical assistance -Preventing active maintenance time -Logistical cost	0: 5/6 people, no coordination 1: 4/5 people 2: 3/4 people 3: 2/3 people 4: 1 person	3
V5	Environment	-Isolation -Exhaust detection system -Presence of high tension cables	0: Very dangerous 1: Dangerous 2: Incredibly dangerous 3: Safe 4: Very safe	3
V6	Tools and equipment for technical assistance	-Weight and volume components for an easy handling -Standard components and tools	0: Too many rare things 1: Many things 2: Too many things 3: Few things 4: Too few things	2
V7	Complexity of the technical assistance	-Possibility of simultaneous maintenance work -Permits to carry out operations	0: Very complicated 1: Complicated 2: Little complicated 3: Regular 4: Very well organized	2

(continued)

Table 12.7 (continued)

Code	Attribute	Criteria	Scale	Value
V8	Documentation	-Convenient instruction and procedure manual for technical assistance	0: No documentation 1: Incomplete 2: Complete but difficult 3: Regular 4: Very good	2
			Level index 3	2,375

12.7.2.6 Variables for a Level 4 Technical Assistance

Table 12.8 Evaluation sheet of variables of Level 4 for technical assistance

Code	Attribute	Criteria	Scale	Value
V1	Accessibility	-Hinges -Doors -Removable shelves	0: Very difficult access 1: Incredibly difficult access 2: Difficult access 3: Normal access 4: Very good access	1
V2	Assembly/ disassembly	-Lock -Welding -Splices -Connections -Edge connectors	0: A lot of difficulties 1: Too many difficulties 2: Some difficulties 3: Good manipulation 4: Very good manipulation	2
V3	Personnel	-Trained people according to the type of work -Training of operators	0: Very bad 1: Bad 2: Too bad 3: Regular 4: Good	3
V4	Technical equipment assistance	-Number of people for technical assistance -Preventing active maintenance time -Logistical cost	0: 5/6 people, no coordination 1: 4/5 people 2: 3/4 people 3: 2/3 people 4: 1 person	2
V5	Environment	-Isolation -Exhaust detection system -Presence of high tension cables	0: Very dangerous 1: Dangerous 2: Incredibly dangerous 3: Safe 4: Very safe	3

(continued)

Table 12.8 (continued)

Code	Attribute	Criteria	Scale	Value
V6	Tools and equipment for technical assistance	-Weight and volume components for an easy handling	0: Too many rare things	1
		-Standard components and tools	1: Many things	
			2: Too many things	
			3: Few things	
			4: Too few things	
V7	Complexity of the technical assistance	-Possibility of simultaneous maintenance work	0: Very complicated	1
		-Permits to carry out operations	1: Complicated	
			2: Little complicated	
			3: Regular	
			4: Very well organized	
V8	Documentation	-Convenient instruction and procedure manual for technical assistance	0: No documentation	2
			1: Incomplete	
			2: Complete but difficult	
			3: Regular	
			4: Very good	
			Level index 4	1,875

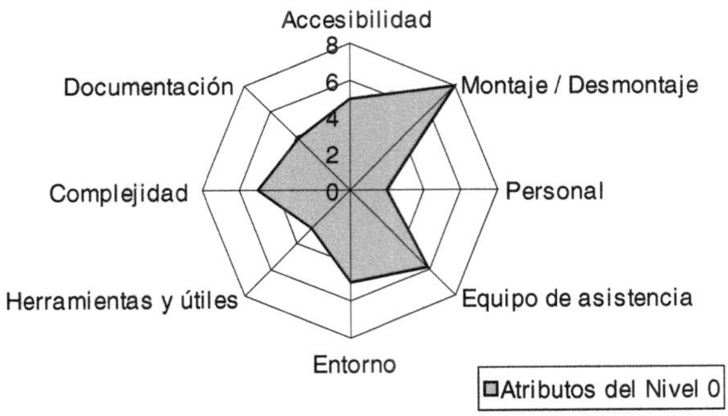

Fig. 12.3 Graph of maintainability index of Level 0

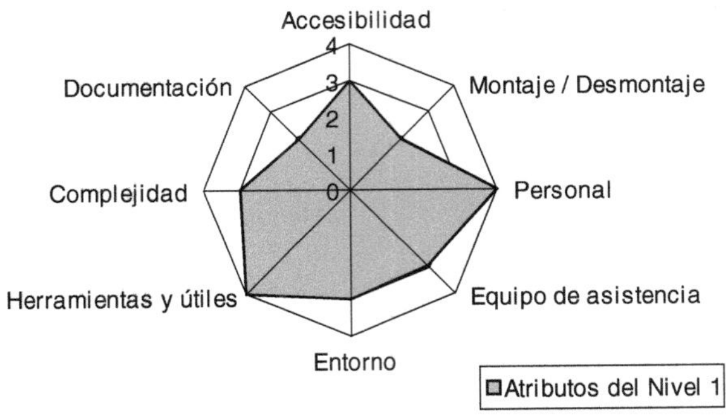

Fig. 12.4 Graph of maintainability index of Level 1

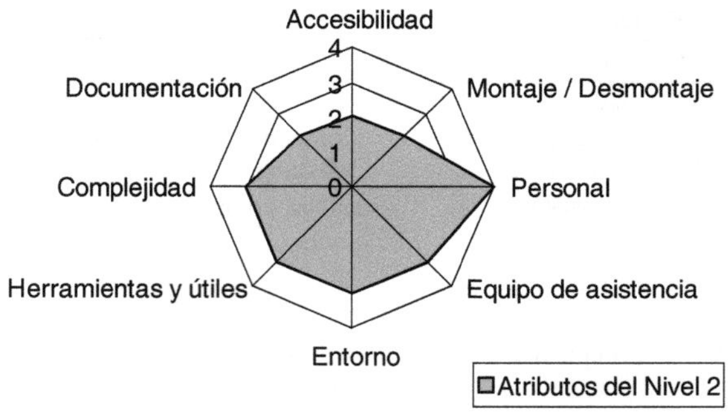

Fig. 12.5 Graph of maintainability index of Level 2

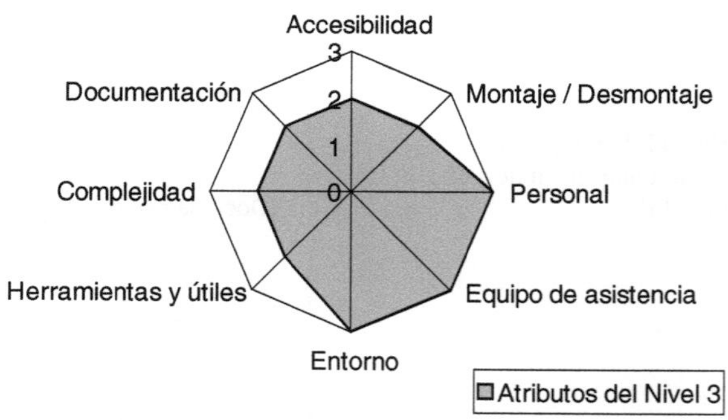

Fig. 12.6 Graph of maintainability index of Level 3

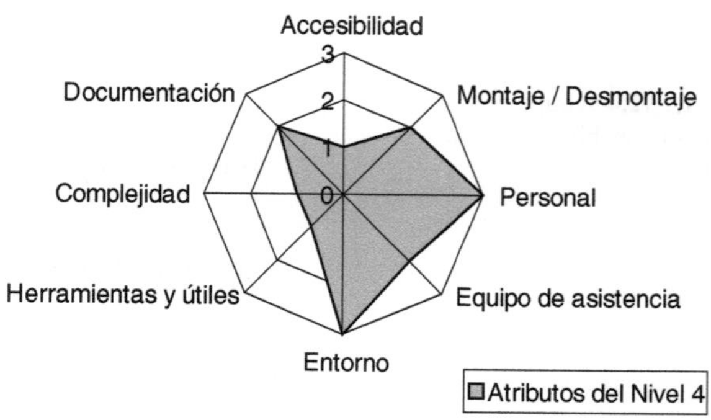

Fig. 12.7 Graph of maintainability index of Level 4

Table 12.9 Summary of results

Code	Attribute	Comments
G1	Simplicity	There is a limited number of items and the number of components is optimized
G2	Identification	All elements are physically identified (pulley, engine, and weighting system), except for the crane's frame of maneuver
G3	Modularity	The separation of different functional units is simple
G4	Tribology	The bearings are lubricated, but there is wear in the rails and the grease is not the adequate for the cables. Also, cable temperature that affects lubrication was not taken into account
G5	Ergonomic	Weight and size of the components is reasonable
G6	Standardization	Hooks, cables, and devices limit the load
G7	Diagnostic ability	There are too many failure indicators and the possibility of parameter monitor is useful for technical assistance
G8	Coordination	There is a good coordination: situation, language, and technologies

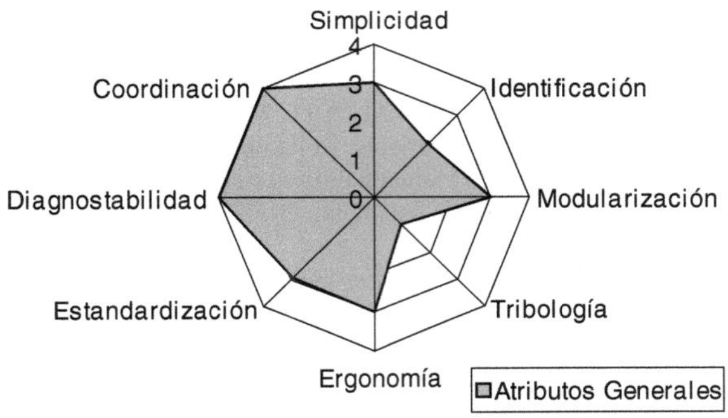

Fig. 12.8 Graph of general attributes

12.7.3 Discussion of the Results

Table 12.9 shows briefly the deduced results from these methodologies. Likewise, Fig. 12.8 shows the general index of the analyzed bridge-crane. Other interesting references to this respect can be found in [12, 13].

References

1. European standard. 2010. EN 13306: Maintenance – Maintenance terminology. CEN/TC 319.
2. Eldon CG. 1990. An Analytic Approach to Performing a Maintainability Demonstration. IEEE Tansactions on Reliability. 39(1).
3. Houshyar A, Ghahramani B. 1997. A practical reliability and maintainability data collection and processing software. Computers & Industrial Engineering. 33(1-2):133-136.

4. Sols A. 2000. Fiabilidad, Mantenibilidad, efectividad: un enfoque sistemático. Madrid: Universidad Pontificia Comillas.
5. Wani MF, Gandhi OP. 2002. Maintainability design and evaluation of mechanical systems based on tribology. Reliability Engineering & System Safety, 77(2):181-188.
6. Chen L, Cai J. 2003. Using Vector Projection Method to evaluate maintainability of mechanical system in design review. Reliability Engineering & System Safety. 81(2):147-154.
7. Smith DJ. 2005. Reliability, maintainability and risk: practical methods for engineers. Amsterdam: Butterworth-Heinemann.
8. Coulibaly A, Houssin R, Mutel B. 2008. Maintainability and safety indicators at design stage for mechanical products. Computers in Industry. 59(5):438-449.
9. Guo B, Jiang P, Xing YY. 2008. A Censored Sequential Posterior Odd Test (SPOT) Method for Verification of the Mean Time To Repair. IEEE Tansactions on Reliability. 57:2.
10. Ben-Daya M, Duffuaa SO, Raouf A, Knezevic J, Ait-Kadi D. 2009. Handbook of Maintenance Management and Engineering. London: Springer-Verlag London.
11. Liu R, Lv C, Zou Y, Zhou D. 2009. The preliminary study on the human-factor evaluation system for maintainability design. International Conference on Intelligent Human-Machine Systems and Cybernetics, IHMSC 2, art. 5336031, pp. 107-112.
12. Lu Z, Sun YC. 2009. Maintainability virtual evaluation method based on fuzzy multiple attribute decision making theory for civil aircraft system. Proceedings of 2009 8th International Conference on Reliability, Maintainability and Safety, ICRMS art. 5270100, pp. 684-689.
13. Abbas B, Javad B, Tore M. 2011. Maintainability analysis considering time-dependent and time-independent covariates. Reliability Engineering and System Safety. 96:210–217
14. UNE 20654:1992. Maintainability requirements. AENOR, Madrid
15. Proyecto de Norma Española. 2011. PNE 151001 del AEN/CTN 151. Mantenimiento. AENOR.
16. Carrión J. 2008. Patente Española solicitada por Motorman radio sistemas y aplicaciones, S.L. Clasificación PCT: B66C13/22. Fecha de Concesión: 17/09/2008

Chapter 13
Dynamic Modeling and Bayesian Networks in the After-sales Service

Nowadays, more developed markets demand more complex products and solutions that are constituted by both tangible and intangible closely related elements. In other words, customers demand more each time of the complementary-determined services affixed to an industrial asset and, consequently, to its respective added value. Recent technical literature has coined a term for this situation: product-service systems (PSS). Maintenance, spare parts supply, warranty extension, etc., or even the dismantling of a system at the end of its life cycle are some examples of services related to the sale of a product that the company must assume as an after-sales service. As we will see in the first part of this chapter, this PSS context increases the potential of the continuous simulation for the support of better decision-making. Specifically for that purpose, a dynamic system model will be developed and applied to evaluate how preventive maintenance contracts, as part of the after-sales service, can affect the general performance of the services, in this particular case, of a manufacturer of agricultural machinery. In the second main part of this chapter, we will study a new approach to after-sales service. This time, it will be carried out at the end of the product's life cycle in order to determine an optimal disassembly plan of a complex industrial asset. Use the Bayesian network theory to achieve this. The best solution is referred to as optimal trajectory. So, a trajectory model will be set out. The model must enable the management of different key factors and specific uncertainties of the disassemble system. After exposing the present question related to disassembly planning, Bayesian nets show the influence diagram allowing the manager to make decisions and proceed to the economical assessment of the different and possible strategies to follow. Among the different factors taken into account, we will also find that the cost related to the warranty of recycled products is included, according to the global trajectory model, the application of warranty costs as variables for decision.

In short, the relevance of this chapter resides not only in the discussion of different gathered results, but also in showing future approaches to analyzing after-sales management in an innovative fashion.

V. González-Prida Díaz and A. Crespo Márquez, *After-sales Service of Engineering Industrial Assets*, DOI: 10.1007/978-3-319-03710-3_13,

13.1 Dynamic Systems Modeling

Traditionally, system modeling has been carried out through mathematical models that endeavor to find analytical solution that would allow predicting system behavior according to a set of parameters and initial conditions. However, there are usually not many simple and analytical solutions for most systems, hence the necessity to use computer simulation models. Simulation models generate a wide range of representative scenarios for a determined model, in which all possible conditions are contemplated (even those that are forbidden or impossible) [1]. A multifaceted, interactive, and dynamic model in structures (as those involved in provision of PSS) will require a strong tool or method that helps to understand the system's own complexity, in order to design better operational policies as well as to orient changes. The modeling of dynamic systems and the simulation of discrete events can be used to model business decisions of the company. For the purposes of this research, it has been decided to apply a dynamic system focus, considering that it is the more adequate method in compound systems [2]. In any case, it also adapts well to those problems associated with continuous processes, where changes of behavior in a nonlineal fashion take place and extensive feedback is allowed.

The main objective of dynamic systems is to understand through the application of quantitative models, the way the systems behave and also to predict the consequences on system performance that occur due to policy changes [3]. On the other hand, the simulation of discrete events represents more often particular processes, not complete systems, and is better at providing a detailed analysis of those systems that take part in lineal processes and discrete models.

Therefore, given the potential of dynamic systems, the model presented in this chapter has been developed to:

1. Comprehend and represent, through the study of connections between service indicators, the nonlineal relations of all involved processes in the provision of a PSS.
2. Evaluate in a quantitative way, how the introduction of new policies for the PSS management significantly affects the company's efficient performance.
3. Allocate appropriate organizational changes to the processes, thanks to the responses (feedback) that the simulation sets out with respect to the changes provided by the provision of the service.

From a technical point of view, the introduction of a new policy involves a change in the exogenous parameters that confirm the dynamic system model and whose value is normally supplied by the company under analysis.

In this particular case, the model has been designed for two specific types of PSS: the provision of maintenance services and spare parts supply. Figure 13.1 shows the elements and the adopted logic to build a dynamic system model.

The dynamic system model has been developed following a general structure and can be used by any company that deals with PSS (especially maintenance and

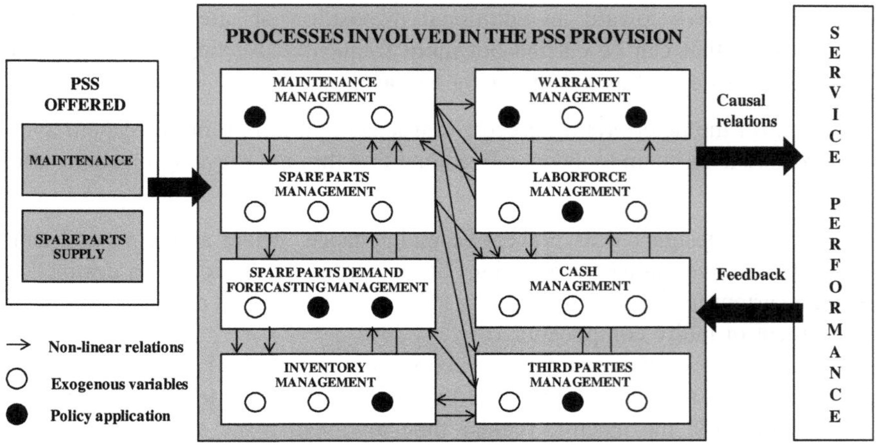

Fig. 13.1 Logic behind the construction of a dynamic system

spare parts supply) as well as to test the impact of the application of a new policy can have in the after-sales service performance. With the aim of proving the potential of the model, this chapter proposes a specific application in a company of the agricultural sector.

13.2 Case of Dynamic Modeling in the Farm Machinery Industry

13.2.1 Study Scenario

This case study regards a worldwide important company allocated in the manufacturing business of agricultural machinery assembly. Together with the production and truck sales, the main market is production, sales, and repair of harvesters. A key question for business improvement and optimizing provision of service on behalf of technical support, in other words, its purpose is to improve after-sales processes in order to increase benefits coming from this business and to retain its customers, so future sales are assured. After-sales activities consist of supplying spare parts and maintenance for customers. An inoperative harvester during harvest season can represent great losses, so the customer demands quick repair assistance, and if possible, not during harvest season. It is important for the customer that this requirement is well assured by the company or their authorized assistance centers.

That is why, the company has embraced a new strategy to reduce the number of maintenance interventions, especially during harvest season, since the services of corrective repair are totally unexpected and hard to tackle. Therefore, it is necessary to move toward an additional disposition of after-sales supportive characteristics that can be carried out more frequently. According to [4], maintenance can be defined as the following actions:

1. To control the deterioration process that leads to a failure in the system and,
2. To restore the system to its operative condition through preventive actions after a failure.

The first is designated as preventive maintenance, where as the second, corrective. Corrective maintenance actions are nonprogrammed activities destined to restore a system from a faulty condition to an operative one. It refers to a repair or replacement of faulty components.

In contrast, preventive maintenance actions are programmed and are carried out to reduce the possibility of failure and to extend the useful life span of the component. Normally, the inactive regular time involved in the application of preventive maintenance activities can represent greater costs to the user that operates the equipment, more than if the equipment is used until the repair is absolutely necessary.

However, it is important to compare both costs, as well as the long-term benefits, and the savings associated with the application of preventive maintenance (for example, reducing inoperative time of the system, a more efficient management of spare parts, replacements inventory, system reliability, etc.). On one part, from a seller perspective, the role of preventive maintenance is becoming more and more important during the warranty period considering that, in general, the buyer pays for preventive maintenance reducing this way for the seller, those failure repair costs generated by corrective maintenance. On the other part, a buyer that does not take into account the impact of the investment in preventive maintenance during warranty and its extension is assuming the total maintenance cost throughout the active life span of an industrial asset. In other words, there is no incentive for the seller in investing its efforts in preventive maintenance during warranty periods, given that the buyer does not pay for those repairs. The only advantage for the seller to perform preventive maintenance is when the additional expense of such maintenance is compensated by a general positive return (for example, customer loyalty).

In short, seller and buyer must evaluate advantages and disadvantages associated with assuming a service contract (refer to Chap. 10) that provides a warranty extension, preventive maintenance, corrective, etc. For these types of questions, the simulation of dynamic systems makes an attempt to give some answers.

13.2.2 Modeling and Results

With the aim of evaluating how the incorporation of preventive maintenance contracts has an impact in the after-sales service performance of the company, the analysis has been carried out on three different scenarios:

- *Escenario Scenario A*: There is no preventive maintenance contracts applied, either in or out of warranty;
- *Escenario Scenario B*: Preventive maintenance contracts are acquired by customers just under a warranty period;
- *Escenario Scenario C*: Preventive maintenance contracts are acquired during the product's whole life cycle, in or out the warranty period.

The analyses that have been carried out, taking into account seller's perspective (manufacturer) and all considerations have been done based on the complete life cycle of the product. The model has been initiated taking into account the installed base in the current company. The simulation time has been established on a monthly period for 30 years in order to analyze the complete life cycle of a harvester, which oscillates around 15–20 years. The model has been based on the following assumptions:

- Each time a failure occurs, it is due to the incorrect operation of a component, and it occurs in all equipment in the market;
- Only one type of replace pieces is considered (in this case in concrete: the rotor);
- The failure rate of spare parts has been estimated as constant and is produced after a certain number of worked hours (or bales produced);
- The customer is provided with a preventive maintenance contract for which it pays a fee that includes the maintenance intervention price and the replaced spare part, in or out of the warranty period;
- Preventive actions are cyclic, they are carried out at predetermined time intervals;
- Preventive and corrective interventions are carried out assuming that the component is restored to a good condition such "as good as new".

With the aim of executing the dynamic model system, exogen parameters have been introduced, such as the demand for sold equipment, warranty time period (one year established), cycle time of a preventive intervention, the failure rate of the product, the price and cost of a preventive and corrective intervention, the unit cost of personnel, the unit cost of the spare parts inventory, etc. These values have been provided by the company.

Following all this, these charts show the main obtained results through a simulation. Figure 13.2 shows the high-level diagram of the modeling.

The fundamental result is that the company actually gains benefits by the introduction of preventive maintenance contracts. In particular, the more frequent the use of preventive maintenance, the greater the benefits it gets from the after-sales service [5, 6].

Limiting preventive maintenance to a warranty period, it can be observed that it is less convenient than when it is extended throughout the vital cycle of the harvesters. Inclusively, in spite of continuing afterward with the same strategy that makes the company incur in greater operational expenses (due to the necessity to

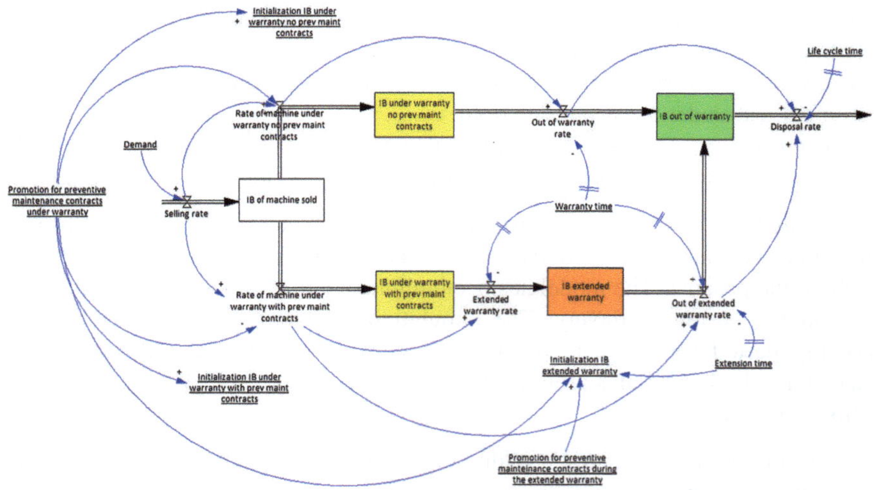

Fig. 13.2 Modeling high-level diagram

carry out interventions both preventive and corrective, and therefore, there is a need to more personnel to execute these interventions), this compensates due to the benefit obtained throughout the product's life cycle (Figs. 13.3 and 13.4). Another interesting result refers to the reduction in spare parts orders. This is because preventive maintenance interventions are regular which reduces the uncertainties with respect to the desired number of spare parts needed. Figure 13.5 illustrates the evolution of accumulated costs in orders of spare part during the product's life cycle.

Finally, it is found interesting that the tendency of costs is derived from personnel inactivity, which accumulates during the life span of the product. As it has been previously mentioned, the maintenance interventions are regularly programmed differently from corrective maintenance; therefore, work time of technical staff can be better planned and optimized.

Figure 13.6 shows how costs of inactive personnel decrease when preventive maintenance interventions increase. This context will also be applied in the Chap. 7 to develop a case study related to the balance scorecard (BSC) and warranty assistance. Future researches can follow the line of applying the results here obtained precisely with a BSC methodology.

In conclusion, other interesting references, apart from the above mentioned, are [7–10], specially pointing out the relevance of the developed work in Ref. [11]. Following this section of this chapter, it is suggested to consult the case study in Chap. 7 where aspects related to BSC are dealt with.

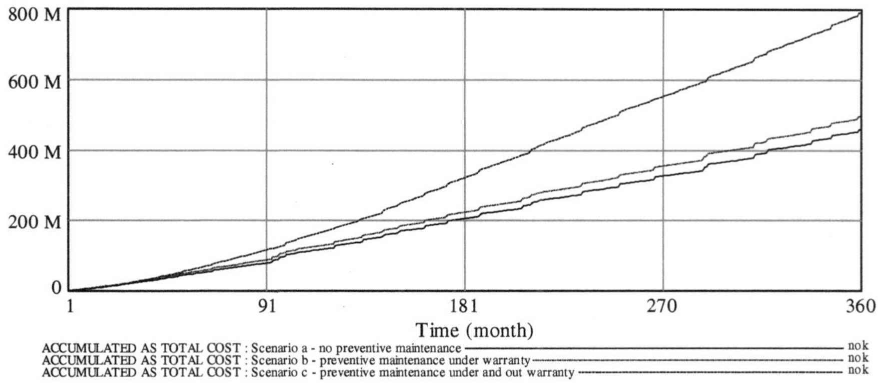

Fig. 13.3 Total life cycle cost for the three scenarios

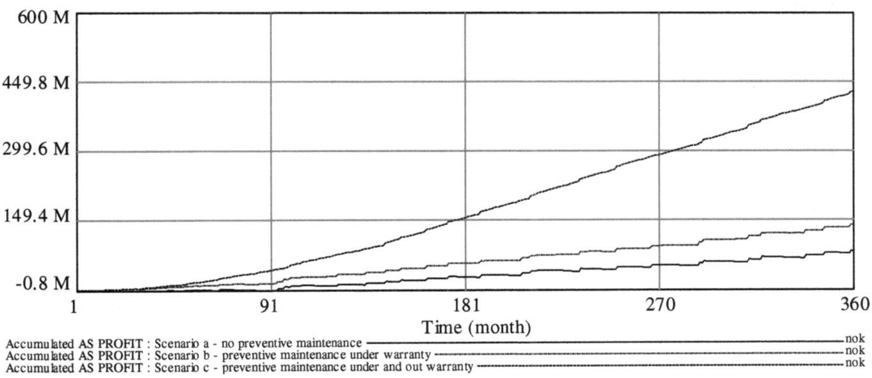

Fig. 13.4 Total life cycle benefit for the three scenarios

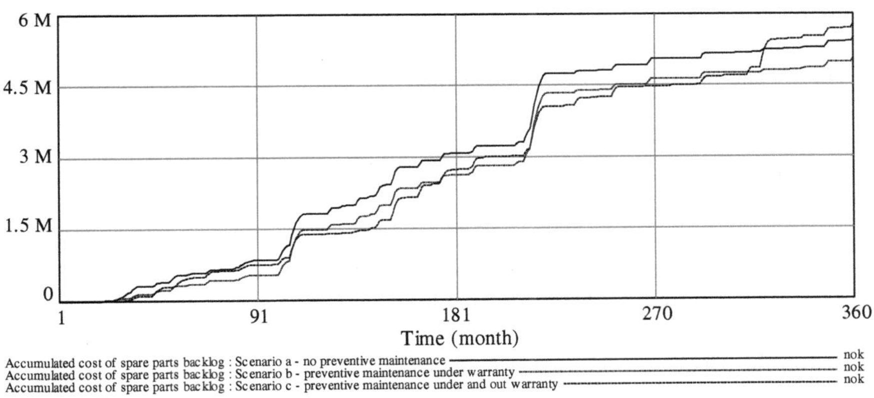

Fig. 13.5 Spare parts cost during the product life span

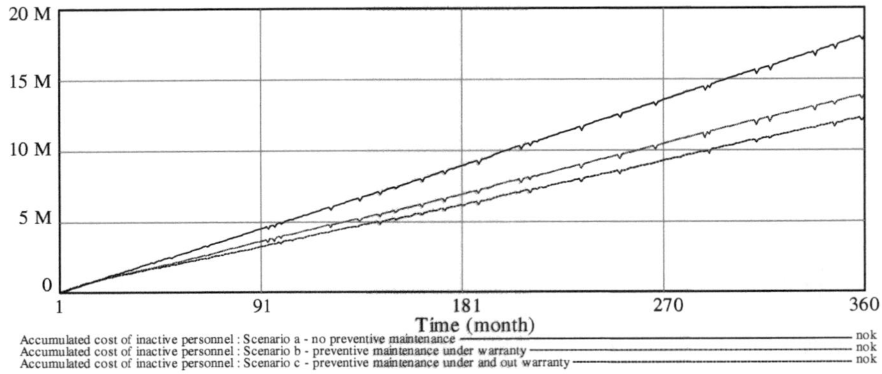

Fig. 13.6 Cost of inactive personnel during the life cycle

13.3 Bayesian Networks in the After-sales Service

The final life stage of complex systems has been an issue of study for several years. Systems like the industrial assets are dealt within this project. This is due, on one part, to the legislative pressures in terms of environmental protection, and on the other, to possible economical benefits that can be obtained by the implementation of recycling solutions to products. It is the responsibility of the designer to integrate recycling limitation by proposing dismantling processes of those systems already from the design stage. The increment of valuable strategies must meet all decision-making problems set out during the system withdrawal stage. Mainly, those valuable products will have to be selected according to technical, economical, and environmental criteria so that facilitates products' dismantling following defined and optimal guidelines. Within this context, three types of decisions have been considered:

1. The first refers to the dismantling level determination, that is to say, the best option for valorization, and for subsets, the best dismantling, or recycling selection.
2. The second refers to the operation sequencing that aims to determine the way of getting products and the logical sequence of operation to get them.
3. Finally, the number of products and their obtaining date must be determined in a given setting.

The support to the dismantling process must be made possible by managing these three previous decisions as well as the establishment of a link between them in order to keep a total control of the strategy. The dismantling trajectory consists in the identification of valuable products in a system at the end of their life cycle, as well as the identification of the channels needed to achieve this and the way to obtain them. In the modeling of dismantling trajectories, some factors must be taken into account as they are of influence in all the set. Among them is warranty, it will consist of, in general, the responsibility of the manufacturer to maintain the integrity of the product and repairing or replacing all faulty components of the industrial asset. There are many investigative works on how this problem is addressed from a state-of-the-art perspective. Some references to that respect are [12–15].

13.3.1 Planning of Disassembly

13.3.1.1 Modeling Framework of the Problem

There are several works in the literature review, which deal with these problems from a determinist context and presenting here a way to relate different focuses. In general, the modeling of dismantling planning involves three main steps. The first step refers to the modeling of the industrial asset structure at the end of its useful life. The objective of such modeling will be to represent the most valuable parts and subparts and the connections between them [16]. The second step regards the modeling of its own dismantling process. The obtained model will represent the different operations that can be carried out in a system in order to get those parts that are valuable. The third step has to do with the search of an optimal sequence between the identified stages in the process model. The purpose is to determine the dismantling level and the sequence jointly. The model must take into account the preferences of the different actors involved in the final phase of life of the system. Classical models focus the problem decision as a lineal program solving it using known algorithms [17].

13.3.1.2 Variables in the Dismantling Process

Problem solving planning entails the use of different models and algorithms. Most of the done on the subject found in the literature review proposes their own method and modeled language. In most cases, the approximations considered do not enable the integration of uncertainties. Our objective here is not to propose an estimation more to solve this problem, but the integration of uncertainties in the dismantling process based on existing models to determine an optimum and solid solution. To achieve this objective, we are using concepts and entities used by distinct focuses later adding the uncertainties, with the help of a modeled language that can cope with those uncertainties. The main entities involved in the resolution of the dismantling planning problem are the ones that follow:

1. *Components*: represent the system composition at the end of their useful life and can correspond to parts, subsets, and/or intermediate dismantling statuses.
2. *Connections*: from a structural point of view, connections complement and represent the articulations and relations between components and subsets.

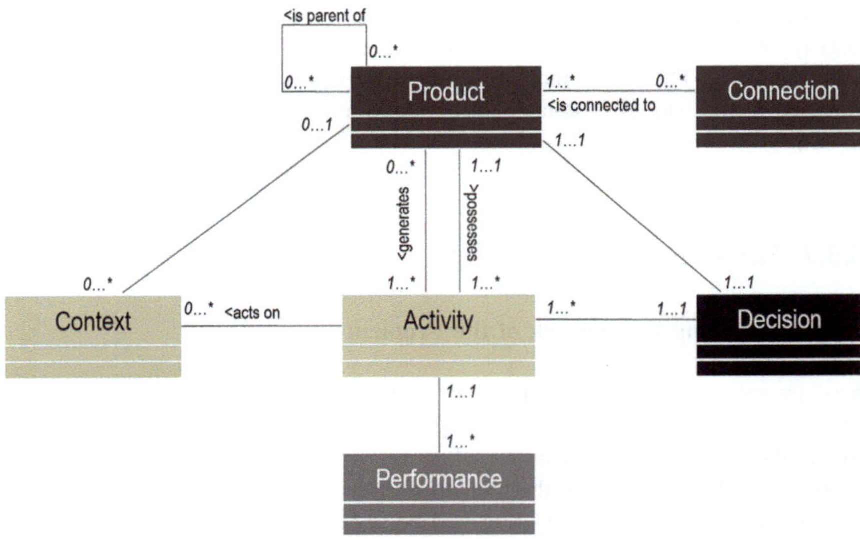

Fig. 13.7 Dismantling variable structure

3. *Variables at the end of their useful life*: related to each component describe the recuperative actions that can be recycling or dismantling operations.
4. *Contextual variables*: attached to components and variables at the end of their life, these model the recycling restrictions as well as other actions in reference to recuperation.
5. *Variables of decision*: related to components and final life variables, these represent the different actions in decision-making and provide a framework for the decision process.
6. *Performance parameters*: attach the variables of final life, describe the consequences of the actions the person responsible for decision-making.

Based on these entities, Fig. 13.7 proposes a modeling generic framework for the planning of the disassembly problem (diagram type UML). In this stage, we have highlighted the steps in the solutions of the problem as well as different variables to be managed by those who make decisions. We are now going to introduce the concept of Bayesian network to tackle uncertainties in some of these variables.

13.3.2 Bayesian Networks in Dismantling Planning

13.3.2.1 Bayesian Networks and Influence Diagrams

The following section proposes the use of Bayesian nets as a mathematical model tool for solving the decision problem: for a determined industrial asset at the end of its useful life, to determine dismantling levels and sequences on the base of a process model, taking into account the uncertainties of such dismantling process. Bayesian nets and their extension to influence diagrams are here used considering that the problems we want to solve present the characteristics that follow [18]:

- They can be represented graphically.
- It is necessary to manage and integrate uncertainties.
- It is necessary to solve an uncertain optimal problem.

The first characteristic listed is important in a multiple-actors' context such as the one here considered. In fact, Bayesian nets and influence diagrams facilitate the comprehension of the problem for all actors through a simple and natural graphical representation. Besides, they permit in unique representation of the interaction between these same actors and the interchange of knowledge. In fact, a Bayesian network is a graph in which knowledge is modeled as a variable and each variable corresponds to a node in the graphic. The arcs directed represent the dependence relation between the variables. The first step in the development of a Bayesian network model consists in obtaining those variables that are not of our interest. The second point corresponds to the purpose of setting out the problem. Bayesian nets permit a statistical inference that consists in determining variable probabilities, hidden, or not evident to the given problem. When all those decision-making managers consider that there are uncertainties in some variables, these can be evaluated using a probability formula. Taking into account, knowledge of all interested parts, these probabilities can be conditional (in other words that depend on other variables) or marginal. Once all uncertainties of the dismantling process have been evaluated, managers must determine the optimal solution in accordance with several criteria. Nodes of decision and utility are added to the Bayesian network becoming an influence diagram.

This models the problem of option selection at the end of the useful life of each component, including the utility of these options. In [19], the authors propose inference algorithms that determine an optimal solution.

13.3.2.2 Decision Problem in the Dismantling Modeling

The following section proposes the use of influence diagrams as tools for decision-making regarding dismantling model problems, and to do that, it is given a model of the process. The disassembly using Petri nets can be used to model such a dismantling process. Petri nets clearly describe precedence restrictions between operations of the dismantling process. The purpose here is to represent all possible operational sequences. The decision problem will be to determine the best sequence according to one or more criteria. Once the behavior of dismantling mechanism had been translated into an influence diagram, the statuses (recovered components, subset, etc.) of the dismantling process will have to be determined according to decision configurations [20]. Dismantling solutions are evaluated through the nodes utilities of the influence diagram. These model the economical performance of the different recuperative actions and dismantling operations.

The influence diagram models permit the integration of utilities in form of a table. There are three types of nodes:

- *Costs of dismantling nodes*: related to the dismantling operation nodes, their value is the in function of the execution of the dismantling operation.
- *Costs of recycling nodes*: evaluate each mode of recycling action, with respect to the linked recycling nodes;
- *Recycling income nodes*: They model the economical flow that is generated when a recycling option is validated.

These nodes with different utilities permit the optimization of a criterion for a dismantling plan selection. This criterion is split for each product in order to select the best operation for each of them. To achieve this, the decision node of each product will indicate the assessment of each option. The decision rule will consist in the selection of the option which maximizes the expected utility of the industrial asset, and this will be designated as the policy at the end of the useful life of the product. The set of all policies will make up the strategy. This will provide the products that have to be analyzed at the end of the useful life of the system, its recycling options and the required dismantling operations. The purpose of the optimization method is in short the strategy for a determined life end of a system.

13.3.3 Integration of Warranty in a Disassembly Model

This section explores a trajectory model that enables all managers, responsible for making decisions, to define the disassembly level in order to increase economical profits. The model uses a Bayesian network that was presented in the *World Congress on Engineering Asset Management* (WCEAM). The originality comes here from the fact of considering warranty costs as factors that can modify decisions in terms of the dismantling level. The problem of the dismantling trajectory is represented by Bayesian nets that allow the representation of the elements in the decision problem.

In general terms, the modeled dismantling system by Bayesian nets is described by the following elements:

1. "Product" nodes that represent the components at the end of the system's useful life and that have one or more options of recycling.
2. "Activity" nodes that represent the dismantling operations or the recycling actions of each product.
3. Arcs that characterize the precedence and exclusion of the relation between activities.
4. Node parameters that make it possible to characterize the progress of the dismantling process.

The variables of decision are attached to every product. They indicate the direction of the dismantling trajectory toward an option (dismantling or recycling), being the specified restrictions by the arcs. The economical parameters are associated with the "activity" nodes through utility nodes. These represent all those costs and incomes that a determined activity can generate, allowing that economical benefits and distinct trajectories to be evaluated and compared. The set of Bayesian network nodes in the dismantling model is designated by *N*. The following subsets characterize the structure of the model:

1. N_P is the set of "product" nodes, being *P* an element of N_P;
2. N_A is the set of "activity" nodes being:
   - N_{AP}^{d} that represents dismantling operations in *P*;
   - N_{AP}^{v} that represents recycling operations in *P*; *A* that represents an elements of N_A;
3. N_U is the set of utility nodes, which are associated with each activity, being U_A the utility node related to the activity modeled by node *A*.

The model represents the completed deconstruction trajectory that the manager or the person in charge of decision-making must identify. The objective is to find an optimal trajectory that enables, each product and status given, to select of the best activity (further than just dismantling, functional, or materials recycling). In this network, *Succ*(*A*) represents the group of subsequent product nodes of a dismantling activity *A*. A certain dismantling policy can be extracted from the global model to evaluate each product separately. This makes it possible to obtain the recursive equation of definition to determine at the same time the optimal dismantling trajectory. The dismantling policies are modeled by nodes of decision associated with each product. These nodes are integrated in the model that follows in Fig. 13.8 (node PL_P).

The product in question is the modeling of node *P* and the modalities of node PL_P that characterized all those possible options that can be selected in the product. The utilities *U*(*Q*), $Q \in Succ(A), A \in N_{\mathrm{AP}}^{d}$ (in other words, *Q* is a component of *P*), represent the evaluation of components of the product generated by each dismantling operation. The utilities in a model in which are associated with each product, such utilities correspond to the optimization results in the policies of product components. Warranty parameters are here introduced through utility nodes related to the functional recycling option, specifying for each dismantling level, the cost of the resulting warranty program as well as the probability (that will be the reliability of the system, subsystem or study component), characterizing all the option of implementing the corresponding actions. That is to say, warranty has been integrated as a decision variable in the recertification process of all parts resulting from the dismantling of an industrial asset at the end of its useful life [21]. Future studies can analyze outcomes obtained from applying this methodology to different scenarios.

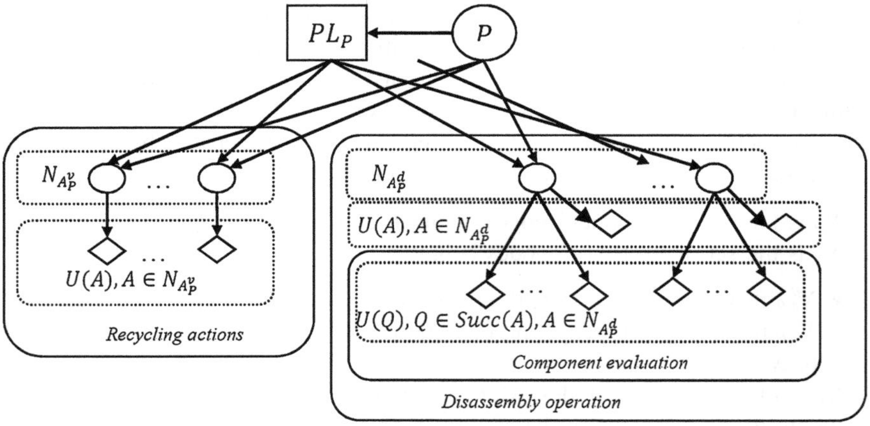

Fig. 13.8 Dismantling policy model

References

1. Crespo Márquez A. 2010. Dynamic modelling for supply chain management – Dealing with front-end and back-end and integration issues. Springer-Verlag, London.
2. Sterman J. 2000. Business dynamics: system thinking and modelling for a complex world. McGraw-Hill.
3. Santos SP, Belton V, Howick S. 2002. Adding value to performance measurement by using system dynamics and multicriteria analysis. International Journal of Operations & Production Management. 22(11):1246–1272.
4. Djamaludin I, Murthy NP, Kim CS. Warranty and preventive maintenance. International Journal of Reliability. Quality and Safety Engineering. 8(2):89–107.
5. Alexander WL, Dayal S, Dempsey JJ, Vander Ark JD. 2002. The secret life of factory service centers. The McKinsey Quarterly. 3:106–115.
6. Gebauer H, Fleisch E, Friedli T. 2005. Overcoming the Service Paradox in Manufacturing Companies. European Management Journal. 23(1):14–26.
7. Baveja SS, Gilbert J, Ledingham D. 2004. From products to services: why it's not so simple. Harvard Management Update. 9(4):3–5.
8. Brandstotter M, Haberl M, Knoth R, Kopacek B, Kopacek P. 2003. IT on demand – towards an environmental conscious service system for Vienna (AT). In: Third International Symposium on Environmentally conscious design and inverse manufacturing – EcoDesign. pp. 799–802.
9. Cohen MA, Agrawal N, Agrawal V. 2006. Winning in the Aftermarket, Harvard Business Review.
10. Oliva R, Kallenberg R. 2003. Managing the transition from products to services. International Journal of Service Industry Management. 14(2):160–172.
11. Legnani E, Cavalieri S, Crespo Márquez A, González-Prida V. 2010. System Dynamics Modeling for Product-Service Systems-A case study in the agri-machine industry, In: Proc. of APMS International Conference on Advances in Production Management Systems, Cernobbio, Como, Italy.
12. Ghodrati B, Akersten PA, Kumar U. 2007. Spare parts estimation and risk assessment conducted at Choghart Iron Ore Mine: A case study. Journal of Quality in Maintenance Engineering. 13(4).

13. Crespo Márquez A. A Structured Approach for the Assessment of System Availability and Reliability Using Continuous Time Monte Carlo Simulation. Proceedings of the 24nd Int. Conf. of the System Dynamics Society. Nijmegen, the Nederlands.
14. Jones JV. Integrated Logistic Support Handbook. McGraw-Hill and Log. Mngt Associates, ISBN 0-07-033079-4.
15. Murthy DNP, Solem O, Roren T. 2004. Product Warranty logistics: Issues and challenges. Euro. Jr. Oper. Res., 156.
16. Moore K, Gungor A, Gupta S. 1998. Disassembly Petri net generation in the presence of XOR precedence relationships, Proceedings of the IEEE International Conference on Systems, Man, and Cybernetics, San Diego, California.
17. Lambert AJD. 1999. Linear Programming in Disassembly/Clustering Sequence Generation. Computers and Industrial Engineering. 36(4):723–738.
18. Godichaud M. 2009. Outils d'aide la décision pour la selection des filières de valorisation des produits de la déconstruction des systèmes en fin de vie : application au domaine aéronautique. PhD. Dissertation, Université de Toulouse, France.
19. Jensen FV, Nielsen TD. 2007. Bayesian network and decision graph. Springer.
20. Godichaud M, Pérès F, Tchangani A. 2009. Disassembly Process Planning Using Bayesian Nets. 4th World Congress on Engineering Asset Management (WCEAM) Athens ISBN 978-1-84996-002-1.
21. Godichaud M, Peres F, González-Prida V, Tchangani A, Crespo Márquez A, Villeneuve E. 2011. Integration of warranty as a decision variable in the process of recertification of parts resulting from end-of-life system dismantling. Quality and Reliability (ICQR). IEEE International Conference. pp: 156–160.

Part VI
Results and Conclusions

Chapter 14
Summary of Results and Conclusions

Throughout this research, we have observed how a proper management framework for after-sales service can help and define new and clearer policies and procedures in the warranty claiming process, technical assistance to the user, spare parts stock, etc., reducing costs, improving the quality of the product, and increasing customer satisfaction. Basically, this research has presented a global reference framework composed by:

- Re-engineering of after-sales processes
- Possibility to outsource technical assistance
- Data analysis
- Analysis, recovery, and collaboration with the supplier
- Repair management
- Replacement planning and implementation of an inventory management
- Logic validation of warranty claims, etc.

Now, we will summarize the conclusions extracted from each of the parts into which this research is divided, to later, respond to the questions brought up at the beginning of this research (Chap. 1) and likewise propose possible future lines for research in this field.

14.1 Summary of Results of this Work and Research

14.1.1 Origins and Background (Part II)

Setting a study scenario (Chap. 2) has made it possible to summarize a business process within a specific framework, like warranty management. The analysis shows how information related to after-sales and recollection services during the useful life of an industrial asset can be advantageously used in decision-making at the time to reduce unnecessary expenses and improve quality of the service and in essence the image of the business to customers. The recollection of data permits us

V. González-Prida Díaz and A. Crespo Márquez, *After-sales Service of Engineering Industrial Assets*, DOI: 10.1007/978-3-319-03710-3_14,

to adjust the necessary parameters to adequately choose between distinct alternatives. Nowadays, the use of software tools can be useful not only to make automated selections, but also to model and simulate business processes with the aim of detecting, for example, their weak points. Together with this scenario, Chap. 3 has been intended to summarize in a few lines the content of some contributions in reference to warranty, its management, information gathering related to objectives, elements, schemes, and comments to this respect, as well as to try to highlight the defined practices from different authors. In the development of this analysis, it was possible to extract steps or desirable stages for a modern and efficient management framework. In this part, it has also been proven how additional costs of a technical assistance service could be reduced through proper strategies in the service and an efficient management of its logistics [1]. In order to adequately manage after-sales costs, businesses must be sure of the length that these costs reach, clearly define all services covered by the warranty or by extension contracts, and sufficiently stipulate detailed specifications in order to provide realistic cost estimations.

14.1.2 Methodology and Resolution Proposal (Part III)

As a consequence of the practice in advanced technologies in manufacturing and adjusted production systems, the nature of the production context has changed enormously in the last few decades. This has allowed companies to manufacture customized products on a massive scale. However, the increasing automation and reduction of inventories have also clearly incremented the pressure on maintenance systems [2]. As it is exposed in Chap. 4, maintenance management models have been created to reduce this pressure on a real necessity of application in the industry. To achieve this, in some cases there is a need to look for an operational guide to reach the organization's own objectives on maintenance. Others, on the other hand, set their final objective as developing an automated system, and some others simply pursue the ability to evaluate the function of maintenance. In all of these cases, the main concern in designing a maintenance management model was the continuous improvement of function performance. Such elements were framed in a new proposed model, which operates within the mechanisms set by the norm ISO 9001:2008 [3]. The new model presents these main characteristics:

- It is a cyclical type.
- It is open to the rest of organizational functions.
- Clearly distinguishes the execution of strategic and operative actions, when these appeared linked.
- Involves management.
- Standardizes evaluation, control, and improvement in efficiency and effectiveness of maintenance.
- Contributes to the generation of documents and records for decision-making.

- It is process focused.
- It is orientated in compliance with a quality management system.

Subsequently, a framework has been proposed in Chap. 5. A framework (as an essential structure of support) necessary to manage a technical assistance service and that suggests methods to improve its organization and decision-making. As it has been successively stated, the proposed framework considers already developed techniques, classifying them in function to their most adequate utilization within a warranty management process. Ideas and tools provided try to solve problems as well as to understand in an easier manner what it is happening in the after-sales field, applying some of the latest tendencies in business management, and also suggesting new ones such as the application of ICTs. Consequently, this chapter provides the reader with an updated revision of important methods, techniques, technologies, that make an organization able to develop different functions related to after-sales management, allowing it to face up to the current complexity in the manufacturing and services fields. In Chap. 6, more aspects can be found that are related to some standards and good practices, as well as the possibility to quantify maturity. All these aspects have been treated in order to take into consideration the importance of quality and its application in after-sales management. This management must assure that the industrial asset meets the needs which it has been designed for, how to complement it with a modern quality management that recognizes the importance of customer satisfaction and the prevention of inspections or reclamations (error prevention costs tend to be much less than error correction costs when there is a customer complaint). After-sales management has to be carried out throughout the life cycle of the product and not just only when it is sold, which involves defining a strategy together with those of technical and commercial type, in order to reach the organization's general objectives [4]. Good practices in after-sales management require taking into account the interactions between the elements that constitute an organization. Some already developed techniques in other fields have been introduced. They can be useful for our subject of study.

14.1.3 Development of the Proposed Stages (Part IV)

14.1.3.1 Effectiveness

As we know, nowadays the manufacturer is obliged by contract to provide warranty assistance to the customer when a product it is sold. Logically, the reduction in such costs will not be the only aspect to be achieved here, but also decisions will have to be global and strategic within the company in order to launch a better and more reliable product in the market, and in the same way, to offer a proper after-sales service for the user. It has been observed in the previously proposed framework that the application of a balanced scorecard is essential to align and harmonize local after-sales services with the global company objectives.

Together with this methodology, Chap. 7 describes an *Analytic Hierarchic Process* that can be useful for complex decisions, permitting also the inclusion of many levels of criteria and sub-criteria. This process can be applied to a wide variety of situations such as strategic planning, resource allocation, business policy selection. Finally, this chapter includes a root cause failure analysis study comparing in respect to a set of criterion, diverse tools commonly used in these analysis and observing their advantages and disadvantages.

14.1.3.2 Efficiency

In Chap. 8, various aspects can be observed regarding efficiency in relation to the improvement of the after-sales management service. Therefore, it is necessary to analyze the *Integrated Logistic Support* applied to a product that is in the market, and how that support facilitates and improves the decision-making process, for example, the selection of important components and with a high complexity. Also, other concepts have been included such as task frequency or the many levels of repairs. Apart from logistic support, other terms such as *Analysis of Risks, Costs and Profits* are also included in the study of after-sales efficiency. Nowadays, concepts and methods of project management theories are applied to many different areas. The intention here is to summarize them and focus them to our research, considering after-sales services as a project with a specific budget and with determined milestones. These tools, in general terms, permit us to control the evolution of a warranty program, to foresee possible new expenses, and, in short, to make proactive decisions that improve, finally, the satisfaction of the customer with the company and with the supplied product.

14.1.3.3 Evaluation and Control

The procedures of assessment and control try to provide, on the one hand, a starting point to recognize in an approximate way, the possible behavior of elements that constitute a piece of equipment, and on the other hand, to extend that analysis to the industrial asset's life cycle. In this way, as the design process becomes more profound and, once in the market, provides experience and a more precise and detailed knowledge of the systems, the initially considered values can be refined and readjusted to reality, providing therefore more accurate conclusions. To achieve this, Chap. 9 presents a formula on *Reliability, Availability, Maintainability, and Safety* that permits us to infer failure frequencies and consequently provide a preliminary estimation about frequency and periodical revisions. These results will also facilitate the outlining of an initial list of recommended spare parts. On the other hand, there is the idea that a product with a relatively long warranty period is more reliable and presents a longer duration than another with a shorter one. That is why the warranty period is often used as an efficient marketing tool by companies competing among themselves. In this chapter, the life cycle costs have been analyzed particularly for the repair study, a formulation based on the non-homogeneous poisson procedure (NHPP) and general renewal process (GRP) has been developed. The different methodologies for the LCCA can make it easier to improve decision-making processes in cases of after-sales services. Specifically, the application of these stochastics models can, for instance, allow the prevision of possible failures in the product, as well as the anticipation of repair expenses that such failures can contribute to the management of products after they are launched in the market. We have seen that the NHPP is a good estimation method of warranty period costs due to the fact that the failure rate, obviously, is going to change as well as the system's operative time, and consequently, it will not be homogenous. On the other hand, GRP is the recommend method for a better adjustment, considering complex repairs of the item.

14.1.3.4 Continuous Improvement

Chapter 10 observes distinct aspects of *Application of ICTs* in reference to a specific case of warranty management process. These technologies can facilitate and improve the decision-making process. In regard to after-sales services, these decisions can be, for example, the possibility to forecast failures in the product and in this way anticipate eventual customer reclamations. Currently, ICTs are being applied in various fields. That is why, the intention of this research to focus these electronic technologies to a new and emerging concept in electronic warranty, in a similar way to the already consolidated term e-maintenance. In a few words and although there are different opinions about it, e-maintenance is understood as the maintenance based on the use of the Internet for the data and information transmission between system to be maintained and the person in charge of making decisions. In the same way, the concept of electronic warranty has been defined as the warranty support that includes management of resources and necessary services that permits proactive decisions in the performance of the technical assistance to the client. These applied electronic technologies for example the control or the diagnosis of reclamations under warranty are, in consequence, key factor to reach high levels of quality, reliability, efficiency, and of course customer's trust. This chapter elaborates on the continuous customer relationship management (CRM), which is closely related to customer's perception on the provided services in terms of quality. Many studies [5] recommend measuring quality costs to analyze and improve the quality for the company and customers. Quality can be measured from the user's point of view through marketing norms and statistical methods based on historical data. These standards focus on the quality perceived from the satisfaction of real or potential customers alike. Nevertheless, without neglecting this aspect, manufacturers are obliged to observe and comply with some rules and guidelines when dealing with consumers. These norms present similarities and differences according to the market we deal with. Recently in Europe, the European norm EN 13269 (guide on the preparation of maintenance contracts) [6] offers an adequate system for making maintenance contracts that can also help writing those contractual documents related to warranty service (or extension), offering a complete view of the contract structure, and discussing different aspects related to their subject and content.

Lastly, the chapter finishes observing the importance of applying a quality methodology, the Six Sigma method, in the improvement of after-sales management as well as to facilitate decision-making. Table 14.1 summarizes the roles according to their objective or purpose, the critical processes where they take part as well as the key activities they carry out. Basically, this section aims to highlight the application of the Six Sigma methodology in order to face the priorities derived from the warranty program and even more, through a solid cycle of continuous improvement, to prevent failure recurrence.

The latest tendencies make an attempt to mix or relate Six Sigma tools with others less consolidated such as Lean philosophy. It was developed by Toyota (originally called TPS or "Toyota Productive System") and has experimented different versions of application by many companies (for instance, the FPS or "Ford Productive System") and currently is producing excellent results when complemented with measuring techniques and Six Sigma analysis (which was developed by Motorola). In spite of detractors of the methodology Lean Six Sigma (LSS), it can be observed that it is a tool with a high level of potential for effectiveness improvement in all types of companies (mainly services and small- and medium-sized companies), especially in countries where, as it is happening in Spain, the implementation of these new management philosophies is in an incipient moment.

14.1.4 Extensions to the Proposed Reference Framework (Part V)

Chapter 11 intends to identify weaknesses, strengths, and improvement opportunities in the after-sales service through the quantification of its intellectual capital. Consequently, this procedure will seek to obtain savings and redistribute expenses, whose combination reflects an improvement in the generated value by customer service management. The importance of this procedure is that it permits us the following points:

- Information is shown for the control and measurement of the organization's maturity in a quantitative way and including intangible costs.
- Facilitates decision-making and its comparison with external agents, competitors, or internationally regulated and accepted values.
- Detects capabilities as well as weaknesses, in order to help make decisions on how to solve them.
- Administrative and management decisions are closer to the client as an improvement in the quality offered by the company and perceived by the user.
- The way, in which the organization concerns itself about its staff and their performance, creating a good impression of security and warranty.

Table 14.1 Summary of roles in a Six Sigma application

Roles	Purpose	Critical processes	Key activities	
Role 1. Understand voice of customer and define priorities	Clearly understands the difference between customer expectation and the current level of quality of the product and translate it into specific requirements to adjust the manufacturer process/product design	Definition of defects and urgent in satisfactions	ECB	Benchmarking segmentation
			EVB	
			Analysis of internal indicators	Inspection study
				Re-contact
		Early detection	Inspection adjustment feedback of the defect generation process	
Role 2. Problem solutions that are non-evident in the work area	Make the abnormality evident at the earliest possible stage	Problem-solving	Critical problem retention and resolution	Reactive role
			Methodology of problem-solving	
Role 3. Assurance of processes and critical quality characteristics	System improvement and key processes and quality characteristic control	Critical variable control	Audits CTQs	Preventive role
			Monitor SCs	
			Reaction a CTQs/SCs out of the expected level	
			Critical process management (WG, supplier, line, sistema, etc.)	
Role 4. Improve the capacity by product design and process	Change product process so that a higher capacity is achieved than the ones that can in the current optimal designing conditions (attempt of product/process design = 0/100)	Align product characteristic with customer expectations	Must wants/AIMs	Proactive role
			PD/PMT Connection/Interface	
			Validation of reduction in cost	
			Participation preproduction	
		Eliminate non-solid processes	Non-solid process follow-up	
			Product changes validation	

This methodology permits us to establish priorities in decision-making based on reliability indicators and facilitating the comparison of the performance of after-sales services according to different geographical areas, technical staff, procedures, operation changes, environmental condition, or in the organization, etc. In short, it is intended with this assessment increase the collective intelligence of the organization through a widening improvement and diffusion of knowledge in a normalized fashion and with the support of information technology. On the other hand, throughout Chap. 12 it has been observed how maintainability is also a key factor for evaluating technical assistance that develops in after-sales services. The object of this chapter has been the definition of a method of index calculation that assesses the maintainability in industrial devices. The indexes that have been developed here intend to show the ease with which technical assistance can be given to an industrial asset, permitting us in this way to improve such tasks in different phases of their life cycle. For that, criteria have been established that characterized the index of maintainability according to these distinct attributes, some are general (that is to say, those that can affect any level such as simplicity, modularity, ergonomics), and others of a variable character (that depend n the level of assistance in question and for example accessibility, the needs of personnel, the working environment). Finally, this fifth part deals with the extensions to the proposed framework. This part serves the purpose of contributing to the research in the field of product-service systems (PSS), which is still a relatively new and not very consolidated subject. In more detail, Chap. 13 develops a dynamic model to quantitatively simulate the services provided by the department of technical assistance and how they can affect the general service performance of the company. In this context, the model has been applied to a specific case study, showing some of the potential outcomes through different scenarios in which a preventive maintenance assistance is requested/hired (or not), in and out the warranty period. This chapter concludes by briefly introducing the Bayesians networks and describing how they can be useful when it comes to determining aspects of a product, for instance, the case of dismantling an industrial asset at the end of its useful life cycle (which can be also an interesting matter in the discussion regarding the after-sales service offered by companies to their customers). The outlined case considers warranty parameters as a known (entrance/input) relating a functional recycling option to the level of disassembly, the cost of the resulting warranty program, as well as the probability and the implementation of the corresponding actions. In other words, it consists simply of observing a warranty management based on the usage of pieces or spare parts previously used, which would link, for example, the areas of reliability or logistical support.

14.1.5 Case Studies

Chapters show a series of cases containing mainly practical and applicable scenarios. They attempt to illustrate with examples the application of the lessons learned in previous theory. Nevertheless, given the fact that it is a chapter that aims to summarize the overall conclusions reached and developed all throughout the research, it is appropriate and pertinent, however, to comment on the results of such case studies. Already introduced in Chap. 1 (in Table 1), are the scenarios, issues to solve and methodology of application in each one of them. Now, Table 14.2 also includes the solutions provided for each one of them. However, it seems far more interesting to know about the processes of reaching certain decisions rather than the outcomes themselves.

14.2 Conclusions Extracted from the Different Parts

In continuation, the results achieved during this work that were proposed in Chap. 1 are shown here, just as in relation to the different contributions, conferences, magazines, and books:

- *1st Question*: **How the different studies, models and approximations… done so far have helped the identification, analysis, elimination and prevention of problems when there has been a failure in an industrial asset?**

The different studies, models, etc., that have been analyzed through contributions, such as [7–11], have facilitated the visibility in as far as the current state-of-the-art situation in which maintenance and warranty management can be found. It is undoubtedly pretentious to consider that this state-of-the-art analysis totally covers the facets related to maintainability, after-sales and its interactions in all fields. However, the reviewed studies and their comparison have sufficiently permitted the ability to differentiate and select forms or manners to prevent the recurrence of problems based on an organizational improvement. Likewise, contributions like [7] have permitted an approximation of a business reality identifying a casuistry of events that tend to arise when managing a warranty assistance program.

- *2nd Question*: **What systems are proposed for after-sales service so that we can the most of the advantages of methodologies that have been successfully applied in other areas?**

The knowledge from technologies successfully implemented in other areas has permitted us to propose in conferences such as [12–17] a generic framework in which different techniques are divided into four sequential phases (effectiveness, efficiency, assessment and control, and continuous improvement) playing a crucial role in the improvement of the organization's maturity concerning after-sales services. Some of these contributions suggest likewise good practice application that improves the quality of the management process.

Table 14.2 Results obtained from the studies discussed in the chapters

Methodology	Study scenario	Issue to solve	Result
Maturity (Chap. 6)	Manufacturer of products that are launched to the market and need warranty services	What's the company's own perception on its after-sales service?	A maturity matrix is obtained where the level of involvement of each department is quantified
BSC (Chap. 7)	Agricultural machinery (harvesters and harvester head) that is sold in the first- and second-hand market	Which after-sales service is the most beneficial?	Sales of new harvesters are the most beneficial in after-sales. Used products tend to exceed objective value
AHP (Chap. 7)	Spare parts management for customized product manufacturing in a determined stage of production	From where is preferable to obtain such spare part in that determined moment?	The best place option to get that spare in this context would be from the warehouse, then cannibalization
Risk and CRM (Chap. 8)	A water distribution company that suspends the supply due to a network failure	What is the cost of keeping or losing a customer after a failure?	In this context, the intangible costs are estimated at 314.31 €
RAMS (Chap. 9)	Electromechanical item in the design phase	Which would be the spare part list best recommended for servicing this warranty?	The welding equipment, rotor set, and stator are the three components which reliability decreases 60 % after a year
LCC (Chap. 9)	Faulty engines are repaired providing a repair warranty	How long would that period of time be to carry out the best repair service?	The minimum value is obtained from the NHPP model, considering different percentages of waiting time until the next failure
Six Sigma (Chap. 10)	An automotive company that repeatedly receives the same type of warranty reclamation	How to deal with a reclamation that affects design?	Instruction sheet for defect corrections, obtaining a 70 % improvement (steel was the critical input of the imprinting)
Intellectual capital (Chap. 11)	Water distribution companies that attend similar areas of customers	Which company presents a better investment of its initial budget?	Company B has a larger investment in intellectual capital (best investment of initial budget)
Maintainability (Chap. 12)	Bridge crane that requires technical assistance	How difficult or easy is to technically assist this item?	Graphics are obtained quantifying attributes and variables of maintainability

- *3rd Question*: **How the after-sales service would be in order to enable us to be more efficient and with a beneficial performance from a business point of view?**

The after-sales services play an important role in companies with rising profits. For that, it is convenient to apply tools and proper methods that are based on a holistic vision of technical assistance to the customer, considering all the areas that are influenced by such service, for example, manufacturing, quality, finance. Both financial and engineering aspects should be taken into account, for which it is necessary to translate technical performance and its economical impact. Contributions made in conventions and conferences and journals such as [18–24] have dealt with these matters presenting applications of BSC or AHP in warranty program assistance or, just describing the management of risks and costs of after-sales service. The presented tools integrate the methods used in the industry and in the general management of projects that identify strengths and avoid weaknesses.

- *4th Question*: **How would the technical and financial performances in after-sales services be measured in order to make an improvement in management possible?**

Even though technical and financial impacts have already been dealt with and elaborated in question 3, this 4th question concludes the proposed cyclical procedure of evaluating such impacts and applying techniques in the area of continuous quality improvement. Some contributions in the field have been made to carry out that analysis. Specialized papers published in magazines such as [25–32]. The direct impact of the after-sales service is, at first, technical, given that it makes an attempt to reduce the number of reclamations or non-conformities of an industrial asset that is already in customer's hands. However, the purpose of the companies is to make an economical profit while maintaining their social and environmental responsibilities; thus, evaluation and control of the economical impacts of technical services become essential. The possibility to control costs is set out by monitoring the constant behavior of the product; this may influence the performance and procedures done in other areas of work within the company different from the after-sales department. In other words, an improvement in the after-sales service foresees savings that directly and holistically have positive repercussions in the rest of the company's organization; beneficial in terms of financial profits; and at the same time, in its own corporate image, giving out a more attractive one to customers. The reliability and life cycle cost analyses of the product launched to the market, as well as the relations with the customer and the legal framework of after-sales contracts, feed the management procedures and make it easier for multidisciplinary equipment (suggested with Six Sigma) to analyze the improvements before and after adopting determined decision or policies. In short, the approximations presented for the assessment of the after-sales service impact, attempt to link technical performance to economical consequences.

- *5th Question*: **What information must be relevant to make decisions regarding the after-sales service and what are the future researches that arise from this line?**

After-sales and maintenance data tend to be stored in the business' different databases, this compilation is usually tedious and time consuming. With the aim of preventing, as much as possible, these inconveniences and contribute at the same time a holistic touch to the subject, it has been defined on the one hand a procedure to evaluate intellectual capital of after-sales service, and on the other hand, the maintainability of the product that permits the comparison of the relevant data for the best performance of the technical staff in particular and of the organization in general. In addition to the previously stated, the management of warranty and after-sales information opens new interesting research lines; an example of these potential lines would be the modeling of dynamic systems of products and services, as well as the application of Bayesian networks in problems where warranty is considered as just another variable for decision-making. All these have been extended in the conferences that follow [33–38].

14.3 Future Lines of Research

The different chapters in this research propose and open different lines of study that can help in the continuation of the analysis of after-sales services and warranty assistance of an organization, and how this management can affect the business holistically, the employees and clients. This way for example, we have begun this research with a generic scenario (Chap. 2) where some other features could be added, such as boundary conditions to consider final products with different reliability, that the cost of the warranty be in terms with the time, different degrees of non-operation, maximum locations at different stages (that is to say, when there is a distribution of products in different destinations), or when the budgetary margins are negative (that is to say, there is a need to prioritize failures to attend, or decide on a reduction in profit margins to have replacements available), etc. Regarding the state-of-the-art analysis (Chap. 3), topics for future researches could be, for example, new economic models for the after-sales service and discussions about its management, etc [39]. In respect to the warranty management models (Chap. 4), an interesting line would consist of the link between the models with specific tools for production systems, or to develop new models of reference for maintenance management that are directly applicable to a chosen organization. In relation to reference framework (Chap. 5), it is possible to propose the application of other methodologies or to more thoroughly study the combination the already mentioned, analyzing from the proposed management framework onwards, information organization and the data necessary to warranty assistance, with the involvement of a Six Sigma team. In reference to the application of best practices of after-sales management of an industrial asset on sale (Chap. 6), it can be interesting to analyze the different viewpoints of different countries and markets where other rules and standards exist. For example, it can be advantageous to know the conceptual difference of this service between the emerging countries (in general producers or manufacturers), with developed countries (in general consumers, and consequently, customers of emerging countries). In addition, another extension to this chapter would be to particularize the quantification of maturity in each stage of the life cycle of the product (pre-launch, launch, and post-launch), taking into account that at each stage there are different necessities. This book attempts to serve as a starting point for future research related to critical analysis. Some key aspects have been presented throughout Chap. 7 with the aim of making the adequate decisions that would enable the company to correctly achieve and reach its global objectives. For that purpose, it is also necessary to pay special attention to the corporate setting in which the companies must their select options and make decisions between alternatives, considering, as one very important aspect, personnel experience in technical service. This research also intends to serve as a base for further researches related to the critical analysis. Other future lines of research that can be opened in this field are for example: related to logistic support and analysis of costs, risks and benefits (Chap. 8), the combination of techniques such as FMECA, RCM with the CRM (Customer Relationship Management); analysis of the advanced mathematical methods with electronic support technologies (electronic surveillance, monitoring, online diagnosis, etc.) with the aim of developing more elaborate models. In other words, as it has been previously mentioned, currently ICTs are being applied to a wide variety of different fields [40]; therefore, future researches can focus on the concept of electronic warranty in the same way as the e-maintenance, etc. (Chap. 10). Besides the application of new technologies in warranty management, the participation of after-sales personnel within the improvement process represents a crucial factor for achieving success. It would be necessary therefore to increase the intellectual capital at higher levels of knowledge, experience, and training (Chap. 11), together with the introduction of new techniques that benefit task performance and personnel participation involved in the effectiveness of the technical service of warranty assistance (Chap. 12). Apart from all mentioned, it is important to highlight that after-sales management appears within the business as a potential area of improvement related to the development of new technologies and the determination of future capabilities in respect to product and service systems (Chap. 13). Alongside the above-mentioned perspectives of future works in this field that try to use the network theory of Bayes to determine risks related to warranty.

References

1. Murthy DNP, Djamaludin I. 2002. New product warranty: A literature review. Int. J. Production Economics 79:231-260.
2. Crespo Márquez A, Benoit I. 2006. Editorial Special issue on eMaintenance. Computers in Industry. 57:473-475.
3. International Std. 2008. ISO 9001:2008 Sistemas De Gestión De La Calidad. AENOR

4. Karim MR, Suzuki K. 2005. Analysis of warranty claim data: a literature review, International Journal of Quality & Reliability Management. 22(7):667-686.
5. Goodman J. 1986. Technical Assistance Research Program (TARP), U.S. Office of Consumer Affairs Study on Complaint Handling in America.
6. European Norm. 2007. UNE-EN 13269:2007 Mantenimiento. Guía para la preparación de contratos de mantenimiento. AENOR
7. González-Prida V, Gómez JF, and Crespo Márquez A. 2009. Case study: Warranty costs estimation according to a defined lifetime distribution of deliverables. 4th World Congress on Engineering Asset Management (WCEAM 2009), Athens, Greece. ISBN: 978-1-84996-002-1, pp. 146-155.
8. González-Prida V, Crespo Márquez A. 2010. Book Review: Reliability Engineering. Warranty Management and Product Manufacture" (By D.N.P. Murthy and W.R. Blischke). Production Planning & Control: The Management of Operations, 1366-5871, 21(7):720-721.
9. González-Prida V, Gómez JF, Crespo Márquez A. 2010. Estado del arte en modelos de garantía: Una revisión práctica al Marco de la gestión de los costes de garantía. Revista Ingeniería y Gestión de Mantenimiento. Vol. Enero/Febrero/Marzo España 2010.
10. González-Prida V, Gómez JF, López M, Crespo Márquez A, & Moreu P. 2010. Warranty cost models State-of-Art: A practical review to the framework of warranty cost management. Reliability, Risk, and Safety. Theory and applications. C. Guedes Soares et al. (Eds), pp. 2051-2059. Taylor & Francis Group, London, ISBN 978-0-415-55509-8.
11. Crespo Márquez A, Moreu P, González-Prida V, Gómez J, Parra C. 2009. The maintenance management framework: A practical view to maintenance management. Taylor & Francis Group, London, ISBN 978-0-415-48513-5, pp. 669-674.
12. González-Prida V, Crespo Márquez A. 2012. A reference framework for the warranty management in industrial assets. Computers in Industry 63:960-971. ISSN: 0166-3615.
13. González-Prida V, Barberá L, Crespo Márquez A, Gómez J. 2012. The Management of a Warranty Assistance Program: A suggestion as Reference Framework. Advances in Safety, Reliability and Risk Management - Bérenguer, Grall & Guedes Soares (eds). Taylor & Francis Group, London, ISBN 978-0-415-68379-1
14. González-Prida V, Parra C, Gómez JF, and Crespo Márquez A. 2010. Reference framework proposal for the management of a warranty program. Euromaintenance. 20th International Industrial Maintenance Congress, from the 12th to the 14th of May in the Verona Conference Centre, Italy, pp 100-105.
15. González-Prida V, Barberá L, Gómez JF, Crespo Márquez A. 2012. Contractual and quality aspects on warranty: best practices for the warranty management and its maturity assessment. International Journal of Quality & Reliability Management. 29(3), ISSN: 0265-671X, Emerald.
16. González-Prida V, Parra C, Crespo Márquez A, Gómez J. 2011. Audit to a specific study scenario according to a reference framework for the improvement of the warranty management. ESREL Conference, Troyes (France). Taylor & Francis Group.
17. Costantino F, De Minicis M, González-Prida V, Crespo Márquez A. 2012. On the use of Quality Function Deployment (QFD) for the identification of risks associated to warranty programs. ESREL Conference, Helsinki (Finland). Taylor & Francis Group.
18. González-Prida V, Gómez JF, Crespo Márquez A. 2011. Practical Applications of AHP for the Improvement of Warranty Management. Journal of Quality in Maintenance Engineering, 17(2):163-182, Emerald Group Publishing Limited, 1355-2511.
19. González-Prida V, Barberá L, Crespo Márquez A, Viveros P, Kristjanpoller F. 2011. Implementing the Balance Scorecard for the Improvement of the Warranty Management: The Strategic Alignment. International Journal of E-Business Development (IJED), 1(1):15-21. ISSN: 2226-7336
20. González-Prida V, Barberá L, Crespo Márquez A, Arata A, Viveros P, Kristjanpoller F. 2011. Aplicación del Cuadro de Mando Integral para la mejora de la gestión de la garantía: la alineación estratégica. Congreso MAPLA 2011, Chile.

21. González-Prida V, Crespo Márquez A, Gómez J. 2010. Practical application of an Analytic Hierarchy Process for the improvement of the warranty management. The 38th ESREDA seminar and 3rd joint ESREDA/ESRA seminar on Advanced Maintenance Modelling, 4-5th may, 2010, hosted by the regional committee of the Hungarian Academy of Science.
22. Barberá L, González-Prida V, Parra C, Crespo Márquez A. 2012. Framework to assess RCM software tools. ESREL Conference, Helsinki (Finland). Taylor & Francis Group.
23. González-Prida V, Pérès F, Crespo Márquez A. 2011. On the risks and costs methodologies applied for the improvement of the warranty management. Journal of Service Science and Management (JSSM). 2(6). ISSN 1940-9907.
24. González-Prida V, Crespo Márquez A, Pérès F, De Minicis M, Tronci M. 2012. Logistic support for the improvement of the warranty management. Advances in Safety, Reliability and Risk Management – Bérenguer, Grall & Guedes Soares (eds). Taylor & Francis Group, London, ISBN 978-0-415-68379-1.
25. González-Prida V, Kristjanpoller F, Arata A, Crespo Márquez A. 2010. Caso de estudio sobre el análisis de la fiabilidad en una planta de triturado de mineral en Chile. Revista Ingeniería y Gestión de Mantenimiento. Vol. Abril/Mayo/Junio 2010 España.
26. González-Prida V, Barberá L, Crespo Márquez A. 2010. Practical application of a RAMS analysis for the improvement of the warranty management. 1st IFAC workshop on Advanced Maintenance Engineering Services and Technology, Lisbon, 1-2 July 2010
27. González-Prida V, Gómez JF, López M, Crespo Márquez A, & Moreu P. 2009. Availability and reliability assessment of industrial complex systems: A practical view applied on a bioethanol plant simulation. Safety, Reliability and Risk Analysis: Theory, Methods and Applications - Martorell et al. (eds), pp. 687-695. Taylor & Francis Group, London, ISBN 978-0-415-48513-5.
28. González-Prida V, Barberá L, Crespo Márquez A, Parra C. 2012. Warranty period calculation after the repair of a complex industrial asset applying non-homogeneous Poisson Process. DYNA Ingeniería e Industria. 87:655-662. ISSN: 0012-7361.
29. González-Prida V, Real C, Barberá L, Crespo Márquez A. 2011. Caso de estudio en el sector automovilístico sobre la aplicación de la metodología 6-Sigma en la gestión de la garantía. Revista Ingeniería y Gestión de Mantenimiento. Vol. Enero/Febrero/Marzo España 2011.
30. González-Prida V, Barberá L, Crespo Márquez A, 2012. New Services Generated by the e-Technology Applied on the Warranty Management. International Journal of E-Business Development (IJED), 2(1):16-22. ISSN: 2226-7336
31. González-Prida V, Barberá L, Crespo Márquez A, Gómez J. 2011. ICT application on the warranty management process. The "e-warranty" concept. ESREL Conference, Troyes (France). Taylor & Francis Group.
32. González-Prida V, Gómez J, Barberá L, Crespo Márquez A. 2012. The customer relationship in the warranty management of Distribution Network Service Providers. ESREL Conference, Helsinki (Finland). Taylor & Francis Group.
33. Gómez J, González-Prida V, Crespo Márquez A, Parra C. 2010. Network maintenance contribution in terms of intellectual capital on distribution network services providers. Reliability, Risk and Safety - Ale, Papazoglou & Zio (Eds). Taylor & Francis Group, London, ISBN 978-0-415-60427-7.
34. Gómez JF, González-Prida V, López M, Crespo Márquez A. 2010. Network maintenance contribution in quality terms on distribution network services providers. Reliability, Risk and Safety: Theory and Applications – Briš, Guedes Soares & Martorell (eds). Taylor & Francis Group, London, ISBN 978-0-415-55509-8
35. Moreu P, González-Prida V, Barberá L, Crespo Márquez A. 2012. A practical method for the maintainability assessment using maintenance indicators and specific attributes. Reliability Engineering and System Safety. 100:84-92. ISSN: 0951-8320.
36. González-Prida V, Moreu P, Crespo Márquez A, Barberá L. 2011. Norma de mantenibilidad: Propuesta para la evaluación de la mantenibilidad en activos industriales. Ingeniería y Gestión del Mantenimiento, N° 74, Jul/Ago/Sept 2011, pp 24-37.

37. Godichaud M, Pérès F, González-Prida V, Tchangani A, Crespo Márquez A, Villeneuve E. 2011. Integration of Warranty as a Decision Variable in The Process of Recertification of Parts Resulting from End-Of-Life System Dismantling. IEEE, Bangkok, International Conference on Quality and Reliability.
38. Legnani E, Cavalieri S, Crespo Márquez A, González-Prida V. 2010. System Dynamics modeling for Product-Service Systems - A case study in the agri-machine industry. APMS, International Conference on Advances in Production Management Systems, Cernobbio, Como, Italy, 11-13 October 2010
39. Ingwald A. 2009. Technologies for better utilisation of production process resources. Doctoral thesis. Växjö University Press. ISBN: 978-91-7636-646-2.
40. Campos J. 2008. ICT tools for e-maintenance. Doctoral thesis. Växjö University Press. ISBN: 978-91-7636-624-0.

MIX
Papier aus verantwortungsvollen Quellen
Paper from responsible sources
FSC® C105338

If you have any concerns about our products,
you can contact us on
ProductSafety@springernature.com

In case Publisher is established outside the EU,
the EU authorized representative is:
Springer Nature Customer Service Center GmbH
Europaplatz 3, 69115 Heidelberg, Germany

Printed by Libri Plureos GmbH
in Hamburg, Germany